To Measure the Sky

An Introduction to Observational Astronomy

With a lively yet rigorous and quantitative approach, Frederick R. Chromey introduces the fundamental topics in optical observational astronomy for undergraduates.

Focussing on the basic principles of light detection, telescope optics, coordinate systems and data analysis, the text introduces students to modern observing techniques and measurements. It approaches cutting-edge technologies such as advanced CCD detectors, integral field spectrometers, and adaptive optics through the physical principles on which they are based, helping students to understand the power of modern space and ground-based telescopes, and the motivations for and limitations of future development. Discussion of statistics and measurement uncertainty enables students to confront the important questions of data quality.

It explains the theoretical foundations for observational practices and reviews essential physics to support students' mastery of the subject. Subject understanding is strengthened through over 120 exercises and problems. Chromey's purposeful structure and clear approach make this an essential resource for all student of observational astronomy.

Frederick R. Chromey is Professor of Astronomy on the Matthew Vassar Junior Chair at Vassar College, and Director of the Vassar College Observatory. He has almost 40 years' experience in observational astronomy research in the optical, radio, and near infrared on stars, gaseous nebulae, and galaxies, and has taught astronomy to undergraduates for 35 years at Brooklyn College and Vassar.

To Measure the Sky
An Introduction to Observational Astronomy

Frederick R. Chromey

Vassar College

CAMBRIDGE
UNIVERSITY PRESS

CAMBRIDGE
UNIVERSITY PRESS

University Printing House, Cambridge CB2 8BS, United Kingdom

Cambridge University Press is part of the University of Cambridge.

It furthers the University's mission by disseminating knowledge in the pursuit of education, learning and research at the highest international levels of excellence.

www.cambridge.org
Information on this title: www.cambridge.org/9780521747684

First published 2010
6th printing 2015

Printed in the United Kingdom by TJ International Ltd, Padstow, Cornwall

A catalog record for this publication is available from the British Library

Library of Congress Cataloguing in Publication data

Chromey, Frederick R., 1944–
To measure the sky : an introduction to observational astronomy /
Frederick R. Chromey.
    p. cm.
Includes bibliographical references and index.
ISBN 978-0-521-76386-8 – ISBN 978-0-521-74768-4 (pbk.)
1. Astronomy–Observations. 2. Astronomy–Technique. I. Title

QB145.C525 2010
522–dc22   2010000052

ISBN 978-0-521-76386-8 Hardback
ISBN 978-0-521-74768-4 Paperback

To Molly

Contents

Preface

There is an old joke: a lawyer, a priest, and an observational astronomer walk into a bar. The bartender turns out to be a visiting extraterrestrial who presents the trio with a complicated-looking black box. The alien first demonstrates that when a bucket ful of garbage is fed into the entrance chute of the box, a small bag of high-quality diamonds and a gallon of pure water appear at its output. Then, assuring the three that the machine is his gift to them, the bartender vanishes.

The lawyer says, "Boys, we're rich! It's the goose that lays the golden egg! We need to form a limited partnership so we can keep this thing secret and share the profits."

The priest says, "No, no, my brothers, we need to take this to the United Nations, so it can benefit all humanity."

"We can decide all that later," the observational astronomer says. "Get me a screwdriver. I need to take this thing apart and see how it works."

This text grew out of 16 years of teaching observational astronomy to undergraduates, where my intent has been partly to satisfy – but mainly to cultivate – my students' need to look inside black boxes. The text introduces the primary tools for making astronomical observations at visible and infrared wavelengths: telescopes, detectors, cameras, and spectrometers, as well as the methods for securing and understanding the quantitative measurements they make. I hope that after this introductory text, none of these tools will remain a completely black box, and that the reader will be ready to use them to pry into other boxes.

The book, then, aims at an audience similar to my students: nominally second- or third-year science majors, but with a sizeable minority containing advanced first-year students, non-science students, and adult amateur astronomers. About three-quarters of those in my classes are *not* bound for graduate school in astronomy or physics, and the text has that set of backgrounds in mind.

I assume my students have little or no preparation in astronomy, but do presume that each has had one year of college-level physics and an introduction to integral and differential calculus. A course in modern physics, although very helpful, is not essential. I make the same assumptions about readers of this book. Since readers' mastery of physics varies, I include reviews of the most relevant physical concepts: optics, atomic structure, and solid-state physics. I also include a brief introduction to elementary statistics. I have written qualitative chapter summaries, but the problems posed at the end of each chapter are all quantitative exercises meant to strengthen and further develop student understanding.

My approach is to be rather thorough on fundamental topics in astronomy, in the belief that individual instructors will supply enrichment in specialized areas as they see fit. I regard as fundamental:

- the interaction between light and matter at the atomic level, both in the case of the formation of the spectrum of an object and in the case of the detection of light by an instrument
- the role of uncertainty in astronomical measurement
- the measurement of position and change of position
- the reference to and bibliography of astronomical objects, particularly with modern Internet-based systems like *simbad* and *ADS*
- the principles of modern telescope design, including space telescopes, extremely large telescopes and adaptive optics systems
- principles of operation of the charge-coupled device (CCD) and other array detectors; photometric and spectroscopic measurements with these arrays
- treatment of digital array data: preprocessing, calibration, background removal, co-addition and signal-to-noise estimation
- the design of modern spectrometers.

The text lends itself to either a one- or two-semester course. I personally use the book for a two-semester sequence, where, in addition to the entire text and its end-of-chapter problems, I incorporate a number of at-the-telescope projects both for individuals and for "research teams" of students. I try to vary the large team projects: these have included a photometric time series of a variable object (in different years an eclipsing exoplanetary system, a Cepheid, and a blazar), an H–R diagram, and spectroscopy of the atmosphere of a Jovian planet. I am mindful that astronomers who teach with this text will have their own special interests in particular objects or techniques, and will have their own limitations and capabilities for student access to telescopes and equipment. My very firm belief, though, is that this book will be most effective if the instructor can devise appropriate exercises that require students to put their hands on actual hardware to measure actual photons from the sky.

To use the text for a one-semester course, the instructor will have to skip some topics. Certainly, if students are well prepared in physics and mathematics, one can dispense with some or all of Chapter 2 (statistics), Chapter 5 (geometrical optics), and Chapter 7 (atomic and solid-state physics), and possibly all detectors (Chapter 8) except the CCD. One would still need to choose between a more thorough treatment of photometry (skipping Chapter 11, on spectrometers), or the inclusion of spectrometry and exclusion of some photometric topics (compressing the early sections of both Chapters 9 and 10).

Compared with other texts, this book has strengths and counterbalancing weaknesses. I have taken some care with the physical and mathematical treatment of basic topics, like detection, uncertainty, optical design, astronomical seeing, and space telescopes, but at the cost of a more descriptive or encyclopedic survey of specialized areas of concern to observers (e.g. little treatment of the details of astrometry or of variable star observing). I believe the book is an excellent fit for courses in which students will do their own optical/infrared observing. Because I confine myself to the optical/infrared, I can develop ideas more systematically, beginning with those that arise from fundamental astronomical questions like position, brightness, and spectrum. But that confinement to a narrow range of the electromagnetic spectrum makes the book less suitable for a more general survey that includes radio or X-ray techniques.

The sheer number of people and institutions contributing to the production of this book makes an adequate acknowledgment of all those to whom I am indebted impossible. Inadequate thanks are better than none at all, and I am deeply grateful to all who helped along the way.

A book requires an audience. The audience I had uppermost in mind was filled with those students brave enough to enroll in my Astronomy 240-340 courses at Vassar College. Over the years, more than a hundred of these students have challenged and rewarded me. All made contributions that found their way into this text, but especially those who asked the hardest questions: Megan Vogelaar Connelly, Liz McGrath, Liz Blanton, Sherri Stephan, David Hasselbacher, Trent Adams, Leslie Sherman, Kate Eberwein, Olivia Johnson, Iulia Deneva, Laura Ruocco, Ben Knowles, Aaron Warren, Jessica Warren, Gabe Lubell, Scott Fleming, Alex Burke, Colin Wilson, Charles Wisotzkey, Peter Robinson, Tom Ferguson, David Vollbach, Jenna Lemonias, Max Marcus, Rachel Wagner-Kaiser, Tim Taber, Max Fagin, and Claire Webb.

I owe particular thanks to Jay Pasachoff, without whose constant encouragement and timely assistance this book would probably not exist. Likewise, Tom Balonek, who introduced me to CCD astronomy, has shared ideas, data, students, and friendship over many years. I am grateful as well to my astronomical colleagues in the Keck Northeast Astronomical Consortium; all provided crucial discussions on how to thrive as an astronomer at a small college, and many, like Tom and Jay, have read or used portions of the manuscript in their observational courses. All parts of the book have benefited from their feedback. I thank every Keckie, but especially Frank Winkler, Eric Jensen, Lee Hawkins, Karen Kwitter, Steve Sousa, Ed Moran, Bill Herbst, Kim McLeod, and Allyson Sheffield.

Debra Elmegreen, my colleague at Vassar, collaborated with me on multiple research projects and on the notable enterprise of building a campus observatory. Much of our joint experience found its way into this volume. Vassar College, financially and communally, has been a superb environment for both my teaching and my practice of astronomy, and deserves my gratitude. My editors at Cambridge University Press have been uniformly helpful and skilled.

My family and friends have had to bear some of the burden of this writing. Clara Bargellini and Gabriel Camera opened their home to me and my laptop during extended visits, and Ann Congelton supplied useful quotations and spirited discussions. I thank my children, Kate and Anthony, who gently remind me that what is best in life is not in a book.

Finally, I thank my wife, Molly Shanley, for just about everything.

Chapter 1
Light

Always the laws of light are the same, but the modes and degrees of seeing vary.
— Henry David Thoreau, *A Week on the Concord and Merrimack Rivers*, 1849

Astronomy is not for the faint of heart. Almost everything it cares for is indescribably remote, tantalizingly untouchable, and invisible in the daytime, when most sensible people do their work. Nevertheless, many — including you, brave reader — have enough curiosity and courage to go about collecting the flimsy evidence that reaches us from the universe outside our atmosphere, and to hope it may hold a message.

This chapter introduces you to astronomical evidence. Some evidence is in the form of material (like meteorites), but most is in the form of light from faraway objects. Accordingly, after a brief consideration of the material evidence, we will examine three theories for describing the behavior of light: light as a wave, light as a quantum entity called a photon, and light as a geometrical ray. The ray picture is simplest, and we use it to introduce some basic ideas like the apparent brightness of a source and how that varies with distance. Most information in astronomy, however, comes from the analysis of how brightness changes with wavelength, so we will next introduce the important idea of spectroscopy. We end with a discussion of the astronomical magnitude system. We begin, however, with a few thoughts on the nature of astronomy as an intellectual enterprise.

1.1 The story

... as I say, the world itself has changed.... For this is the great secret, which was known by all educated men in our day: that by what men think, we create the world around us, daily new.
— Marion Zimmer Bradley, *The Mists of Avalon*, 1982

Astronomers are storytellers. They spin tales of the universe and of its important parts. Sometimes they envision landscapes of another place, like the roiling liquid-metal core of the planet Jupiter. Sometimes they describe another time, like the era before Earth when dense buds of gas first flowered into stars, and a darkening Universe filled with the sudden blooms of galaxies. Often the stories solve mysteries or illuminate something commonplace or account for something monstrous: How is it that stars shine, age, or explode? Some of the best stories tread the same ground as myth: What threw up the mountains of the Moon? How did the skin of our Earth come to teem with life? Sometimes there are fantasies: What would happen if a comet hit the Earth? Sometimes there are prophecies: How will the Universe end?

Like all stories, creation of astronomical tales demands imagination. Like all storytellers, astronomers are restricted in their creations by many conventions of language as well as by the characters and plots already in the literature. Astronomers are no less a product of their upbringing, heritage, and society than any other crafts people. Astronomers, however, think their stories are special, that they hold a larger dose of "truth" about the universe than any others. Clearly, the subject matter of astronomy – the Universe and its important parts – does not belong only to astronomers. Many others speak with authority about just these things: theologians, philosophers, and poets, for example. Is there some characteristic of astronomers, besides arrogance, that sets them apart from these others? Which story about the origin of the Moon, for example, is the truer: the astronomical story about a collision 4500 million years ago between the proto-Earth and a somewhat smaller proto-planet, or the mythological story about the birth of the Sumerian/Babylonian deity Nanna-Sin (a rather formidable fellow who had a beard of lapis-lazuli and rode a winged bull)?

This question of which is the "truer" story is not an idle one. Over the centuries, people have discovered (by being proved wrong) that it is very difficult to have a commonsense understanding of what the whole Universe and its most important parts are like. Common sense just isn't up to the task. For that reason, as Morgan le Fey tells us in *The Mists of Avalon,* created stories about the Universe themselves actually *create* the Universe the listener lives in. The real Universe (like most scientists, you and I behave as if there is one) is not silent, but whispers very softly to the storytellers. Many whispers go unheard, so that real Universe is probably very different from the one you read about today in any book that claims to tell its story. People, nevertheless, must act. Most recognize that the bases for their actions are fallible stories, and they must therefore select the most trustworthy stories that they can find.

Most of you won't have to be convinced that it is better to talk about colliding planets than about Nanna-Sin if your aim is to understand the Moon or perhaps plan a visit. Still, it is useful to ask the question: What is it, if anything, that makes astronomical stories a more reliable basis for action, and in that sense more truthful or factual than any others? Only one thing, I think: *discipline.* Astronomers feel an obligation to tell their story with great care, following a rather strict, scientific, discipline.

Scientists, philosophers, and sociologists have written about what it is that makes science different from other human endeavors. There is much discussion and disagreement about the necessity of making scientific stories "broad and deep and simple", about the centrality of paradigms, the importance of predictions, the strength or relevance of motivations, and the inevitability of conformity to social norms and professional hierarchies.

But most of this literature agrees on the perhaps obvious point that a scientist, in creating a story (scientists usually call them "theories") about, say, the Moon, must pay a great deal of attention to all the relevant evidence. A scientist, unlike a science-fiction writer, may only fashion a theory that cannot be shown to violate that evidence.

This is a book about how to identify and collect relevant evidence in astronomy.

1.2 The evidence: astronomical data

> [Holmes said] I have no data yet. It is a capital mistake to theorize before one has data. Insensibly one begins to twist facts to suit theories, instead of theories to suit facts
>
> — Arthur Conan Doyle, *The Adventures of Sherlock Holmes*, 1892

> Facts are not pure and unsullied bits of information; culture also influences what we see and how we see it. Theories moreover are not inexorable inductions from facts. The most creative theories are often imaginative visions imposed upon facts;...
>
> — Stephen Jay Gould, *The Mismeasure of Man*, 1981

A few fortunate astronomers investigate cosmic rays or the Solar System. All other astronomers must construct stories about objects with which they can have no direct contact, things like stars and galaxies that can't be manipulated, isolated, or made the subject of experiment. This sets astronomers apart from most other scientists, who can thump on, cut up, and pour chemicals over their objects of study. In this sense, astronomy is a lot more like paleontology than it is like physics. Trying to tell the story of a galaxy is like trying to reconstruct a dinosaur from bits of fossilized bone. We will never have the galaxy or dinosaur in our laboratory, and must do guesswork based on flimsy, secondhand evidence. To study any astronomical object we depend on intermediaries, entities that travel from the objects to us. There are two categories of intermediaries – particles with mass, and those without. First briefly consider the massive particles, since detailed discussion of them is beyond the scope of this book.

1.2.1 Particles with mass

Cosmic rays are microscopic particles that arrive at Earth with extraordinarily high energies. *Primary cosmic rays* are mostly high-speed atomic nuclei, mainly hydrogen (84%) and helium (14%). The remainder consists of heavier nuclei, electrons, and positrons. Some primary cosmic rays are produced in solar flares, but many, including those of highest energies, come from outside the Solar System. About 6000 cosmic rays strike each square meter of the Earth's upper atmosphere every second. Since all these particles move at a large fraction of the speed of light, they carry a great deal of kinetic energy. A convenient unit for measuring particle energies is the electron volt (eV):

$$1 \text{ eV} = 1.602 \times 10^{-19} \text{ joules}$$

Primary cosmic rays have energies ranging from 10^6 to 10^{20} eV, with relative abundance declining with increasing energy. The mean energy is around 10 GeV $= 10^{10}$ eV. At relativistic velocities, the relation between speed, v, and total energy, E, is

$$E = \frac{mc^2}{\sqrt{1 - v^2/c^2}}$$

Here m is the rest mass of the particle and c is the speed of light. For reference, the rest mass of the proton (actually, the product mc^2) is 0.93 GeV. The highest-energy cosmic rays have energies far greater than any attainable in laboratory particle accelerators. Although supernova explosions are suspected to be the source of some or all of the higher-energy primary cosmic rays, the exact mechanism for their production remains mysterious.

Secondary cosmic rays are particles produced by collisions between the primaries and particles in the upper atmosphere – generally more that 50 km above the surface. Total energy is conserved in the collision, so the kinetic energy of the primary can be converted into the rest-mass of new particles, and studies of the secondaries gives some information about the primaries. Typically, a cosmic-ray collision produces many fragments, including pieces of the target nucleus, individual nucleons, and electrons, as well as particles not present before the collision: positrons, gamma rays, and a variety of more unusual short-lived particles like kaons. In fact, cosmic-ray experiments were the first to detect pions, muons, and positrons.

Detection of both primary and secondary cosmic rays relies on methods developed for laboratory particle physics. Detectors include cloud and spark chambers, Geiger and scintillation counters, flash tubes, and various solid-state devices. Detection of primaries requires placement of a detector above the bulk of the Earth's atmosphere, and only secondary cosmic rays can be studied directly from the Earth's surface. Since a shower of secondary particles generally spreads over an area of many square kilometers by the time it reaches sea level, cosmic-ray studies often utilize arrays of detectors. Typical arrays consist of many tens or hundreds of individual detectors linked to a central coordinating computer. Even very dense arrays, however, can only sample a small fraction of the total number of secondaries in a shower.

Neutrinos are particles produced in nuclear reactions involving the weak nuclear force. They are believed to have tiny rest masses (the best measurements to date are uncertain but suggest something like 0.05 eV). They may very well be the most numerous particles in the Universe. Many theories predict intense production of neutrinos in the early stages of the Universe, and the nuclear reactions believed to power all stars produce a significant amount of energy in the form of neutrinos. In addition, on Earth, a flux of high-energy "atmospheric" neutrinos is generated in cosmic-ray secondary showers.

Since neutrinos interact with ordinary matter only through the weak force, they can penetrate great distances through dense material. The Earth and the Sun, for example, are essentially transparent to them. Neutrinos can nonetheless be detected: the trick is to build a detector so massive that a significant number of neutrino reactions will occur within it. Further, the detector must also be shielded from secondary cosmic rays, which can masquerade as neutrinos. About a half-dozen such "neutrino telescopes" have been built underground.

For example, the Super-Kamiokande instrument is a 50 000-ton tank of water located 1 km underground in a zinc mine 125 miles west of Tokyo. The water acts as both the target for neutrinos and as the detecting medium for the products of the neutrino reactions. Reaction products emit light observed and analyzed by photodetectors on the walls of the tank.

Neutrinos have been detected unambiguously from only two astronomical objects: the Sun and a nearby supernova in the Large Magellanic Cloud, SN 1987A. These are promising results. Observations of solar neutrinos, for example, provide an opportunity to test the details of theories of stellar structure and energy production.

Meteorites are macroscopic samples of solid material derived primarily from our Solar System's asteroid belt, although there are a few objects that originate from the surfaces of the Moon and Mars. Since they survive passage through the Earth's atmosphere and collision with its surface, meteorites can be subjected to physical and chemical laboratory analysis. Some meteorites have remained virtually unchanged since the time of the formation of the Solar System, while others have endured various degrees of processing. All, however, provide precious clues about the origin, age, and history of the Solar System. For example, the age of the Solar System (4.56 Gyr) is computed from radioisotopic abundances in meteorites, and the inferred original high abundance of radioactive aluminum-26 in the oldest mineral inclusions in some meteorites suggests an association between a supernova, which would produce the isotope, and the events immediately preceding the formation of our planetary system.

Exploration of the Solar System by human spacecraft began with landings on the Moon in the 1960s and 1970s. Probes have returned samples – Apollo and Luna spacecraft brought back several hundred kilograms of rock from the Moon. Humans and their mechanical surrogates have examined remote surfaces *in situ*. The many landers on Mars, the Venera craft on Venus, and the Huygens lander on Titan, for example, made intrusive measurements and conducted controlled experiments.

1.2.2 Massless particles

Gravitons, theorized particles corresponding to gravity waves, have only been detected indirectly through the behavior of binary neutron stars. Graviton detectors designed to sense the local distortion of space-time caused by a passing gravity wave have been constructed, but have not yet detected waves from an astronomical source.

Photons are particles of light that can interact with all astronomical objects.[1] Light, in the form of visible rays as well as invisible rays like radio and X-rays, has historically constituted the most important channel of astronomical information. This book is about using that channel to investigate the Universe.

1.3 Models for the behavior of light

Some (not astronomers!) regard astronomy as applied physics. There is some justification for this, since astronomers, to help tell some astronomical story, persistently drag out theories proposed by physicists. Physics and astronomy differ partly because astronomers are interested in telling the story of an object, whereas physicists are interested in uncovering the most fundamental rules of the material world. Astronomers tend to find physics useful but sterile; physicists tend to find astronomy messy and mired in detail. We now invoke physics, to ponder the question: how does light behave? More specifically, what properties of light are important in making meaningful astronomical observations and predictions?

1.3.1 Electromagnetic waves

> ... we may be allowed to infer, that homogeneous light, at certain equal
> distances in the direction of its motion, is possessed of opposite qualities,
> capable of neutralizing or destroying each other, and extinguishing the light,
> where they happen to be united; ...
>
> – Thomas Young, *Philosophical Transactions, The Bakerian Lecture*, 1804

---

[1] Maybe not. There is strong evidence for the existence of very large quantities of "dark" matter in the Universe. This matter seems to exert gravitational force, but is the source of no detectable light. It is unclear whether the dark matter is normal stuff that is well hidden, or unusual stuff that can't give off or absorb light. Even more striking is the evidence for the presence of "dark energy" – a pressure-like effect in space itself which contains energy whose mass equivalent is even greater than that of visible and dark matter combined.

Electromagnetic waves are a model for the behavior of light which we know to be incorrect (*incomplete* is perhaps a better term). Nevertheless, the wave theory of light describes much of its behavior with precision, and introduces a lot of vocabulary that you should master. Christian Huygens,[2] in his 1678 book, *Traité de la Lumière*, summarized his earlier findings that visible light is best regarded as a *wave* phenomena, and made the first serious arguments for this point of view. Isaac Newton, his younger contemporary, opposed Huygens' wave hypothesis and argued that light was composed of tiny solid particles.

A **wave** is a disturbance that propagates through space. If some property of the environment (say, the level of the water in your bathtub) is disturbed at one place (perhaps by a splash), a wave is present if that disturbance moves continuously from place to place in the environment (ripples from one end of your bathtub to the other, for example). Material particles, like bullets or ping-pong balls, also propagate from place to place. Waves and particles share many characteristic behaviors – both can **reflect**, (change directions at an interface) **refract** (change speed in response to a change in the transmitting medium), and can carry energy from place to place. However, waves exhibit two characteristic behaviors not shared by particles:

Diffraction – the ability to bend around obstacles. A water wave entering a narrow opening, for example, will travel not only in the "shadow" of the opening but will spread in all directions on the far side.

Interference – an ability to combine with other waves in predictable ways. Two water waves can, for example, destructively interfere if they combine so that the troughs of one always coincide with the peaks of the other.

Although Huygens knew that light exhibited the properties of diffraction and interference, he unfortunately did not discuss them in his book. Newton's reputation was such that his view prevailed until the early part of the nineteenth century, when Thomas Young and Augustin Fresnel were able to show how Huygen's wave idea could explain diffraction and interference. Soon the evidence for waves proved irresistible.

Well-behaved waves exhibit certain measurable qualities – amplitude, wavelength, frequency, and wave speed – and physicists in the generation following Fresnel were able to measure these quantities for visible light waves. Since light was a wave, and since waves are disturbances that propagate, it was

---

[2] Huygens (1629–1695), a Dutch natural philosopher and major figure in seventeenth-century science, had an early interest in lens grinding. He discovered the rings of Saturn and its large satellite, Titan, in 1655–1656, with a refracting telescope of his manufacture. At about the same time, he invented the pendulum clock, and formulated a theory of elastic bodies. He developed his wave theory of light later in his career, after he moved from The Hague to the more cosmopolitan environment of Paris. Near the end of his life, he wrote a treatise on the possibility of extraterrestrial life.

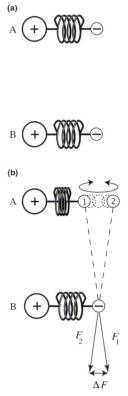

Fig. 1.1 Acceleration of an electron produces a wave. (a) Undisturbed atoms in a source (A) and a receiver (B). Each atom consists of an electron attached to a nucleus by some force, which we represent as a spring. In (b) of the figure, the source electron has been disturbed, and oscillates between positions (1) and (2). The electron at B experiences a force that changes from F_1 to F_2 in the course of A's oscillation. The difference, ΔF, is the amplitude of the changing part of the electric force seen by B.

natural to ask: "What 'stuff' does a light wave disturb?" In one of the major triumphs of nineteenth century physics, James Clerk Maxwell proposed an answer in 1873.

Maxwell (1831–1879), a Scot, is a major figure in the history of physics, comparable to Newton and Einstein. His doctoral thesis demonstrated that the rings of Saturn (discovered by Huygens) must be made of many small solid particles in order to be gravitationally stable. He conceived the kinetic theory of gases in 1866 (Ludwig Boltzmann did similar work independently), and transformed thermodynamics into a science based on statistics rather than determinism. His most important achievement was the mathematical formulation of the laws of electricity and magnetism in the form of four partial differential equations. Published in 1873, **Maxwell's equations** completely accounted for separate electric and magnetic phenomena and also demonstrated the connection between the two forces. Maxwell's work is the culmination of classical physics, and its limits led both to the theory of relativity and the theory of quantum mechanics.

Maxwell proposed that light disturbs **electric and magnetic fields**. The following example illustrates his idea.

Consider a single electron, electron A. It is attached to the rest of the atom by means of a spring, and is sitting still. (The spring is just a mechanical model for the electrostatic attraction that holds the electron to the nucleus.) This pair of charges, the negative electron and the positive ion, is a dipole. A second electron, electron B, is also attached to the rest of the atom by a spring, but this second dipole is at some distance from A. Electron A repels B, and B's position in its atom is in part determined by the location of A. The two atoms are sketched in Figure 1.1a. Now to make a wave: set electron A vibrating on its spring. Electron B must respond to this vibration, since the force it feels is changing direction. It moves in a way that will echo the motion of A. Figure 1.1b shows the changing electric force on B as A moves through a cycle of its vibration.

The disturbance of dipole A has propagated to B in a way that suggests a wave is operating. Electron B behaves like an object floating in your bathtub that moves in response to the rising and falling level of a water wave.

In trying to imagine the actual thing that a vibrating dipole disturbs, you might envision the water in a bathtub, and imagine an entity that fills space continuously around the electrons, the way a fluid would, so a disturbance caused by moving one electron can propagate from place to place. The physicist Michael Faraday[3]

---

[3] Michael Faraday (1791–1867), considered by many the greatest experimentalist in history, began his career as a bookbinder with minimal formal education. His amateur interest in chemistry led to a position in the laboratory of the renowned chemist, Sir Humphrey Davy, at the Royal Institution in London. Faraday continued work as a chemist for most of his productive life, but conducted an impressive series of experiments in electromagnetism in the period 1834–1855. His ideas, although largely rejected by physicists on the continent, eventually formed the empirical basis for Maxwell's theory of electromagnetism.

supplied the very useful idea of a *field* – an abstract *entity* (not a material fluid at all) created by charged particles that permeates space and gives other charged particles instructions about what force they should experience. In this conception, electron B consults the local field in order to decide how to move. Shaking (accelerating) the electron at A distorts the field in its vicinity, and this distortion propagates to vast distances, just like the ripples from a rock dropped into a calm and infinite ocean.

The details of propagating a field disturbance turned out to be a little complicated. Hans Christian Oerstead and Andre Marie Ampère in 1820 had shown experimentally that a changing electric field, such as the one generated by an accelerated electron, produces a magnetic field. Acting on his intuition of an underlying unity in physical forces, Faraday performed experiments that confirmed his guess that a changing magnetic field must in turn generate an electric field. Maxwell had the genius to realize that his equations implied that the electric and magnetic field changes in a vibrating dipole would support one another, and produce a wave-like self-propagating disturbance. Change the electric field, and you thereby create a magnetic field, which then creates a different electric field, which creates a magnetic field, and so on, forever. Thus, it is proper to speak of the waves produced by an accelerated charged particle as *electromagnetic*. Figure 1.2 shows a schematic version of an electromagnetic wave. The changes in the two fields, electric and magnetic, vary at right angles to one another and the direction of propagation is at right angles to both.

Thus, a disturbance in the electric field does indeed seem to produce a wave. Is this *electromagnetic* wave the same thing as the *light* wave we see with our eyes?

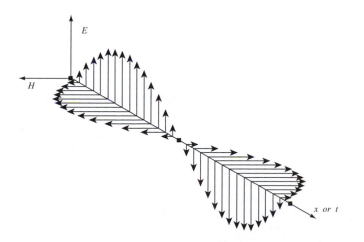

Fig. 1.2 A plane polarized electromagnetic wave. The electric and magnetic field strengths are drawn as vectors that vary in both space and time. The illustrated waves are said to be plane-polarized because all electric vectors are confined to the same plane.

From his four equations – the laws of electric and magnetic force – Maxwell derived the speed of any electromagnetic wave, which, in a vacuum, turned out to depend only on constants:

$$c = (\varepsilon\mu)^{-\frac{1}{2}}$$

Here ε and μ are well-known constants that describe the strengths of the electric and magnetic forces. (They are, respectively, the electric permittivity and magnetic permeability of the vacuum.) When he entered the experimental values for ε and μ in the above equation, Maxwell computed the electromagnetic wave speed, which turned out to be numerically identical to the speed of light, a quantity that had been experimentally measured with improving precision over the preceding century. This equality of the predicted speed of electromagnetic waves and the known speed of light really was quite a convincing argument that light waves and electromagnetic waves were the same thing. Maxwell had shown that three different entities, electricity, magnetism, and light, were really one.

Other predictions based on Maxwell's theory further strengthened this view of the nature of light. For one thing, one can note that for any well-behaved wave the speed of the wave is the product of its frequency and wavelength:

$$c = \lambda v$$

There is only one speed that electromagnetic waves can have in a vacuum; therefore there should be a one-dimensional classification of electromagnetic waves (the **electromagnetic spectrum**). In this spectrum, each wave is characterized only by its particular wavelength (or frequency, which is just c/λ). Table 1.1 gives the names for various portions or **bands** of the electromagnetic spectrum.

Maxwell's wave theory of light very accurately describes the way light behaves in many situations. In summary, the theory says:

Table 1.1. *The electromagnetic spectrum. Region boundaries are not well-defined, so there is some overlap. Subdivisions are based in part on distinct detection methods*

Band	Wavelength range	Frequency range	Subdivisions (long λ–short λ)
Radio	>1 mm	< 300 GHz	VLF-AM-VHF-UHF
Microwave	0.1 mm–3 cm	100 MHz–3000 GHz	Millimeter–submillimeter
Infrared	700 nm–1 mm	3×10^{11}–4×10^{14} Hz	Far–Middle–Near
Visible	300 nm–800 nm	4×10^{14}–1×10^{15} Hz	Red–Blue
Ultraviolet	10 nm–400 nm	7×10^{14}–3×10^{16} Hz	Near–Extreme
X-ray	0.001 nm–10 nm	3×10^{16}–3×10^{20} Hz	Soft–Hard
Gamma ray	<0.1 nm	$>3 \times 10^{18}$ Hz	Soft–Hard

1. Light exhibits all the properties of classical, well-behaved waves, namely:
   - reflection at interfaces
   - refraction upon changes in the medium
   - diffraction around edges
   - interference with other light waves
   - polarization in a particular direction (plane of vibration of the electric vector).
2. A light wave can have any wavelength, selected from a range from zero to infinity. The range of possible wavelengths constitutes the electromagnetic spectrum.
3. A light wave travels in a straight line at speed c in a vacuum. Travel in other media is slower and subject to refraction and absorption.
4. A light wave carries energy whose magnitude depends on the squares of the amplitudes of the electric and magnetic waves.

In 1873, Maxwell predicted the existence of electromagnetic waves outside the visible range and by1888, Heinrich Hertz had demonstrated the production of radio waves based on Maxwell's principles. Radio waves traveled at the speed of light and exhibited all the familiar wave properties like reflection and interference. This experimental confirmation convinced physicists that Maxwell had really discovered the secret of light. Humanity had made a tremendous leap in understanding reality. This leap to new heights, however, soon revealed that Maxwell had discovered only a part of the secret.

1.3.2 Quantum mechanics and light

It is very important to know that light behaves like particles, especially for those of you who have gone to school, where you were probably told something about light behaving like waves. I'm telling you the way it does behave — like particles.

— Richard Feynman, *QED*, 1985

Towards the end of the nineteenth century, physicists realized that electromagnetic theory could not account for certain behaviors of light. The theory that eventually replaced it, **quantum mechanics**, tells a different story about light. In the quantum story, light possesses the properties of a particle as well as the wave-like properties described by Maxwell's theory. Quantum mechanics insists that there are situations in which we cannot think of light as a wave, but must think of it as a collection of particles, like bullets shot out of the source at the speed of light. These particles are termed **photons**. Each photon "contains" a particular amount of energy, E, that depends on the frequency it possesses when it exhibits its wave-like properties:

$$E = h\nu = \frac{hc}{\lambda}$$

Here h is Planck's constant (6.626×10^{-34} J s) and ν is the frequency of the wave. Thus a single radio photon (low frequency) contains a small amount of energy, and a single gamma-ray photon (high frequency) contains a lot.

The quantum theory of light gives an elegant and successful picture of the interaction between light and matter. In this view, atoms no longer have electrons bound to nuclei by springs or (what is equivalent in classical physics) electric fields. Electrons in an atom have certain permitted energy states described by a wave function – in this theory, everything, including electrons, has a wave as well as a particle nature. The generation or absorption of light by atoms involves the electron changing from one of these permitted states to another. Energy is conserved because the energy lost when an atom makes the transition from a higher to a lower state is exactly matched by the energy of the photon emitted. In summary, the quantum mechanical theory says:

1. Light exhibits all the properties described in the wave theory in situations where wave properties are measured.
2. Light behaves, in other circumstances, as if it were composed of massless particles called photons, each containing an amount of energy equal to its frequency times Planck's constant.
3. The interaction between light and matter involves creation and destruction of individual photons and the corresponding changes of energy states of charged particles (usually electrons).

We will make great use of the quantum theory in later chapters, but for now our needs are more modest.

1.3.3 A geometric approximation: light rays

Since the quantum picture of light is as close as we can get to the real nature of light, you might think quantum mechanics would be the only theory worth considering. However, except in simple situations, application of the theory demands complex and lengthy computation. Fortunately, it is often possible to ignore much of what we know about light, and use a very rudimentary picture which pays attention to only those few properties of light necessary to understand much of the information brought to us by photons from out there. In this geometric approximation, we treat light as if it traveled in "rays" or streams that obey the laws of reflection and refraction as described by geometrical optics. It is sometimes helpful to imagine a collection of such rays as the paths taken by a stream of photons.

Exactly how we picture a stream of photons will vary from case to case. Sometimes it is essential to recognize the discrete nature of the particles. In this case, think of the light ray as the straight path that a photon follows. We might then think of astronomical measurements as acts of *counting* and classifying the individual photons as they hit our detector like sparse raindrops tapping on a tin roof.

On the other hand, there will be circumstances where it is profitable to ignore the lumpy nature of the photon stream – to assume it contains so many photons that the stream behaves like a smooth fluid. In this case, we think of astronomical measurements as recording smoothly varying quantities. In this case it is like measuring the volume of rain that falls on Spain in a year: we might be aware that the rain arrived as discrete drops, but it is safe to ignore the fact.

We will adopt this simplified ray picture for much of the discussion that follows, adjusting our awareness of the discreet nature of the photon stream or its wave properties as circumstances warrant. For the rest of this chapter, we use the ray picture to discuss two of the basic measurements important in astronomy:

> **photometry** measures the amount of energy arriving from a source;
> **spectrometry** measures the distribution of photons with wavelength.

Besides photometry and spectroscopy, the other general categories of measurement are ***imaging*** and ***astrometry***, which are concerned with the appearance and positions of objects in the sky; and ***polarimetry***, which is concerned with the polarization of light from the source. Incidentally, the word "wavelength" does not mean we are going to think deeply about the wave theory just yet. It will be sufficient to think of wavelength as a property of a light ray that can be measured – by where the photon winds up when sent through a spectrograph, for example.

1.4 Measurements of light rays

> Twinkle, twinkle, little star,
> Flux says just how bright you are.
>   – Anonymous, *c.* 1980

1.4.1 Luminosity and brightness

Astronomers have to construct the story of a distant object using only the tiny whisper of electromagnetic radiation it sends us. We define the (electromagnetic) luminosity, L, as the total amount of energy that leaves the surface of the source per unit time in the form of photons. Energy per unit time is called power, so we can measure L in physicists' units for power (SI units): joules per second or watts. Alternatively, it might be useful to compare the object with the Sun, and we then might measure the luminosity in solar units:

$$L = \textbf{Luminosity} = \text{Energy per unit time emitted by the entire source}$$

$$L_\odot = \text{Luminosity of the Sun} = 3.825 \times 10^{26} \text{ W}$$

The luminosity of a source is an important clue about its nature. One way to measure luminosity is to surround the source completely with a box or bag of perfect energy-absorbing material, then use an "energy gauge" to measure the

Fig. 1.3 Measuring
luminosity by
intercepting all the power
from a source.

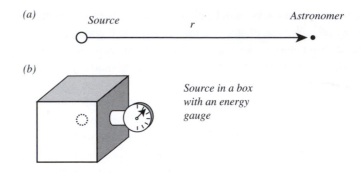

total amount of energy intercepted by this enclosure during some time interval.
Figure 1.3 illustrates the method. Luminosity is the amount of energy absorbed
divided by the time interval over which the energy accumulates. The astronomer, however, cannot measure luminosity in this way. She is too distant from
the source to put it in a box, even in the unlikely case she has a big enough box.

Fortunately, there is a quantity related to luminosity, called the **apparent
brightness** of the source, which is much easier to measure.

Measuring apparent brightness is a local operation. The astronomer holds up
a scrap of perfectly absorbing material of known area so that its surface is
perpendicular to the line of sight to the source. She measures how much energy
from the source accumulates in this material in a known time interval. Apparent
brightness, F, is defined as the total energy per unit time per unit area that arrives
from the source:

$$F = \frac{E}{tA}$$

This quantity F is usually known as the **flux** or the **flux density** in the
astronomical literature. In the physics literature, the same quantity is usually
called the **irradiance** (or, in studies restricted to visual light, the **illuminance**.)
To make matters not only complex but also confusing, what astronomers call
luminosity, L, physicists call the **radiant flux**.

Whatever one calls it, F will have units of power per unit area, or $W\,m^{-2}$. For
example, the average flux from the Sun at the top of the Earth's atmosphere (the
apparent brightness of the Sun) is about $1370\ W\,m^{-2}$, a quantity known as the
solar constant.

1.4.2 The inverse square law of brightness

Refer to Figure 1.4 to derive the relationship between the flux from a source and
the source's luminosity. We choose to determine the flux by measuring the

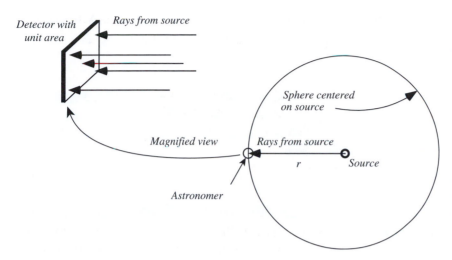

Fig. 1.4 Measuring the apparent brightness of a source that is at distance r. The astronomer locally detects the power reaching a unit area perpendicular to the direction of the source. If the source is isotropic and there is no intervening absorber, then its luminosity is equal to the apparent brightness multiplied by the area of a sphere of radius r.

power intercepted by the surface of a very large sphere of radius r centered on the source. The astronomer is on the surface of this sphere. Since this surface is everywhere perpendicular to the line of sight to the source, this is an acceptable way to measure brightness: simply divide the intercepted power by the total area of the sphere. But surrounding the source with a sphere is also like putting it into a very large box, as in Figure 1.3. The total power absorbed by the large sphere must be the luminosity, L, of the source. We assume that there is nothing located between the source and the spherical surface that absorbs light – no dark cloud or planet. The brightness, averaged over the whole sphere, then, is

$$\langle F \rangle = \frac{L}{4\pi r^2}$$

Now make the additional assumption that the radiation from the source is isotropic (the same in all directions). Then the *average* brightness is the same as the brightness measured locally, using a small surface:

$$F = \frac{L}{4\pi r^2} \tag{1.1}$$

Both assumptions, isotropy and the absence of absorption, can be violated in reality. Nevertheless, in its simple form, Equation (1.1) not only represents one of the fundamental relationships in astronomy, it also reveals one of the central problems in our science.

The problem is that the left-hand side of Equation (1.1) is a quantity that can be determined by direct observation, so potentially could be known to great accuracy. The expression on the right-hand side, however, contains two unknowns, luminosity and distance. Without further information these two cannot be disentangled. This is a frustration – you can't say, for example, how much power a quasar is producing without knowing its distance, and you can't know its distance unless you know how much power it is producing. A fundamental problem in astronomy is determining the third dimension.

1.4.3 Surface brightness

One observable quantity that does not depend on the distance of a source is its surface brightness on the sky. Consider the simple case of a uniform spherical source of radius a and luminosity L. On the surface of the source, the amount of power leaving a unit area is

$$s = \frac{L}{4\pi a^2}$$

Note that s, the power emitted per unit surface area of the source, has the same dimensions, $W\,m^{-2}$, as F, the apparent brightness seen by a distant observer. The two are very different quantities, however. The value of s is characteristic only of the object itself, whereas F changes with distance. Now, suppose that the object has a detectable angular size – that our telescope can resolve it, that is, distinguish it from a point source. The **solid angle**, in **steradians**, subtended by a spherical source of radius a and distance r is (for $a \ll r$)

$$\Omega \cong \frac{\pi a^2}{r^2}\,[\text{steradians}]$$

Now we write down σ, the apparent **surface brightness** of the source on the sky, that is, the flux that arrives per unit solid angle of the source. In our example, it is both easily measured as well as **independent of distance**:

$$\sigma = \frac{F}{\Omega} = \frac{s}{\pi}$$

A more careful analysis of non-spherical, non-uniform emitting surfaces supports the conclusion that, for objects that can be resolved, σ is invariant with distance. Ordinary optical telescopes can measure Ω with accuracy only if it has a value larger than a few square seconds of arc (about 10^{-10} steradians), mainly because of turbulence in the Earth's atmosphere. Space telescopes and ground-based systems with adaptive optics can resolve solid angles perhaps a hundred times smaller. Unfortunately, the majority of even the nearest stars have angular sizes too small (diameters of a few milli-arcsec) to resolve with present instruments, so that for them Ω (and therefore σ) cannot be measured directly. Astronomers do routinely measure σ for "extended objects" like planets, gaseous nebulae, and galaxies and find these values immensely useful.

1.5 Spectra

If a question on an astronomy exam starts with the phrase "how do we know…" then the answer is probably "spectrometry".

– Anonymous, *c.* 1950

Astronomers usually learn most about a source not from its flux, surface brightness, or luminosity, but from its spectrum: the way in which light is distributed with wavelength. Measuring its luminosity is like glancing at the cover of a book about the source – with luck, you might read the title. Measuring its spectrum is like opening the book and skimming a few chapters, chapters that might explain the source's chemical composition, pressure, density, temperature, rotation speed, or radial velocity. Although evidence in astronomy is usually meager, the most satisfying and eloquent evidence is spectroscopic.

1.5.1 Monochromatic flux

Consider measuring the flux from of a source in the usual fashion, with a set-up like the one in Figure 1.4. Arrange our perfect absorber to absorb only photons that have frequencies between v and $v + dv$, where dv is an infinitesimally small frequency interval. Write the result of this measurement as $F(v, v + dv)$. As with all infinitesimal quantities, you should keep in mind that dv and $F(v, v + dv)$ are the *limits* of finite quantities called Δv and $F(v, v + \Delta v)$. We then define ***monochromatic flux*** or ***monochromatic brightness*** as

$$f_v = \frac{F(v, v + dv)}{dv} = \lim_{\Delta v \to 0} \frac{F(v, v + \Delta v)}{\Delta v} \qquad (1.2)$$

The complete function, f_v, running over all frequencies (or even over a limited range of frequencies), is called the ***spectrum*** of the object. It has units $[\mathrm{W\,m^{-2}\,Hz^{-1}}]$. The extreme right-hand side of Equation (1.2) reminds us that f_v is the limiting value of the ratio as the quantity Δv [and correspondingly $F(v, v + \Delta v)$] become indefinitely small. For measurement, Δv must have a finite size, since $F(v, v + dv)$ must be large enough to register on a detector. If Δv is large, the detailed wiggles and jumps in the spectrum will be smoothed out, and one is said to have measured a ***low-resolution spectrum***. Likewise, a ***high-resolution spectrum*** will more faithfully show the details of the limiting function, f_v.

If we choose the wavelength as the important characteristic of light, we can define a different monochromatic brightness, f_λ. Symbolize the flux between wavelengths λ and $\lambda + d\lambda$ as $F(\lambda, \lambda + d\lambda)$ and write

$$f_\lambda = \frac{F(\lambda, \lambda + d\lambda)}{d\lambda}$$

Although the functions f_v and f_λ are each called the spectrum, they differ from one another in numerical value and overall appearance for the same object. Figure 1.5 shows schematic low-resolution spectra of the bright star, Vega, plotted over the same range of wavelengths, first as f_λ, then as f_v.

Light

Fig. 1.5 Two forms of the ultraviolet and visible outside-the-atmosphere spectrum of Vega. The two curves convey the same information, but have very different shapes. Units on the vertical axes are arbitrary.

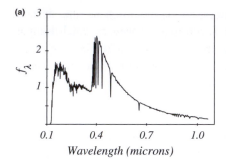

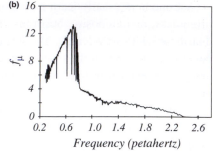

1.5.2 Flux within a band

Less is more.

– Robert Browning, *Andrea del Sarto* (1855), often quoted by L. Mies van der Rohe

An **ideal bolometer** is a detector that responds to all wavelengths with perfect efficiency. In a unit time, a bolometer would record every photon reaching it from a source, regardless of wavelength. We could symbolize the **bolometric flux** thereby recorded as the integral:

$$F_{\text{bol}} = \int_0^\infty f_\lambda d\lambda$$

Real bolometers operate by monitoring the temperature of a highly absorbing (i.e. black) object of low thermal mass. They are imperfect in part because it is difficult to design an object that is "black" at all wavelengths. More commonly, practical instruments for measuring brightness can only detect light within a limited range of wavelengths or frequencies. Suppose a detector registers light between wavelengths λ_1 and λ_2, and nothing outside this range. In the notation of the previous section, we might then write the flux in the 1, 2 pass-band as:

$$F(\lambda_1, \lambda_2) = \int_{\lambda_1}^{\lambda_2} f_\lambda \, d\lambda = F(v_2, v_1) = \int_{v_2}^{v_1} f_v \, dv$$

Usually, the situation is even more complex. In addition to having a sensitivity to light that "cuts on" at one wavelength, and "cuts off" at another, practical detectors vary in detecting efficiency over this range. If $R_A(\lambda)$ is the fraction of the incident flux of wavelength λ that is eventually detected by instrument A, then the flux actually recorded by such a system might be represented as

$$F_A = \int_0^\infty R_A(\lambda) f_\lambda \, d\lambda \tag{1.3}$$

The function $R_A(\lambda)$ may be imposed in part by the environment rather than by the instrument. The Earth's atmosphere, for example, is (imperfectly) transparent only in the visible and near-infrared pass-band between about 0.32 and 1 micron (extending in restricted bands to 25 μm at very dry, high-altitude sites), and in the microwave–radio pass-band between about 0.5 millimeters and 50 meters.

Astronomers often intentionally restrict the range of a detector's sensitivity, and employ filters to control the form of the efficiency function, R_A, particularly to define cut-on and cut-off wavelengths. Standard filters are useful for several reasons. First, a well-defined band makes it easier for different astronomers to compare measurements. Second, a filter can block troublesome wavelengths, ones where the background is bright, perhaps, or where atmospheric transmission is low. Finally, comparison of two or more different band-pass fluxes for the same source is akin to measuring a very low-resolution spectrum, and thus can provide some of the information, like temperature or chemical composition, that is conveyed by its spectrum.

Hundreds of bands have found use in astronomy. Table 1.2 lists the broad band filters (i.e. filters where the bandwidth, $\Delta\lambda = \lambda_2 - \lambda_1$, is large) that are most commonly encountered in the visible–near-infrared window. Standardization of bands is less common in radio and high-energy observations.

Table 1.2. *Common broad band-passes in the visible (UBVRI), near-infrared (JHKLM), and mid-infrared (NQ). Chapter 10 discusses standard bands in greater detail*

Name	λ_c (μm)	Width (μm)	Rationale
U	0.365	0.068	Ultraviolet
B	0.44	0.098	Blue
V	0.55	0.089	Visual
R	0.70	0.22	Red
I	0.90	0.24	Infrared
J	1.25	0.38	
H	1.63	0.31	
K	2.2	0.48	
L	3.4	0.70	
M	5.0	1.123	
N	10.2	4.31	
Q	21.0	8	

1.5.3 Spectrum analysis

> [With regard to stars] . . . we would never know how to study by any means their
> chemical composition . . . In a word, our positive knowledge with respect to stars
> is necessarily limited solely to geometrical and mechanical phenomena . . .
>
> — Auguste Comte, *Cours de Philosophie Positive* II, 19th Lesson, 1835

> . . . I made some observations which disclose an unexpected explanation of the
> origin of Fraunhofer's lines, and authorize conclusions therefrom respecting
> the material constitution of the atmosphere of the sun, and perhaps also of that of
> the brighter fixed stars.
>
> — Gustav R. Kirchhoff, *Letter to the Academy of Science at Berlin*, 1859

Astronomers are fond of juxtaposing Comte's pronouncement about the impossibility of knowing the chemistry of stars with Kirchhoff's breakthrough a generation later. Me too. Comte deserves better, since he wrote quite thoughtfully about the philosophy of science, and would certainly have been among the first to applaud the powerful new techniques of spectrum analysis developed later in the century. Nevertheless, the failure of his dictum about what is knowable is a caution against pomposity for all.

Had science been quicker to investigate spectra, Comte might have been spared posthumous deflation. In 1666, Newton observed the dispersion of visible "white" sunlight into its component colors by glass prisms, but subsequent applications of spectroscopy were very slow to develop. It was not until the start of the nineteenth century that William Herschel and Johann Wilhelm Ritter used spectrometers to demonstrate the presence of invisible electromagnetic waves beyond the red and violet edges of the visible spectrum. The English physicist William Wollaston, in 1802, noted the presence of dark *lines* in the visible solar spectrum.

The Fraunhofer spectrum. Shortly thereafter (*c.* 1812), Joseph von Fraunhofer (1787–1826), using a much superior spectroscope, unaware of Wollaston's work, produced an extensive map of the solar absorption lines.

Of humble birth, Fraunhofer began his career as an apprentice at a glass-making factory located in an abandoned monastery in Benediktbeuern, outside Munich. By talent and fate (he survived a serious industrial accident) he advanced quickly in the firm. The business became quite successful and famous because of a secret process for making large blanks of high-quality crown and flint glass, which had important military and civil uses. Observing the solar spectrum with the ultimate goal of improving optical instruments, Fraunhofer pursued what he believed to be his discovery of solar absorption lines with characteristic enthusiasm and thoroughness. By 1814, he had given precise positions for 350 lines and approximate positions for another 225 fainter lines (see Figure 1.6). By 1823, Fraunhofer was reporting on the spectra of bright stars and planets, although his main attention was devoted

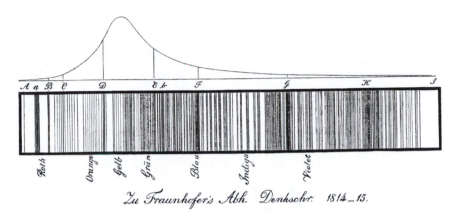

Fig. 1.6 A much-reduced reproduction of one of Fraunhofer's drawings of the solar spectrum. Frequency increases to the right, and the stronger absorption lines are labeled with his designations. Modern designations differ slightly.

to producing high-quality optical instruments. Fraunhofer died from tuberculosis at the age of 39. One can only speculate on the development of astrophysics had Fraunhofer been able to remain active for another quarter century.

Fraunhofer's lines are narrow wavelength bands where the value of function f_ν drops almost discontinuously, then rises back to the previous "continuum" (see Figures 1.6 and 1.7). The term "line" arises because in visual spectroscopy one actually examines the image of a narrow slit at each frequency. If the intensity is unusually low at a particular frequency, then the image of the slit there looks like a dark line. Fraunhofer designated the ten most prominent of these dark lines in the solar spectrum with letters (Figure 1.6, and Appendix B2). He noted that the two dark lines he labeled with the letter D occurred at wavelengths identical to the two bright emission lines produced by a candle flame. (Emission lines are narrow wavelength regions where the value of function f_λ increases almost discontinuously, then drops back down to the continuum; see Figure 1.7.) Soon several observers noted that the bright D lines, which occur in the yellow part of the spectrum at wavelengths of 589.0 and 589.6 nanometers, always arise from the presence of sodium in a flame. At about this same time, still others noted that heated solids, unlike the gases in flames, produce **continuous spectra** (no bright or dark lines – again, see Figure 1.7). Several researchers (John Herschel, William Henry Fox Talbot, David Brewster) in the 1820s and 1830s suggested that there was a connection between the composition of an object and its spectrum, but none could describe it precisely.

The Kirchhoff–Bunsen results. Spectroscopy languished for the thirty years following Fraunhofer's death in 1826. Then, in Heidelberg in 1859, physicist Gustav Kirchhoff and chemist Robert Bunsen performed a crucial experiment. They passed a beam of sunlight through a sodium flame, perhaps expecting the dark D lines to be filled in by the bright lines from the flame in the resulting spectrum. What they observed instead was that the dark D lines became darker

Fig. 1.7 Three types of spectra and the situations that produce them. (a) A solid, liquid, or dense gas produces a continuous spectrum. (b) An emission-line spectrum is produced by rarefied gas like a flame or a spark. (c) An absorption-line spectrum is produced when a source with a continuous spectrum is viewed through a rarefied gas.

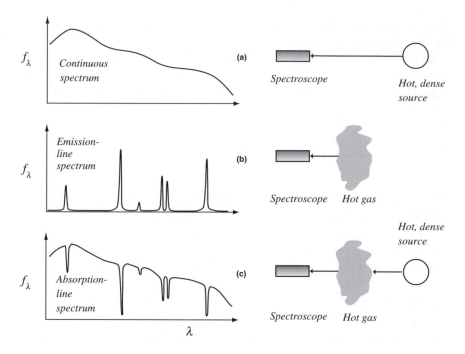

still. Kirchhoff reasoned that the hot gas in the flame had both absorbing and emitting properties at the D wavelengths, but that the absorption became more apparent as more light to be absorbed was supplied, whereas the emitting properties remained constant. This suggested that absorption lines would always be seen in situations like that sketched on the right of Figure 1.7c, so long as a sufficiently bright source was observed through a gas. If the background source were too weak or altogether absent, then the situation sketched in Figure 1.7b would hold and emission lines would appear.

Kirchhoff, moreover, proposed an explanation of all the Fraunhofer lines: The Sun consists of a bright source – a very dense gas, as it turns out – that emits a continuous spectrum. A low-density, gaseous atmosphere surrounds this dense region. As the light from the continuous source passes through the atmosphere, the atmosphere absorbs those wavelengths characteristic of its chemical composition. One could then conclude, for example, that the Fraunhofer D lines demonstrated the presence of sodium in the solar atmosphere. Identification of other chemicals in the solar atmosphere became a matter of obtaining an emission-line "fingerprint" from a laboratory flame or spark spectrum, then searching for the corresponding absorption line or lines at identical wavelengths in the Fraunhofer spectrum. Kirchhoff and Bunsen quickly confirmed the presence of potassium, iron, and calcium; and the absence (or very low abundance) of lithium in the solar atmosphere. The Kirchhoff–Bunsen results were not limited to the Sun. The spectra of almost all stars turned out to be absorption spectra, and it was easy to identify many of the lines present.

Quantitative chemical analysis of solar and stellar atmospheres became possible in the 1940s, after the development of astrophysics in the early twentieth century. At that time astronomers showed that most stars were composed of hydrogen and helium in a roughly 12 to 1 ratio by number, with small additions of other elements. However, the early qualitative results of Kirchhoff and Bunsen had already demonstrated to the world that stars were made of ordinary matter, and that one could hope to learn their exact composition by spectrometry. By the 1860s they had replaced the "truth" that stars were inherently unknowable with the new "truth" that stars were made of ordinary stuff.

Blackbody spectra. In 1860, Kirchhoff discussed the ratio of absorption to emission in hot objects by first considering the behavior of a perfect absorber, an object that would absorb all light falling on its surface. He called such an object a *blackbody*, since it by definition would reflect nothing. Blackbodies, however, must emit (otherwise their temperatures would always increase from absorbing ambient radiation). One can construct a simple blackbody by drilling a small hole into a uniform oven. The hole is the blackbody. The black walls of the oven will always absorb light entering the hole, so that the hole is a perfect absorber. The spectrum of the light *emitted* by the hole will depend on the temperature of the oven (and, it turns out, on nothing else). The blackbody spectrum is usually a good approximation to the spectrum emitted by any solid, liquid, or dense gas. (Low-density gases produce line spectra.)

In 1878, Josef Stefan found experimentally that the surface brightness of a blackbody (total power emitted per unit area) depends only on the fourth power of its temperature, and, in 1884, Ludwig Boltzmann supplied a theoretical understanding of this relation. **The *Stefan–Boltzmann law*** is

$$s = \sigma T^4$$

where σ = the Stefan–Boltzmann constant = 5.6696×10^{-8} W m^{-2} K^{-4}.

Laboratory studies of blackbodies at about this time showed that although their spectra change with temperature, all have a similar shape: a smooth curve with one maximum (see Figure 1.8). This peak in the monochromatic flux curve (either f_λ or f_ν) shifts to shorter wavelengths with increasing temperature, following a relation called *Wein's displacement law* (1893). Wein's law states that for f_λ

$$T\lambda_{\text{MAX}} = 2.8979 \times 10^{-3} \text{ m} \cdot \text{K}$$

or equivalently for f_ν

$$\frac{T}{\nu_{\text{MAX}}} = 1.7344 \times 10^{-11} \text{ Hz}^{-1} \text{ K}$$

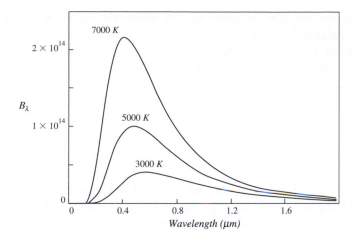

Thus blackbodies are never black: the color of a glowing objects shifts from red to yellow to blue as it is heated.

Max Planck presented the actual functional form for the blackbody spectrum at a Physical Society meeting in Berlin in 1900. His subsequent attempt to supply a theoretical understanding of the empirical "***Planck function***" led him to introduce the quantum hypothesis – that energy can only radiate in discrete packets. Later work by Einstein and Bohr eventually showed the significance of this hypothesis as a fundamental principle of quantum mechanics.

The Planck function gives the specific intensity, that is, the monochromatic flux per unit solid angle, usually symbolized as $B(v, T)$ or $B_v(T)$. The total power emitted by a unit surface blackbody over all angles is just $s(v,T) = \pi B(v,T)$. The Planck function is

$$B(v, T) = \frac{2hv^3}{c^2} \frac{1}{\left(\exp\left(\frac{hv}{kT}\right) - 1\right)}$$

$$B(\lambda, T) = \frac{2hc^2}{\lambda^5} \frac{1}{\left(\exp\left(\frac{hc}{\lambda kT}\right) - 1\right)}$$

For astronomers, the Planck function is especially significant because it shows exactly how the underlying continuous spectrum of a dense object depends on temperature and wavelength. Since the shape of the spectrum is observable, this means that one can deduce the temperature of any object with a "Planck-like" spectrum. Figure 1.8 shows the Planck function for several temperatures. Note that even if the position of the peak of the spectrum cannot be observed, the slope of the spectrum gives a measure of the temperature.

At long wavelengths, it is useful to note the ***Rayleigh–Jeans approximation*** to the tail of the Planck function:

$$B(\lambda, T) = \frac{2ckT}{\lambda^4}$$

$$B(v, T) = 2kT\frac{v^2}{c^2}$$

1.5.4 Spectra of stars

When astronomers first examined the absorption line spectra of large numbers of stars other than the Sun, they did not fully understand what they saw. Flooded with puzzling but presumably significant observations, most scientists have the (good) impulse to look for similarities and patterns: to sort the large number of observations into a small number of classes. Astronomers did their initial sorting of spectra into classes on the basis of the overall simplicity of the pattern of lines, assigning the simplest to class A, next simplest to B and so on through the alphabet. Only after a great number of stars had been so classified from photographs[4] (in the production of the Henry Draper Catalog – see Chapter 4) did astronomers come to understand, through the new science of astrophysics, that a great variety in stellar spectra arises mainly from temperature differences.

There is an important secondary effect due to surface gravity, as well as some subtle effects due to variations in chemical abundance. The chemical differences usually involve only the minor constituents – the elements other than hydrogen and helium.

The spectral type of a star, then, is basically an indication of its ***effective temperature*** – that is, the temperature of the blackbody that would produce the same amount of radiation per unit surface area as the star does. If the spectral type is sufficiently precise, it might also indicate the surface gravity or relative diameter or luminosity (if two stars have the same temperature and mass, the one with the larger diameter has the lower surface gravity as well as the higher luminosity). For reference, Table 1.3 lists the modern spectral classes and corresponding effective temperatures. The ***spectral type*** of a star consists of three designations: (a) a letter, indicating the general temperature class, (b) a decimal subclass number between 0 and 9 refining the temperature estimate (0 indicating the hottest and 9.9 the coolest subclass), and (c) a Roman numeral indicating the relative surface gravity or luminosity class. Luminosity class I (the ***supergiants***) is most luminous, III (the ***giants***) is intermediate, and V (the ***dwarves***) is least

---

[4] Antonia Maury suggested in 1897 that the correct sequence of types should be O through M (the first seven in Table 1.3 – although Maury used a different notation scheme). Annie Cannon, in 1901, justified this order on the basis of continuity, not temperature. Cannon's system solidified the modern notation and was quickly adopted by the astronomical community. In 1921, Megh Nad Saha used atomic theory to explain the Cannon sequence as one of stellar temperature.

Table 1.3. *Modern spectral classes in order of decreasing temperature. The L and T classes are recent additions. Some objects of class L and all of class T are not true stars but brown dwarves. Temperatures marked with a colon are uncertain*

Type	Temperature range, K	Main characteristic of absorption line spectra
O	$> 30\,000$	Ionized He lines
B	30 000–9800	Neutral He lines, strengthening neutral H
A	9800–7200	Strong neutral H, weak ionized metals
F	7200–6000	H weaker, ionized Ca strong, strong ionized and neutral metals
G	6000–5200	Ionized Ca strong, very strong neutral and ionized metals
K	5200–3900	Very strong neutral metals, CH and CN bands
M	3900–2100:	Strong TiO bands, some neutral Ca
L	2100:–1500:	Strong metal hydride molecules, neutral Na, K, Cs
T	< 1500:	Methane bands, neutral K, weak water

luminous. Dwarves are by far the most common luminosity class. The Sun, for example, has spectral type G2 V.

There are some rarely observed spectral types that are not listed in Table 1.3. A few exhibit unusual chemical abundances, like *carbon stars* (spectral types R and N) and *Wolf–Rayet stars* (type W) and stars with strong zirconium oxide bands (type S). White dwarves (spectral types DA, DB, and several others) are small, dense objects of low luminosity that have completely exhausted their supply of nuclear fuel and represent endpoints of stellar evolution. *Brown dwarves*, which have spectral types L and T, are stars that do not have sufficient mass to initiate full-scale thermonuclear reactions. We observe relatively few white dwarves and stars in classes L and T because their low luminosities make them hard to detect. All three groups are probably quite abundant in the galaxy. Stars of types W, R, N, and S, in contrast, are luminous and intrinsically rare.

1.6 Magnitudes

1.6.1 Apparent magnitudes

When Hipparchus of Rhodes (*c.* 190–120 BC), arguably the greatest astronomer in the Hellenistic school, published his catalog of 600 stars, he included an estimate of the brightness of each – our quantity *F*. Strictly, what Hipparchus and all visual observers estimate is F_{vis}, the flux in the visual band-pass, the band corresponding to the response of the human eye. The eye has two different response functions, corresponding to two different types of receptor cells – rods and cones. At high levels of illumination, only the cones operate (*photopic vision*), and the eye is relatively sensitive to red light. At low light levels (*scotopic vision*) only the rods operate, and sensitivity shifts to the blue. Greatest sensitivity is at about 555 nm (yellow) for cones and 505 nm (green) for rods. Except for extreme red wavelengths, scotopic vision is more sensitive than photopic and most closely corresponds to the Hipparchus system. See Appendix B3.

Hipparchus cataloged brightness by assigning each star to one of six classes, the first class (or first *magnitude*) being the brightest, the sixth class the faintest. The choice of six classes, rather than some other number – ten, for example – is a curious one, and may be tied to earlier Babylonian mysticism, which held six to be a significant number. For the next two millennia, astronomers perpetuated this system, eventually extending it to fainter stars at higher magnitudes: magnitudes 7, 8, 9, etc. could only be seen with a telescope. With the introduction of photometers in the nineteenth century, William Pogson (*c*. AD 1856) discovered that Hipparchus's classes were in fact approximately a geometric progression in F, with each class two or three times fainter than the preceding. Pogson proposed regularizing the system so that *a magnitude difference of 5 corresponds to a brightness ratio of* 100:1, a proposal eventually adopted by international agreement early in the twentieth century.

Astronomers who observe in the visual and near infrared persist in using this system. It has advantages: for example, all astronomical bodies have apparent magnitudes that fall in the restricted and easy-to-comprehend range of about −26 (the Sun) to +30 (the faintest telescopic objects). However, when Hipparchus and Pogson assigned the more positive magnitudes to the fainter objects, they were asking for trouble. Avoid the trouble and remember that *smaller* (more negative) magnitudes mean *brighter* objects. Those who work at other wavelengths are less burdened by tradition and use less confusing (but sometimes less convenient) units. Such units linearly relate to the apparent brightness, F, or to the monochromatic brightness, f_λ. In radio astronomy, for example, one often encounters the *jansky* (1 Jy = 10^{-26} W m^{-2} Hz^{-1}) as a unit for f_ν.

The relationship between apparent magnitude, m, and brightness, F, is:

$$m = -2.5 \log_{10}(F) + K \qquad (1.4)$$

The constant K is often chosen so that modern measurements agree, more or less, with the older catalogs, all the way back to Hipparchus. For example, the bright star, Vega, has $m \approx 0$ in the modern magnitude system. If the flux in Equation (1.4) is the total or bolometric flux (see Section 1.4) then the magnitude defined is called the *apparent bolometric magnitude*, m_{bol}. Most practical measurements are made in a restricted band-pass, but the modern definition of such a band-pass magnitude remains as in Equation (1.4), even to the extent that K is often chosen so that Vega has $m \approx 0$ in any band. This standardization has many practical advantages, but is potentially confusing, since the function f_λ is not at all flat for Vega (see Figure 1.5) or any other star, nor is the value of the integral in Equation (1.3) similar for different bands. In most practical systems, the constant K is specified by defining the values of m for some set of **standard stars**. Absolute calibration of such a system, so that magnitudes can be converted into energy units, requires comparison of at least one of the standard stars to a source of known brightness, like a blackbody at a stable temperature.

Consideration of Equation (1.4) leads us to write an equation for the *magnitude difference* between two sources as

$$\Delta m = m_1 - m_2 = -2.5 \log_{10} \left(\frac{F_1}{F_2} \right) \qquad (1.5)$$

This equation holds for both bolometric and band-pass magnitudes. It should be clear from Equation (1.5) that once you define the magnitudes of a set of standard stars, measuring the magnitude of an unknown is a matter of measuring a flux ratio between the standard and the unknown. Magnitudes are thus almost always measured in a **differential** fashion without conversion from detector response to absolute energy units. It is instructive to invert Equation (1.5), and write a formula for the flux ratio as a function of magnitude difference:

$$\frac{F_1}{F_2} = 10^{-0.4\Delta m} = 10^{-0.4(m_1 - m_2)}$$

A word about notation: you can write magnitudes measured in a band-pass in two ways, by (1) using the band-pass name as a subscript to the letter "m", or (2) by using the name itself as the symbol. So for example, the B band apparent magnitudes of a certain star could be written as $m_B = 5.67$ or as $B = 5.67$ and its visual band magnitude written as $m_V = V = 4.56$.

1.6.2 Absolute magnitudes

The magnitude system can also be used to express the luminosity of a source. The **absolute magnitude** of a source is defined to be the apparent magnitude (either bolometric or band-pass) that the source would have if it were at the standard distance of 10 parsecs in empty space (1 parsec = 3.086×10^{21} meters – see Chapter 3). The relation between the apparent and absolute magnitudes of the same object is

$$m - M = 5 \log(r) - 5 \qquad (1.6)$$

where M is the absolute magnitude and r is the actual distance to the source in parsecs. The quantity $(m - M)$ on the left-hand side of Equation (1.6) depends only on distance, and is called the **distance modulus** of the source. You should recognize this equation as the equivalent of the inverse square law relation between apparent brightness and luminosity (Equation 1.2). Equation (1.6) must be modified if the source is not isotropic or if there is absorption along the path between it and the observer.

To symbolize the absolute magnitude in a band-pass, use the band name as a subscript to the symbol M. The Sun, for example, has absolute magnitudes:

$$M_B = 5.48$$
$$M_V = 4.83$$
$$M_{bol} = 4.75$$

1.6.3 Measuring brightness or apparent magnitude from images

We will consider in detail how to go about measuring the apparent magnitude of a source in Chapter 10. However, it is helpful to have to have a simplified description of what is involved. Imagine a special detector that will take pictures of the sky through a telescope — a visible-light image of stars, similar to what you might obtain with a black-and-white digital camera.

Our picture, shown in Figure 1.9, is composed of a grid of many little square elements called *pic*ture *el*ements or **pixels**. Each pixel stores a number that is proportional to the energy that reaches it during the exposure. Figure 1.9 displays this data by mimicking a photographic negative: each pixel location is painted a shade of gray, with the pixels that store the largest numbers painted darkest.

The image does *not* show the surfaces of the stars. Notice that the images of the stars are neither uniform nor hard-edged — they are most intense in the center and fade out over several pixels. Several effects cause this — the finite resolving power of any lens and scattering of photons by molecules and particles in the air, for example, or the scattering of photons from pixel to pixel within the detector itself. The diameter of a star image in the picture has to do with the strength of the blurring effect and the choice of grayscale mapping — not with the physical

Fig. 1.9 A digital image. The width of each star image, despite appearances, is the same. Width is measured as the width at the brightness level equal to half the peak brightness. Bright stars have a higher peak and their half-power levels are at a higher brightness or gray level.

size of the star. Despite appearances, the size of each star image is actually the same when scaled by peak brightness.

Suppose we manage to take a single picture with our camera that records an image of the star Vega as well as that of some other star whose brightness we wish to measure. To compute the brightness of a star, we add up the energies it deposits in each pixel in its image. If E_{xy} is the energy recorded in pixel x, y due to light rays *from the star*, then the brightness of the star will be

$$F = \frac{1}{tA} \sum_{x,y} E_{xy}$$

where t is the exposure time in seconds and A is the area of the camera lens. Measuring F is just a matter of adding up the E_{xy}'s.

Of course things are not quite so simple. A major problem stems from the fact that the detector isn't smart enough to distinguish between light rays coming from the star and light rays coming from any other source in the same general direction. A faint star or galaxy nearly in the same line of sight as the star, or a moonlit terrestrial dust grain or air molecule floating in front of the star, can make unwelcome additions to the signal. All such sources contribute *background* light rays that reach the same pixels as the star's light. In addition, the detector itself may contribute a background that has nothing to do with the sky. Therefore, the actual signal, S_{xy}, recorded by pixel x, y will be the sum of the signal from the star, E_{xy}, and that from the background, B_{xy}, or

$$E_{xy} = S_{xy} - B_{xy}$$

The task then is to determine B_{xy} so we can subtract it from each E_{xy}. You can do this by measuring the energy reaching some pixels near but not within a star image, and taking some appropriate average (call it B). Then assume that every-where in the star image, B_{xy} is equal to B. Sometimes this is a good assumption, sometimes not so good. Granting the assumption, then the brightness of the star is

$$F_{\text{star}} = \frac{1}{tA} \sum_{x,y} \left[S_{xy} - B \right]_{\text{star}}$$

The apparent magnitude difference between the star and Vega is

$$m_{\text{star}} - m_{\text{Vega}} = -2.5 \log_{10} \frac{F_{\text{star}}}{F_{\text{Vega}}} = -2.5 \log_{10} \left\{ \frac{\sum (S_{xy} - B)_{\text{star}}}{\sum (S_{xy} - B)_{\text{Vega}}} \right\} \qquad (1.7)$$

Notice that Equation (1.7) contains no reference to the exposure time, t, nor to the area of the camera lens, A. Also notice that since a ratio is involved, the pixels need not record units of energy – anything proportional to energy will do. Furthermore, although we have used Vega as a standard star in this example, it would be possible to solve Equation (1.7) for m_{star} without using Vega, by employing either of two different strategies:

Differential photometry. Replace Vega with any other star whose magnitude is known that happens to be in the same detector field as the unknown, then employ Equation (1.7). Atmospheric absorption effects should be nearly identical for both the standard and the unknown if they are in the same image.

All-sky photometry. Take two images, one of the star to be measured, the second of a standard star, keeping conditions as similar as possible in the two exposures. Ground-based all-sky photometry can be difficult in the optical-NIR window, because you must look through different paths in the air to observe the standard and program stars. Any variability in the atmosphere (e.g. clouds) defeats the technique. In the radio window, clouds are less problematic and the all-sky technique is usually appropriate. In space, of course, there are no atmospheric effects, and all-sky photometry is usually appropriate at every wavelength.

Summary

- Astronomers can gather information about objects outside the Earth from multiple channels: neutrinos, meteorites, cosmic rays, and gravitational waves; however, the vast majority of information arrives in the form of electromagnetic radiation.
- The wave theory formulated by Maxwell envisions light as a self-propagation transverse disturbance in the electric and magnetic fields.
- A light wave is characterized by wavelength or frequency $\lambda = c/v$, and exhibits the properties of diffraction and interference.
- Quantum mechanics improves on the wave theory, and describes light as a stream of photons, massless particles that can exhibit wave properties.
- A simple but useful description of light postulates a set of geometric rays that carry luminous energy, from a source of a particular luminosity to an observer who locally measures an apparent brightness, flux, or irradiance.
- The spectrum of an object gives its brightness as a function of wavelength or frequency.
- The Kirchhoff–Bunsen rules specify the circumstances under which an object produces an emission line, absorption line, or continuous spectrum.
- Line spectra contain information about (among other things) chemical composition. However, although based on patterns of absorption lines, the spectral types of stars depend primarily on stellar temperatures.
- A blackbody, described by Planck's law, Wien's law, and the Stefan Boltzmann law, emits a continuous spectrum whose shape depends only on its temperature.
- The astronomical magnitude system uses apparent and absolute magnitudes to quantify brightness measurements on a logarithmic scale.

(continued)

Summary (*cont.*)

- Photometry and spectroscopy are basic astronomical measurements. They often depend on direct comparison to standard objects, and the use of standard band-passes.
- Important constants and formulae:

$$1 \, \text{eV} = 1.602 \times 10^{-19} \, \text{J}$$

$$h = \text{Planck's constant} = 6.626 \times 10^{-34} \, \text{J s}$$

$$1 \, \text{Jansky} = 10^{-26} \, \text{W m}^{-2} \, \text{Hz}^{-1}$$

$$L_\odot = 3.87 \times 10^{26} \, \text{W}$$

Solar constant (flux at the top of the atmosphere) = 1370 W m^{-2}

σ = Stefan–Boltzmann constant = 5.6696×10^{-8} W m^{-2} K^{-4}

k = Boltzmann constant = 1.3806×10^{-23} J K^{-1} = 8.6174×10^{-5} eV K^{-1}

Energy of a photon: $E = h\nu$

Inverse square law of light: $F = \dfrac{L}{4\pi r^2}$

Stefan–Boltzmann law: $s = \sigma T^4$

Wein's law: $T\lambda_{\text{MAX}} = 2.8979 \times 10^{-3}$ m $\cdot$ K

Magnitude difference and flux:

$$\Delta m = m_1 - m_2 = -2.5 \log_{10}\left(\frac{F_1}{F_2}\right)$$

$$\frac{F_1}{F_2} = 10^{-0.4\Delta m} = 10^{-0.4(m_1 - m_2)}$$

Distance modulus:

$$m - M = 5\log(r) - 5$$

Exercises

> 'Therefore,' I said, 'by the use of problems, as in geometry, we shall also pursue astronomy, ... by really taking part in astronomy we are going to convert the prudence by nature in the soul from uselessness to usefulness.'
>
> – Socrates, in Plato, *The Republic,* Book VII, 530b (*c.* 360 BC)

1. Propose a definition of astronomy that distinguishes it from other sciences like physics and geology.

2. What wavelength photon would you need to:
   (a) ionize a hydrogen atom (ionization energy = 13.6 eV)
   (b) dissociate the molecular bond in a diatomic hydrogen molecule (dissociation energy = 4.48 eV)
   (c) dissociate a carbon monoxide molecule (dissociation energy = 11.02 eV)

3. What are the units of the monochromatic brightness, f_λ?

4. What is the value of the ratio f_λ/f_ν for any source?

5. (a) Define $A(T)$ to be the average slope of the Planck function, $B(\lambda,T)$, between 400 nm (violet) and 600 nm (yellow). In other words:

$$A(T) = [B(600, T) - B(400, T)]/200\,\text{nm}$$

Use a spreadsheet to compute and plot $A(T)$, over the temperature range 2000 K to 30 000 K (typical of most stars). What happens to A at very large and very small temperatures?

(b) Compute and plot the function

$$C(T) = \log[B(400, T)] - [\log B(600, T)]$$

over the same range of temperature. Again, what happens to C at the extremes of temperature?

(c) Comment on the usefulness of A and C as indicators of the temperature of a blackbody.

6. A certain radio source has a monochromatic flux density of 1 Jy at a frequency of 1 MHz. What is the corresponding flux density in photon number? (How many photons arrive per m^2 in one second with frequencies between 1 000 000 Hz and 1 000 001 Hz?)

7. The bolometric flux from a star with $m_{\text{bol}} = 0$ is about 2.65×10^{-8} W m^{-2} outside the Earth's atmosphere.

(a) Compute the value of the constant K in Equation (1.4) for bolometric magnitudes.

(b) Compute the bolometric magnitude of an object with a total flux of:

(i) one solar constant

(ii) 1.0 W m^{-2}

8. The monochromatic flux at the center of the B band-pass (440 nm) for a certain star is 375 Jy. If this star has a blue magnitude of $m_B = 4.71$, what is the monochromatic flux, in Jy, at 440 nm for:

(a) a star with $m_B = 8.33$

(b) a star with $m_B = -0.32$

9. A double star has two components of equal brightness, each with a magnitude of 8.34. If these stars are so close together that they appear to be one object, what is the apparent magnitude of the combined object?

10. A gaseous nebula has an average surface brightness of 17.77 magnitudes per square second of arc.

(a) If the nebula has an angular area of 145 square arcsec, what is its total apparent magnitude?

(b) If the nebula were moved to twice its original distance, what would happen to its angular area, total apparent magnitude, and surface brightness?

11. At maximum light, type Ia supernovae are believed to have an absolute visual magnitude of −19.60. A supernova in the Pigpen Galaxy is observed to reach

apparent visual magnitude 13.25 at its brightest. Compute the distance to the Pigpen Galaxy.

12. Derive the distance modulus relation in Equation (1.6) from the inverse square law relation in Equation (1.1).

13. Show that, for small values of Δm, the difference in magnitude is approximately equal to the fractional difference in brightness, that is

$$\Delta m \approx -\frac{\Delta F}{F}$$

     Hint: consider the derivative of m with respect to F.

14. An astronomer is performing synthetic aperture photometry on a single unknown star and standard star (review Section 1.6.3) in the same field. The data frame is in the figure below. The unknown star is the fainter one. If the magnitude of the standard is 9.000, compute the magnitude of the unknown.

     Actual data numbers are listed for the frame in the table. Assume these are proportional to the number of photons counted in each pixel, and that the band-pass is narrow enough that all photons can be assumed to have the same energy. Remember that photometrically both star images have the same size.

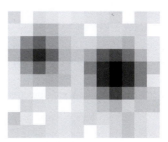

34	16	26	33	37	22	25	25	29	19	28	25
22	20	44	34	22	26	14	30	30	20	19	17
31	70	98	66	37	25	35	36	39	39	23	20
34	99	229	107	38	28	46	102	159	93	37	22
33	67	103	67	36	32	69	240	393	248	69	30
22	33	34	29	36	24	65	241	363	244	68	24
28	22	17	16	32	24	46	85	157	84	42	22
18	25	27	26	17	18	30	29	35	24	30	27
32	23	16	29	25	24	30	28	20	35	22	23
28	28	28	24	26	26	17	19	30	35	30	26

Chapter 2
Uncertainty

Errare humanum est.

– Anonymous Latin saying

Upon foundations of evidence, astronomers erect splendid narratives about the lives of stars, the anatomy of galaxies or the evolution of the Universe. Inaccurate or imprecise evidence weakens the foundation and imperils the astronomical story it supports. Wrong ideas and theories are vital to science, which normally works by proving many, many ideas to be incorrect until only one remains. Wrong data, on the other hand, are deadly.

As an astronomer you need to know how far to trust the data you have, or how much observing you need to do to achieve a particular level of trust. This chapter describes the formal distinction between accuracy and precision in measurement, and methods for estimating both. It then introduces the concepts of a population, a sample of a population, and the statistical descriptions of each. Any characteristic of a population (e.g. the masses of stars) can be described by a probability distribution (e.g. low-mass stars are more probable than high-mass stars) so we next will consider a few probability distributions important in astronomical measurements. Finally, armed with new statistical expertise, we revisit the question of estimating uncertainty, both in the case of an individual measurement, as well as the case in which multiple measurements combine to produce a single result.

2.1 Accuracy and precision

In common speech, we often do not distinguish between these two terms, but we will see that it is very useful to attach very different meanings to them. An example will help.

2.1.1 An example

In the distant future, a very smart theoretical astrophysicist determines that the star Malificus might soon implode to form a black hole, and in the process destroy all life on its two inhabited planets. Careful computations show that if

Malificus is fainter than magnitude 14.190 by July 24, as seen from the observing station orbiting Pluto, then the implosion will not take place and its planets will be spared. The Galactic government is prepared to spend the ten thousand trillion dollars necessary to evacuate the doomed populations, but needs to know if the effort is really called for; it funds some astronomical research. Four astronomers and a demigod each set up experiments on the Pluto station to measure the apparent magnitude of Malificus.

The demigod performs photometry with divine perfection, obtaining a result of 14.123 010 (all the remaining digits are zeros). The truth, therefore, is that Malificus is brighter than the limit and will explode. The four astronomers, in contrast, are only human, and, fearing error, repeat their measurements – five times each. I'll refer to a single one of these five as a "trial." Table 2.1 lists the results of each trial, and Figure 2.1 illustrates them.

2.1.2 Accuracy and systematic error

In our example, we are fortunate a demigod participates, so we feel perfectly confident to tell the government, sorry, Malificus is doomed, and those additional taxes are necessary. The **accuracy** of a measurement describes (usually numerically) how close it is to the "true" value. The demigod measures with perfect accuracy.

Table 2.1. *Results of trials by four astronomers. The values for σ and s are computed from Equations (2.3) and (2.1), respectively*

Astronomer	A	B	C	D
Trial 1	14.115	14.495	14.386	14.2
Trial 2	14.073	14.559	14.322	14.2
Trial 3	14.137	14.566	14.187	14.2
Trial 4	14.161	14.537	14.085	14.2
Trial 5	14.109	14.503	13.970	14.2
Mean	14.119	14.532	14.190	14.2
Deviation from truth	−0.004	+0.409	+0.067	+0.077
Spread	0.088	0.071	0.418	0
σ	0.033	0.032	0.174	0
s	0.029	0.029	0.156	0
Uncertainty of the mean	0.013	0.013	0.070	(0.05)
Interpretation	Evacuate	Stay	Uncertain	Uncertain
Accuracy?	Accurate	Inaccurate	Accurate	Inaccurate
Precision?	Precise	Precise	Imprecise	Imprecise

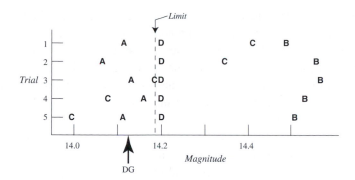

Fig. 2.1 Apparent magnitude measurements. The arrow points to the demigod's result, which is the true value. The dotted line marks the critical limit.

What is the accuracy of the human results? First, decide what we mean by "a result": since each astronomer made five trials, we choose a single value that summarizes these five measurements. In this example, each astronomer chooses to compute the mean – or average – of the five, a reasonable choice. (We will see there are others.) Table 2.1 lists the mean values from each astronomer – "results" that summarize the five trials each has made.

Since we know how much each result deviates from the truth, we could express its accuracy with a sentence like: "Astronomer B's result is relatively inaccurate," or, more specifically: "The result of Astronomer A is 0.004 magnitude smaller than the true value." Statements of this kind are easy to make because the demigod tells us the truth, but in the real Universe, how could you determine the "true" value, and hence the accuracy? In science, after all, the whole point is to discover values that are unknown at the start, and no demigods work at observatories.

How, then, can we judge accuracy? The alternative to divinity is variety. We can only repeat measurements using different devices, assumptions, strategies, and observers, and then check for general agreements (and disagreements) among the results. We suspect a particular set-up of inaccuracy if it disagrees with all other experiments. For example, the result of Astronomer B differs appreciably from those of his colleagues. Even in the absence of the demigod result, we would guess that B's result is the least accurate. In the absence of the demigod, in the real world, the best we can hope for is a good *estimate* of accuracy.

If a particular set-up always produces consistent (estimated) inaccuracies, if its result is always biased by about the same amount, then we say it produces a **systematic error**. Although Astronomer B's trials do not have identical outcomes, they all tend to be much too large, and are thus subject to a systematic error of around +0.4 magnitude. Systematic errors are due to some instrumental or procedural fault, or some mistake in modeling the phenomena under investigation. Astronomer B, for example, used the wrong magnitude for the standard star in his measurements. He could not improve his measurement just by repeating it — making more trials would give the same general result, and B would continue to recommend against evacuation.

In a second example of inaccuracy, suppose the astrophysicist who computed the critical value of $V = m_V = 14.19$ had made a mistake because he neglected the effect of the spin of Malificus. Then even perfectly accurate measurements of brightness would result in false (inaccurate) predictions about the collapse of the star.

2.1.3 Precision and random error

Precision differs from accuracy. The **precision** of a measurement describes how well or with what certainty a particular result is known, without regard to its truth. Precision denotes the ability to be very specific about the exact value of the measurement itself. A large number of legitimately significant digits in the numerical value, for example, indicates high precision. Because of the possibility of systematic error, of course, high precision does not mean high accuracy.

Poor precision *does* imply a great likelihood of poor accuracy. You might console yourself with the possibility that an imprecise result *could* be accurate, but the Universe seldom rewards that sort of optimism. You would do better to regard precision as setting a limit on the accuracy expected. Do not expect accuracy better than your precision, and do not be shocked when, because of systematic error, it is a lot worse.

Unlike accuracy, precision is often easy to quantify without divine assistance. You can determine the precision numerically by examining the degree to which multiple trials agree with one another. If the outcome of one trial differs from the outcome of the next in an unpredictable fashion, the scattering is said to arise from **stochastic**, **accidental**, or **random error**. (If the outcome of one trial differs from the next in a *predictable* fashion, then one should look for some kind of systematic error or some unrecognized physical effect.) The term random "error" is unfortunate, since it suggests some sort of mistake or failure, whereas you should really think of it as a scattering of values due to the uncertainty inherent in the measuring process itself. Random error limits precision, and therefore limits accuracy.

To quantify random error, you could examine the **spread** in values for a finite number of trials:

spread = largest trial result − smallest trial result

The spread will tend to be larger for experiments with the largest random error and lowest precision. A better description of the scatter or "dispersion" of a set of N trials, $\{x_1, x_2, \ldots x_N\}$, would depend on all N values. One useful statistic of this sort is the **estimated standard deviation**, s:

$$s = \sqrt{\frac{1}{N-1} \sum_{i=1}^{N} (x_i - \bar{x})^2} \qquad (2.1)$$

We examine Equation (2.1) more carefully in later sections of this chapter. The values for s and for the spread in our example are in Table 2.1. These confirm the subjective impression from Figure 2.1 – in relative terms, the results of the astronomers are as follows:

A is precise and accurate;

B is precise but inaccurate;

C is imprecise and accurate (to the degree expected from the precision);

D is a special imprecise case, discussed below.

The basic statistical techniques for coping with random error and estimating the resulting uncertainty are the subjects of this chapter. A large volume of literature deals with more advanced topics in the statistical treatment of data dominated by stochastic error – a good introduction is the book by Bevington (1969). Although most techniques apply only to stochastic error, in reality, systematic error is usually the more serious limitation to good astronomy.

Techniques for detecting and coping with systematic error are varied and indirect, and therefore difficult to discuss at an elementary level. Sometimes, one is aware of systematic error only after reconciling different methods for determining the same parameter. This is the case with Astronomer B, whose result differs from the others by more than the measured stochastic error. Sometimes, what appears to be stochastic variation turns out to be a systematic effect. This might be the case with astronomer C, whose trial values decrease with time, suggesting perhaps some change in the instrument or environment. Although it is difficult to recognize systematic error, the fact that it is the consequence of some sort of mistake means that it is often possible to correct the mistake and improve accuracy.

Stochastic error and systematic error both contribute to the uncertainty of a particular result. That result is useless until the size of its uncertainty is known.

2.1.4 Uncertainty

> . . . one discovers over the years that as a rule of thumb accidental errors are twice as large as observers indicated, and systematic errors may be five times larger than indicated.
>
> – C. Jaschek, Error, *Bias and Uncertainties in Astronomy* (1990)

In our Malificus example, the recommendation that each astronomer makes depends on two things – the numerical value of the result and the uncertainty the astronomer attaches to its accuracy. Astronomer A recommends evacuation because (a) her result is below the cut-off by 0.07 magnitude, and (b) the uncertainty she feels is small because her random error, as measured by s, is small compared to 0.07 and because she assumes her systematic error is also small. The assumption of a small systematic error is based mostly on A's confidence that she "knows what she is doing" and hasn't made a mistake. Later, when she is aware that both C and D agree with her result, she can be even more sanguine about this. Astronomical literature sometimes makes a distinction between *internal error*, which is the uncertainty computed from the scatter of trials, and *external error*, which is the total uncertainty, including systematic effects.

Astronomer A should quote a numerical value for her *uncertainty*. If u is the uncertainty of a result, r, then the probability that the true value is between $r + u$ and $r - u$ is 1/2.

Statistical theory (see below) says that under certain broad conditions, the *uncertainty of the mean* of N values is something like $s/\sqrt{N}$. Thus, the uncertainty imposed by random error (the internal error) alone for A is about 0.013. The additional uncertainty due to systematic error is harder to quantify. The astronomer should consider such things as the accuracy of the standard star magnitudes and the stability of her photometer. In the end, she might feel that her result is uncertain (external error) by 0.03 magnitudes. She concludes the chances are much greater than 50% that the limit is passed, and thus must recommend evacuation in good conscience.

Astronomer B goes through the same analysis as A, and recommends *against* evacuation with even greater (why?) conviction. Since quadrillions of dollars and billions of lives are at stake, it would be criminal for A and B not to confront their disagreement. They must compare methods and assumptions and try to determine which (if either) of them has the accurate result.

Astronomer C shouldn't make a recommendation because his uncertainty is so large. He can't rule out the need for an evacuation, nor can he say that one is necessary. We might think C's measurements are so imprecise that they are useless, but this is not so. Astronomer C's precision is sufficient to cast doubt on B's result (but not good enough to confirm A's). The astronomers thus should first concentrate on B's experimental method in their search for the source of their disagreement. Astronomer C should also be suspicious of his relatively large random error compared to the others. This may represent the genuine accidental errors that limit his particular method, or it may result from a systematic effect that he could correct.

2.1.5 Digitizing effects

What about Astronomer D, who performed five trials that gave identical results? Astronomer D made her measurements with a digital light meter that only reads to the nearest 0.2 magnitude, and this digitization is responsible her very uniform data.

From the above discussion, it might seem that since her scatter is zero, then D's measurement is perfectly precise. This is misguided, because it ignores what D knows about her precision: rounding off every measurement produces uncertainty. She reasons that in the absence of random errors, there is a nearly 100% chance that the true value is within ± 0.1 of her measurement, and there is a nearly 50% chance that the true value lies within 0.05 magnitudes of her measurement. Thus, D would report an uncertainty of around ± 0.05 magnitude. This is a case where a known systematic error (digitization) limits precision, and where stochastic error is so small compared to the systematic effect that it cannot be investigated at all.

You can control digitization effects, usually by building a more expensive instrument. If you can arrange for the digitization effects to be smaller than the expected random error, then your measurements will exhibit a stochastic scatter and you can ignore digitization when estimating your precision.

2.1.6 Significant digits

One way to indicate the uncertainty in a measurement is to retain only those digits that are warranted by the uncertainty, with the remaining insignificant digits rounded off. In general, only one digit with "considerable" uncertainty (more than ± 1) should be retained. For example, Astronomer C had measured a value of 14.194 with an uncertainty of at least $0.156/\sqrt{5} = 0.070$. Astronomer C would realize that the last digit "4" has no significance whatever, the digit "1" is uncertain by almost ± 1, so the digit "9", which has considerable uncertainty, is the last that should be retained. Astronomer C should quote his result as 14.19.

Astronomer A, with the result 14.119, should likewise recognize that her digit "1" in the hundredths place is uncertain by more than ± 1, so she should round off her result to 14.12.

It is also very good practice to quote the actual uncertainty. Usually one or two digits in the estimate of uncertainty are all that are significant. The first three astronomers might publish (internal errors):

> A's result: 14.12 ± 0.013
> B's result: 14.53 ± 0.013
> C's result: 14.19 ± 0.07

Note that Astronomers A and C retain the same number of significant digits, even though A's result is much more certain than C's. Astronomer B, who is unaware of his large systematic error, estimates his (internal) uncertainty in good faith, but nevertheless illustrates Jaschek's rule of thumb about underestimates of systematic error.

2.2 Populations and samples

> As some day it may happen that a victim must be found,
> I've got a little list – I've got a little list
>
> — W.S. Gilbert, *The Mikado*, Act I, 1885

In the Malificus problem, our fictional astronomers used simple statistical computations to estimate both brightness and precision. We now treat more systematically the statistical analysis of observational data of all kinds, and begin with the concept of a population.

Consider the problem of determining a parameter (e.g. the brightness of a star) by making several measurements under nearly identical circumstances. We define the **population** under investigation as the hypothetical set of all possible measurements that could be made with an experiment substantially identical to our own. We then imagine that we make our actual measurements by drawing a finite **sample** (five trials, say) from this much larger population. Some populations are indefinitely large, or are so large that taking a sample is the only practical method for investigating the population. Some populations are finite in size, and sometimes are small enough to be sampled completely. Table 2.2 gives some examples of populations and samples.

Table 2.2. *Populations and samples. Samples can be more or less representative of the population from which they are drawn*

Population	Sample	Better sample
1000 colored marbles mixed in a container: 500 red, 499 blue, 1 purple	5 marbles drawn at random from the container	50 marbles drawn at random
The luminosities of each star in the Milky Way galaxy (about 10^{11} values)	The luminosities of each of the nearest 100 stars (100 values)	The luminosities of 100 stars at random locations in the galaxy (100 values)
The weights of every person on Earth	The weights of each person in this room	The weights of 100 people drawn from random locations on Earth
The outcomes of all possible experiments in which one counts the number of photons that arrive at your detector during one second from the star Malificus	The outcome of 1 such experiment	The outcomes of 100 such experiments

2.2.1 Descriptive statistics of a finite population

> But all evolutionary biologists know that variation itself is nature's only
> irreducible essence. Variation is the hard reality, not a set of imperfect measures
> for a central tendency
>
> — Stephen Jay Gould, The Median Isn't the Message, *Bully for Brotosaurus*, 1991

Imagine a small, finite population, which is a set of M values or members, $\{x_1, x_2, \ldots, x_M\}$. The list of salaries of the employees in a small business, like the ones in Table 2.3, would be an example. We can define some statistics that summarize or describe the population as a whole.

Measures of the central value. If every value in the population is known, a familiar descriptive statistic, the **population mean**, is just

$$\mu = \frac{1}{M} \sum_{i=1}^{M} x_i$$

Two additional statistics also measure the *central* or *representative* value of the population. The **median**, or midpoint, is the value that divides the population exactly in half: just as many members have values above as have values below the median. If $n(E)$ is the number of members of a population with a particular characteristic, E, then the median, $\mu_{1/2}$, satisfies

$$n\left(x_i \leq \mu_{1/2}\right) = n\left(x_i \geq \mu_{1/2}\right) \approx \frac{M}{2}$$

Compared to the mean, the median is a bit more difficult to compute if M is large, since you have to sort the list of values. In the pre-sorted list in Table 2.3, we can see by inspection that the median salary is $30,000, quite a bit different from the mean ($300,000). The third statistic is the **mode**, which is the most common or most frequent value. In the example, the mode is clearly $15,000, the salary of the four astronomers. In a sample in which there are no identical values, you can still compute the mode by sorting the values into bins, and then searching for the bin with the most members. Symbolically, if μ_{max} is the mode, then

$$n(x_i = \mu_{max}) > n(x_i = y, y \neq \mu_{max})$$

Table 2.3. *Employee salaries at Astroploitcom*

Job title (number of employees)	Salary in thousands of dollars
President (1)	2000
Vice president (1)	500
Programmer (3)	30
Astronomer (4)	15

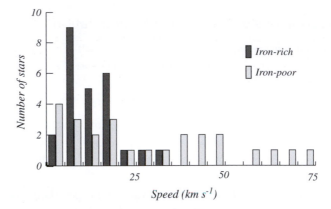

Fig. 2.2 Histogram of the data in Table 2.4.

Which measure of the central value is the "correct" one? The mean, median, and mode all legitimately produce a central value. Which one is most relevant depends on the question being asked. In the example in 2.3, if you were interested in balancing the corporate accounts, then the mean would be most useful. If you are interested in organizing a workers' union, the mode might be more interesting.

Measures of dispersion. How scattered are the members of a population? Are values clustered tightly around the central value, or are many members significantly different from one another? Table 2.4 gives the speeds of stars in the direction perpendicular to the Galactic plane. Two populations differ in their chemical compositions: one set of 25 stars (Group A) contains the nearby solar-type stars that most closely match the Sun in iron abundance. Group A is relatively iron-rich. A second group, B, contains the 25 nearby solar-type stars that have the lowest known abundances of iron in their atmospheres. Figure 2.2 summarizes the table with a histogram. Clearly, the central value of the speed is different for the two populations. Group B stars, on average, zoom through the plane at a higher speed than do members of Group A.

A second difference between these populations – how spread out they are – concerns us here. The individual values in Group B are more dispersed than those in Group A. Figure 2.2 illustrates this difference decently, but we want a compact and quantitative expression for it. To compute such a statistic, we first examine the **deviation** of each member from the population mean:

$$\text{deviation from the mean} = (x_i - \mu)$$

Those values of x_i that differ most from μ will have the largest deviations. The definition of μ insures that the average of all the deviations will be zero (positive deviations will exactly balance negative deviations), so the average deviation is an uninteresting statistic. The average of all the *squares of the deviations*, in contrast, must be a positive number. This is called the **population variance**:

Table 2.4. *Speeds perpendicular to the Galactic plane, in km s^{-1}, for 50 nearby solar type stars*

Group A: 25 Iron-rich stars					Group B: 25 Iron-poor stars				
0.5	7.1	9.2	14.6	18.8	0.3	7.9	16.8	35.9	48.3
1.1	7.5	10.7	15.2	19.6	0.4	10.0	18.1	38.8	55.5
5.5	7.8	12.0	16.1	24.2	2.5	10.8	23.1	42.2	61.2
5.6	7.9	14.3	17.1	26.6	4.2	14.5	26.0	42.3	67.2
6.9	8.1	14.5	18.0	32.3	6.1	15.5	32.1	46.6	76.6

$$\sigma^2 = \frac{1}{M} \sum_{i=1}^{M} (x_i - \mu)^2 = \frac{1}{M} \sum_{i=1}^{M} x_i^2 - \mu^2 \tag{2.2}$$

The variance tracks the dispersion nicely – the more spread out the population members are, the larger the variance. Because the deviations enter Equation (2.2) as quadratics, the variance is especially sensitive to population members with large deviations from the mean.

The square root of the population variance is called the **standard deviation of the population**:

$$\sigma = \sqrt{\frac{1}{M} \sum_{i=1}^{M} (x_i - \mu)^2} \tag{2.3}$$

σ has the same dimensions as the population values themselves. For example, the variance of Group A in 2.4 is 57.25 km^2 s^{-2}, and the standard deviation is 7.57 km s^{-1}. The standard deviation is usually the statistic employed to measure population spread. The mean of Group A is 12.85 km s^{-1} and examination of Figure 2.2 or the table shows that a deviation of $\sigma = 7.57$ km s^{-1} is "typical" for a member of this population.

2.2.2 Estimating population statistics

Many populations are so large that it is impractical to tabulate all members. This is the situation with most astronomical measurements. In this case, the strategy is to *estimate* the descriptive statistics for the population from a small sample. We chose the sample so that it *represents* the larger population. For example, a sample of five or ten trials at measuring the brightness of the star Malificus represents the population that contains *all* possible equivalent measurements of its brightness. Most scientific measurements are usually treated as samples of a much larger population of possible measurements.

A similar situation arises if we have a physical population that is finite but very large (the masses of every star in the Milky Way, for example). We can discover the characteristics of this population by taking a sample that has only N members (masses for 50 stars picked at random in the Milky Way).

In any sampling operation, we estimate the population mean from the sample mean, $\bar{x}$. All other things being equal, we believe a larger sample will give a better estimate. In this sense, the population mean is the limiting value of the sample mean, and the sample mean is the best estimator of the population mean. If the sample has N members:

$$\mu = \lim_{N \to \infty} \frac{1}{N} \sum_{i=1}^{N} x_i = \lim_{N \to \infty} \bar{x}$$

$$\mu \approx \bar{x}$$

To estimate the population variance from a sample, the best statistic is s^2, the **sample variance** computed with *(N − 1)* weighting.

$$s^2 = \frac{1}{N-1} \sum_{i=1}^{N} (x_i - \bar{x})^2 \tag{2.4}$$

$$\sigma^2 \approx s^2$$

The $(N-1)^{-1}$ factor in Equation (2.4) (instead of just N^{-1}) arises because $\bar{x}$ is an *estimate* of the population mean, and is not μ itself. The difference is perhaps clearest in the case where $N = 2$. For such small N, it is likely that $\bar{x} \neq \mu$. Then the definition of $\bar{x}$ guarantees that

$$\sum_{i=1}^{2} (x_i - \mu)^2 \geq \sum_{i=1}^{2} (x_i - \bar{x})^2$$

which suggests that N^{-1} weighting will consistently *underestimate* the population variance. In the limit of large N, the two expressions are equivalent:

$$\sigma^2 = \lim_{N \to \infty} \frac{1}{N} \sum_{i=1}^{N} (x_i - \mu)^2 = \lim_{N \to \infty} \frac{1}{N} \sum_{i=1}^{N} (x_i - \bar{x})^2 = \lim_{N \to \infty} \frac{1}{N-1} \sum_{i=1}^{N} (x_i - \bar{x})^2$$

$$= \lim_{N \to \infty} s^2$$

Proof that Equation (2.4) is the best estimate of σ^2 can be found in elementary references on statistics. The square root of s^2 is called the **standard deviation of the sample**. Since most astronomical measurements are samples of a population, the dispersion of the population is usually estimated as

$$s = \sqrt{\frac{1}{N-1} \sum_{i=1}^{N} (x_i - \bar{x})^2} \approx \sigma \tag{2.5}$$

which is the expression introduced at the beginning of the chapter (Equation (2.1)).

The terminology for s and σ can be confusing. It is unfortunately common to shorten the name for s to just "the standard deviation," and to represent it with the symbol σ. You, the reader, must then discern from the context whether the statistic is an estimate (i.e. *s,* computed from a sample by Equation (2.5)) or a complete description (i.e. σ, computed from the population by Equation (2.3)).

2.3 Probability distributions

> The most important questions of life are, for the most part, really only problems of probability.
>
> — Pierre-Simon Laplace, *A Philosophical Essay on Probabilities*

2.3.1 The random variable

Since scientific measurements generally only sample a population, we consider the construction of a sample a little more carefully. Assume we have a large population, Q. For example, suppose we have a jar full of small metal spheres of differing diameters, and wish to sample those diameters in a representative fashion. Imagine doing this by stirring up the contents, reaching into the jar without looking, and measuring the diameter of the sphere selected. This operation is a trial, and its result is a diameter, x. We call x a **random variable** – its value depends not at all on the selection method (we hope). Although the value of x is unpredictable, there clearly is a function that describes how likely it is to obtain a particular value for x in a single trial. This function, P_Q, is called the **probability distribution** of x in Q. In the case where x can take on any value over a continuous range, we define:

$$P_Q(x)\mathrm{d}x = \text{the probability that the result of a single trial will}$$
$$\text{have a value between } x \text{ and } x + \mathrm{d}x$$

Sometimes a random variable is restricted to a discrete set of possible values. In our example, this would be the case if a ball-bearing factory that only made spheres with diameters that were integral multiples of 1 mm manufactured all the spheres. In this case, the definition of the probability distribution function has to be a little different:

$$P_Q(x_j) = \text{the probability that the result of a single trial will have}$$
$$\text{a value } x_j, \text{where } j = 1, 2, 3, \ldots$$

For our example, P_Q might look like Figure 2.3, where (a) shows a continuous distribution in which any diameter over a continuous range is possible. Plot (b) shows a discrete distribution with only six possible sizes.

In experimental situations, we sometimes know or suspect something about the probability distribution before conducting any quantitative trials. We might, for example, look into our jar of spheres and get the impression that "there seem to be only two general sizes, large and small." Knowing something about the expected distribution before making a set of trials can be helpful in designing the experiment and in analyzing the data. Nature, in fact, favors a small number of distributions. Two particular probability distributions arise so often in astronomy that they will repay special attention.

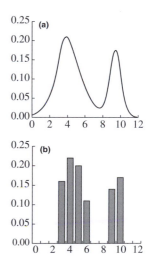

Fig. 2.3 Probability distributions of the diameters of spheres, in millimeters. (a) A continuous distribution; (b) a discrete distribution, in which only six sizes are present.

2.3.2 The Poisson distribution

The Poisson[1] distribution describes a population encountered in certain count-ing experiments. These are cases in which the random variable, x, is the number of events counted in a unit time: the number of raindrops hitting a tin roof in 1 second, the number of photons hitting a light meter in 10 seconds, or the number of nuclear decays in an hour. For counting experiments where non-correlated events occur at an average rate, μ, the probability of counting x events in a single trial is

$$P_p(x, \mu) = \frac{\mu^x}{x!} e^{-\mu}$$

Here, $P_p(x, \mu)$ is the Poisson distribution. For example, If you are listening to raindrops on the roof in a steady rain, and on average hear 3.25 per second, then $P_p(0, 3.25)$ is the probability that you will hear zero drops in the next one-second interval. Of course, $P_p(x, \mu)$ is a discrete distribution, with x restricted to non-negative integer values (you can never hear 0.266 drops, nor could you hear -1 drops). Figure 2.4 illustrates the Poisson distribution for three different values of μ. Notice that as μ increases, so does the dispersion of the distribution. An important property of the Poisson distribution, in fact, is that its variance is exactly equal to its mean:

$$\sigma^2 = \mu$$

This behavior has very important consequences for planning and analyzing experiments. For example, suppose you count the number of photons, N, that arrive at your detector in t seconds. If you count N things in a single trial, you can estimate that the average result of a single trial of length t seconds will be a count of $\mu \approx \bar{x} = N$ photons. How uncertain is this result? The uncertainty in a result can be judged by the standard deviation of the population from which the measurement is drawn. So, assuming Poisson statistics apply, the uncertainty of the measurement should be $\sigma = \sqrt{\mu} \approx \sqrt{N}$. The uncertainty in the rate *in units of counts per second* would be $\sigma/t \approx \sqrt{N}/t$. The fractional uncertainty is:

$$\textit{Fractional uncertainty in counting N events} = \frac{\sigma}{\mu} \approx \frac{1}{\sqrt{N}}$$

[1] Siméon-Denis Poisson (1781–1840) in youth resisted his family's attempts to educate him in medicine and the law. After several failures in finding an occupation that suited him, he became aware of his uncanny aptitude for solving puzzles, and embarked on a very prolific career in mathematics, becoming Laplace's favorite pupil. Poisson worked at a prodigious rate, both in mathematics and in public service in France. Given his rather undirected youth, it is ironic that he characterized his later life with his favorite phrase: "La vie, c'est le travail."

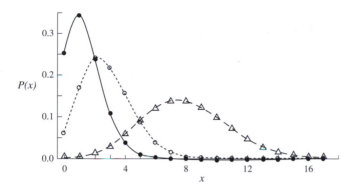

Fig. 2.4 The Poisson distribution for values of $\mu = 1.4$ (filled circles), $\mu = 2.8$ (open circles), and $\mu = 8.0$ (open triangles). Note that only the plotted symbols have meaning as probabilities. The curves merely assist the eye in distinguishing the three distributions.

Thus, to decrease the uncertainty in your estimate of the photon arrival rate μ/t, you should increase the number of photons you count (by increasing either the exposure time or the size of your telescope). To cut uncertainty in half, for example, increase the exposure time by a factor of 4.

2.3.3 The Gaussian, or normal, distribution

The Gaussian[2], or normal, distribution is the most important continuous distribution in the statistical analysis of data. Empirically, it seems to describe the distribution of trials for a very large number of different experiments. Even in situations where the population itself is not described by a Gaussian (e.g. Figure 2.3) estimates of the summary statistics of the population (e.g. the mean) are described by a Gaussian.

If a population has a Gaussian distribution, then in a single trial the probability that x will have a value between x and $x + dx$ is

$$P_G(x, \mu, \sigma)dx = \frac{dx}{\sigma\sqrt{2\pi}} \exp\left[-\frac{1}{2}\left(\frac{x-\mu}{\sigma}\right)^2\right] \qquad (2.6)$$

Figure 2.5 illustrates this distribution, a shape sometimes called a ***bell curve***. In Equation (2.6), μ and σ are the mean and standard deviation of the distribution, and they are independent of one another (unlike the Poisson distribution). We will find it useful to describe the dispersion of a Gaussian by specifying its

---

[2] Karl Friedrich Gauss (1777–1855) was a child prodigy who grew to dominate mathematics during his lifetime. He made several important contributions to geometry and number theory in his early 20s, after rediscovering many theorems because he did not have access to a good mathematics library. In January 1801 the astronomer Piazzi discovered Ceres, the first minor planet, but the object was soon lost. Gauss immediately used a new method and only three recorded observations to compute the orbit of the lost object. His predicted positions led to the recovery of Ceres, fame, and eventually a permanent position at Göttingen Observatory. At Göttingen, Gauss made important contributions to differential geometry and to many areas of physics, and was involved in the invention of the telegraph.

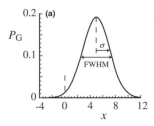

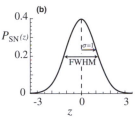

Fig. 2.5 (a) A Gaussian distribution with a mean of 5 and a standard deviation of 2.1. The curve peaks at $P = 1.2$, and the FWHM is drawn at half this level. In contrast to the Poisson distribution, note that P is defined for negative values of x. (b) The standard normal distribution.

full width at half-maximum (**FWHM**), that is, the separation in x between the two points where

$$P_G(x, \mu, \sigma) = \frac{1}{2} P_G(\mu, \mu, \sigma)$$

The FWHM is proportional to σ:

$$\mathrm{FWHM}_{\mathrm{Gaussian}} = 2.354\sigma$$

The dispersion of a distribution determines how likely it is that a single sample will turn out to be close to the population mean. One measure of dispersion, then, is the *probable error*, or P.E. By definition, a single trial has a 50% probability of lying closer to the mean than the P.E., that is

$$\text{Probability that } (|x - \mu| \leq \text{P.E.}) = 1/2$$

The P.E. for a Gaussian distribution is directly proportional to its standard deviation:

$$(\text{P.E.})_{\mathrm{GAUSSIAN}} = 0.6745\sigma = 0.2865(\text{FWHM})$$

2.3.4 The standard normal distribution

$P_G(x, \mu, \sigma)$ is difficult to tabulate since its value depends not only on the variable, x, but also on the two additional parameters, μ and σ. This prompts us to define a new random variable:

$$z = \frac{x - \mu}{\sigma}$$
$$dz = \sigma^{-1} dx \qquad (2.7)$$

After substitution in (2.6), this gives:

$$P_{SN}(z) = P_G(z, 0, 1) = \frac{1}{\sqrt{2\pi}} \exp\left[-\frac{z^2}{2}\right] \qquad (2.8)$$

Equation (2.8), the Gaussian distribution with zero mean and unit variance, is tabulated in Appendix C1. You can extract values for a distribution with a specific μ and σ from the table through Equations (2.7).

2.3.5 Other distributions

Many other distributions describe populations in nature. We will not discuss these here, but only remind you of their existence. You have probably encountered some of these in everyday life. A uniform distribution, for example, describes a set of equally likely outcomes, as when the value of the random variable, x, is the outcome of the roll of a single die. Other distributions are important in elementary physics. The Maxwell distribution, for example, describes the probability that in a gas of temperature T a randomly selected particle will have energy E.

2.3.6 Mean and variance of a distribution

Once the distribution is known, it is a simple matter to compute the population mean. Suppose $P(x, \mu, \sigma)$ is a continuous distribution, and $P'(x_i, \mu', \sigma')$ is a discrete distribution. The mean of each population is:

$$\mu = \frac{\int_{-\infty}^{\infty} xP dx}{\int_{-\infty}^{\infty} P dx} = \frac{1}{N} \int_{-\infty}^{\infty} xP dx \qquad (2.9)$$

and

$$\mu' = \frac{\sum\limits_{i=-\infty}^{\infty} x_i P'(x_i, \mu', \sigma')}{\sum\limits_{i=-\infty}^{\infty} P'(x_i, \mu', \sigma')} = \frac{1}{N'} \sum\limits_{i=-\infty}^{\infty} x_i P'(x_i, \mu', \sigma') \qquad (2.10)$$

If P' and P are, respectively, the probability and the probability density, then the terms in the denominators of the above equations (the "normalizations," N and N') should equal one.

Population variance (and standard deviation) also is easy to compute from the distribution

$$\sigma^2 = \frac{1}{N} \int_{-\infty}^{\infty} (x - \mu)^2 P dx = \frac{1}{N} \int_{-\infty}^{\infty} x^2 P dx - \mu^2 \qquad (2.11)$$

and

$$\sigma'^2 = \frac{1}{N'} \sum\limits_{i=-\infty}^{\infty} (x_i - \mu')^2 P'(x_i, \mu', \sigma') = \frac{1}{N'} \sum\limits_{i=-\infty}^{\infty} x_i^2 P'(x_i, \mu', \sigma') - \mu'^2 \qquad (2.12)$$

2.4 Estimating uncertainty

We can now address the central issue of this chapter: How do you estimate the uncertainty of a particular quantitative measurement? You now recognize most measurements result from sampling the very large population of all possible measurements. We consider a very common situation: a scientist samples a population by making n measurements, and computes the mean of the sample. He knows that this sample mean is the best guess for the population mean. The question is: How good is this guess? How close to the population mean is the sample mean? What uncertainty should he attach to it?

2.4.1 The Central Limit Theorem

Return to the example of a very large population of metal spheres that have a distribution of diameters as illustrated by Figure 2.3a. This distribution is clearly

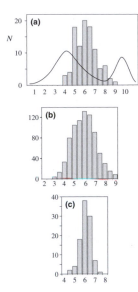

Fig. 2.6 (a) The distribution of a sample of 100 trials of the random variable $\bar{x}_5$. The solid curve is the distribution of the individual x values. Distribution (b) is for a sample of 800 trials of the random variable $\bar{x}_5$. This distribution is approximately Gaussian with a standard deviation of 1.13. Distribution (c) is the same as (a), except the random variable is $\bar{x}_{20}$. Its standard deviation is 0.54.

not Gaussian. Nevertheless, properties of the Gaussian are relevant even for this distribution. Consider the problem of estimating the average size of a sphere. Suppose we ask Dora, our cheerful assistant, to conduct a prototypical experiment: select five spheres at random, measure them and compute the average diameter. The result of such an experiment is a new random variable, $\bar{x}_5$, which is an estimate of the mean of the entire non-Gaussian population of spheres. Dora is a tireless worker. She does not stop with just five measurements, but enthusiastically conducts many experiments, pulling out many spheres at random, five at a time, and tabulating many different values for $\bar{x}_5$. When we finally get her to stop measuring, Dora becomes curious about the distribution of her tabulated values. She plots the histograms shown in Figures 2.6a and 2.6b, the results for 100 and 800 determinations of $\bar{x}_5$ respectively.

"Looks like a Gaussian," says Dora. "In fact, the more experiments I do, the more the distribution of $\bar{x}_5$ looks like a Gaussian. This is curious, because the original distribution of diameters (the solid curve in Figure 2.6a) was not Gaussian."

Dora is correct. Suppose that $P(x)$ is the probability distribution for random variable x, where $P(x)$ is characterized by mean μ and variance σ^2, but otherwise can have any form whatsoever. In our example, $P(x)$ is the bimodal function plotted in Figure 2.3(a). The **Central Limit Theorem** states that if $\{x_1, x_2, \ldots, x_n\}$ is a sequence of n independent random variables drawn from P, then as n becomes large, the distribution of the variables

$$\bar{x}_n = \frac{1}{n} \sum_{i=1}^{n} x_i$$

will approach a Gaussian distribution with mean μ and variance σ^2/n

To illustrate this last statement, Dora computes the values of a new random variable $\bar{x}_{20}$, which is the mean of 20 individual xs. The distribution of 100 $\bar{x}_{20}$s is shown in Figure 2.6c. As expected, the new distribution has about one-half the dispersion of the one for the $\bar{x}_5$s.

Since so many measurements in science are averages of individual experiments, the Central Limit Theorem means that the Gaussian distribution will be central to the analysis of experimental results. In addition, the conclusion that the variance of the average is proportional to $1/n$ relates directly to the problem of estimating uncertainty. Since s, the estimated standard deviation, is the best guess for σ, we should estimate $\sigma_\mu(n)$, the standard deviation of $\bar{x}_n$, the mean, as

$$\sigma_\mu(n) = \frac{s}{\sqrt{n}} \qquad (2.13)$$

Here, s is computed according to Equation (2.5) from the scatter in the n individual measurements. It is common to simply quote the value of $\sigma_\mu(n)$ as the uncertainty in a measurement. The interpretation of this number is clear because the Central Limit Theorem implies that $\sigma_\mu(n)$ is the standard deviation of an approximately *Gaussian* distribution – one then knows, for example, that there is a 50% probability that $\bar{x}_n$ is within $0.6745\sigma_\mu(n)$ of the "true" value, μ.

2.4.2 Reducing uncertainty

The Central Limit Theorem, which applies to all distributions, as well as the elementary properties of the Poisson distribution, which applies to counting experiments, both suggest that the way to reduce the uncertainty (and increase both precision and possibly accuracy) in any estimate of the population mean is *repetition*. Either increase the number of trials, or increase the number of things counted. (In astronomy, where a trial often involves counting photons, the two sometimes amount to the same thing.) If N is either the number of repetitions, or the number of things counted, then the basic rule is:

$$\text{relative uncertainty} \propto \frac{1}{\sqrt{N}} \tag{2.14}$$

This implies that success, or at least experimental accuracy, in astronomy involves making N large. This means having a large telescope (i.e. able to collect many photons) for a long time (i.e. able to conduct many measurements). You should keep a number of very important cautions in mind while pondering the importance of Equation (2.14).

- The cost of improved accuracy is high. To decrease uncertainty by a factor of 100, for example, you have to increase the number of experiments (the amount of telescope time, or the area of its light-gathering element) by a factor of 10,000. At some point the cost becomes too high.
- Equation (2.14) only works for experiments or observations that are completely independent of one another and sample a stationary population. In real life, this need not be the case: for example, one measurement can have an influence on another by sensitizing or desensitizing a detector, or the brightness of an object can change with time. In such cases, the validity of (2.14) is limited.
- Equation (2.13) only describes uncertainties introduced by scatter in the parent population. You should always treat this as the very minimum possible uncertainty. Systematic errors will make an additional contribution, and often the dominant one.

2.5 Propagation of uncertainty

2.5.1 Combining several variables

We consider first the special case where the quantity of interest is the sum or difference of more than one measured quantity. For example, in differential photometry, you are often interested in the magnitude difference

$$\Delta m = m_1 - m_2$$

Here m_1 is the measured instrumental magnitude of a standard or comparison object, and m_2 is the instrumental magnitude of an unknown object. Clearly, the uncertainty in Δm depends on the uncertainties in both m_1 and m_2. If these uncertainties are known to be σ_1 and σ_2 then the uncertainty in Δm is given by

$$\sigma^2 = \sigma_1^2 + \sigma_2^2 \tag{2.15}$$

The above formula could be stated: "the variance of a sum (or difference) is the sum of the variances." Equation (2.15) could also be restated by saying that the uncertainties "add in quadrature."

Equation (2.15) suggests that a magnitude difference, Δm, will be more uncertain than the individual magnitudes. This is true only with respect to random errors, however. For example, if a detector is badly calibrated and always reads out energy values that are too high by 20%, then the values of individual fluxes and magnitudes it yields will be very inaccurate. However, systematic errors of this kind will often cancel each other if a flux ratio or magnitude difference is computed. It is a very common strategy to use such *differential measurements* as a way of reducing systematic errors.

A second special case of combining uncertainties concerns products or ratios of measured quantities. If, for example, one were interested in the ratio between two fluxes, $F_1 \pm \sigma_1$ and $F_2 \pm \sigma_2$

$$R = \frac{F_1}{F_2}$$

then the uncertainty in R is given by:

$$\left(\frac{\sigma_R}{R}\right)^2 = \left(\frac{\sigma_1}{F_1}\right)^2 + \left(\frac{\sigma_2}{F_2}\right)^2 \tag{2.16}$$

One might restate this equation by saying that for a product (or ratio) the *relative uncertainties* of the factors add in quadrature.

2.5.2 General rule

In general, if a quantity, G, is a function of n variables, $G = G(x_1, x_2, x_3, \ldots, x_n)$, and each variable has uncertainty (or standard deviation) $\sigma_1, \sigma_2, \sigma_3, \ldots, \sigma_n$, then the variance in G is given by:

$$\sigma_G^2 = \sum_{i=1}^{n} \left(\frac{\partial G}{\partial x_i}\right)^2 \sigma_{x_i}^2 + \text{covar} \tag{2.17}$$

Here the term "covar" measures the effect of correlated deviations. We assume this term is zero. You should be able to verify that Equations (2.15) and (2.16) follow from this expression.

2.5.3 Several measurements of a single variable

Suppose three different methods for determining the distance to the center of our Galaxy yield values 8.0 ± 0.3, 7.8 ± 0.7 and 8.25 ± 0.20 kiloparsecs. What is the best combined estimate of the distance, and what is its uncertainty? The general rule in this case is that if measurements $y_1, y_2, y_3, \ldots y_n$ have associated uncertainties $\sigma_1, \sigma_2, \sigma_3, \ldots, \sigma_n$, then the best estimate of the central value is

$$y_c = \sigma_c^2 \sum_{i=1}^{n} \left(y_i / \sigma_i^2 \right) \tag{2.18}$$

where the uncertainty of the combined quantity is

$$\sigma_c^2 = \frac{1}{\sum\limits_{i=1}^{n} (1/\sigma_i^2)}$$

If we define the weight of a measurement to be

$$w_i = \frac{1}{\sigma_i^2}$$

Equation (2.18) becomes

$$y_c = \frac{\sum\limits_{i=1}^{n} y_i w_i}{\sum\limits_{i=1}^{n} w_i} \tag{2.19}$$

Computed in this way, y_c is called the **weighted mean**. Although the weights, $\{w_1, w_2, \ldots, w_n\}$, should be proportional to the reciprocal of the variance of each of the y valuess, they can often be assigned by rather subjective considerations like the "reliability" of different astronomers, instruments, or methods. However assigned, each w_i represents the relative probability that a particular observation will yield the "correct" value for the quantity in question.

To complete our example, the best estimate for the distance to the center of the Galaxy from the above data would assign weights 11.1, 2.0 and 25 to the three measurements, resulting in a weighted mean of

$$y_c = \frac{1}{11.1 + 2 + 25} [11.1(8) + 2(7.8) + 25(8.25)] = 8.15 \text{ kpc}$$

and an uncertainty of

$$(11.1 + 2 + 25)^{-1/2} = 0.16 \text{ kpc}$$

Notice that the uncertainty of the combined result is less than the uncertainty of even the best of the individual results, and that of the three measurements, the one with the very large uncertainty (7.8 ± 0.7 kpc) has little influence on the weighted mean.

2.6 Additional topics

Several topics in elementary statistics are important in the analysis of data but are beyond the scope of this introduction. The ***chi-square (χ^2) statistic*** measures the deviation between experimental measurements and their theoretically expected values (e.g. from an assumed population distribution). Tests based on this statistic can assign a probability to the truth of the theoretical assumptions. ***Least-square fitting*** methods minimize the χ^2 statistic in the case of an assumed functional fit to experimental data (e.g. brightness as a function of time, color as a function of brightness, . . .).

You can find elementary discussions of these and many more topics in either Bevington (1969) or Lyons (1991). The books by Jaschek and Murtagh (1990) and by Wall and Jenkins (2003) give more advanced treatments of topics particularly relevant to astronomy.

Summary

Precision, but not ***accuracy***, can be estimated from the scatter in measurements.

Standard deviation is the square root of the ***variance***. For a ***population***:

$$\sigma^2 = \frac{1}{M} \sum_{i=1}^{M} (x_i - \mu)^2 = \frac{1}{M} \sum_{i=1}^{M} x_i^2 - \mu^2$$

From a ***sample***, the best estimate of the population variance is

$$s^2 = \frac{1}{N-1} \sum_{i=1}^{N} (x_i - \bar{x})^2$$

Probability distributions describe the expected values of a random variable drawn from a parent population. The ***Poisson distribution*** describes measurements made by counting uncorrelated events like the arrival of photons. For measurements following the Poisson distribution,

$$\sigma_{\text{Poisson}}^2 = \mu_{\text{Poisson}}$$

The ***Gaussian distribution*** describes many populations whose values have a smooth and symmetric distribution. The ***Central Limit Theorem*** contends that the mean of n random samples drawn from a population of mean μ and variance σ^2 will take on a Gaussian distribution in the limit of large n. This is a distribution whose mean approaches μ and whose variance approaches σ^2/n. For any distribution, the uncertainty (standard deviation) of the mean of n measurements of the variable x is

$$\sigma_\mu(n) = \frac{s}{\sqrt{n}}$$

The variance of a function of several uncorrelated variables, each with its own variance, is given by

$$\sigma_G^2 = \sum_{i=1}^{n} \left(\frac{\partial G}{\partial x_i}\right)^2 \sigma_{x_i}^2 + \text{covar}$$

For measurements of unequal variance, the **weighted mean** is

$$y_c = \sigma_c^2 \sum_{i=1}^{n} \left(y_i/\sigma_i^2\right)$$

and the combined variance is

$$1/\sigma_c^2 = \sum_{i=1}^{n} \left(1/\sigma_i^2\right)$$

Exercises

1. There are some situations in which it is impossible to compute the mean value for a set of data. Consider this example. The ten crew members of the starship *Nostromo* are all exposed to an alien virus at the same time. The virus causes the deaths of nine of the crew at the following times, in days, after exposure:

$$1.2, 1.8, 2.1, 2.4, 2.6, 2.9, 3.3, 4.0, 5.4$$

The tenth crew member is still alive after 9 days, but is infected with the virus. Based only on this data:

   (a) Why can't you compute the "average survival time" for victims of the *Nostromo* virus?

   (b) What is the "expected survival time"? (A victim has a 50–50 chance of surviving this long.) Justify your computation of this number.

2. An experimenter makes eleven measurements of a physical quantity, X, that can only take on integer values. The measurements are

$$0, 1, 2, 3, 4, 5, 6, 7, 8, 9, 10$$

   (a) Estimate the mean, median, variance (treating the set as a sample of a population) and standard deviation of this set of measurements.

   (b) The same experimenter makes a new set of 25 measurements of X, and finds the that the values

$$0, 1, 2, 3, 4, 5, 6, 7, 8, 9, 10$$

occur

$$0, 1, 2, 3, 4, 5, 4, 3, 2, 1, \text{and } 0$$

times respectively. Again, estimate the mean, median, variance and standard deviation of this set of measurements.

3. Describe your best guess as to the parent distributions of the samples given in questions 2(a) and 2(b).

4. Assume the parent distribution for the measurements in problem 2(b) is actually a Poisson distribution. (a) Explain how you would estimate the Poisson parameter, μ. From the resulting function, $P_P(X, \mu)$, compute (b) the probability of obtaining a value of exactly zero in a single measurement, and (c) the probability of obtaining a value of exactly 11.

5. Now assume the parent distribution for the measurements in problem 2(b) is actually a Gaussian distribution. From your estimate of $P_G(x, \mu, \sigma)$, again compute (a) the probability of obtaining a value of exactly zero in a single measurement, as well as (b) the probability of obtaining a value of exactly 11. You will have to decide what "exactly 0" etc. means in this context.

6. Which distribution, Poisson or Gaussian, do you think is the better fit to the data in problem 2(b)? Suggest a method for quantifying the goodness of each fit. (Hint: see Section 2.6.)

7. Compute the mean, median, and standard deviation of the population for group B in Table 2.4.

8. An astronomer wishes to make a photon-counting measurement of a star's brightness that has a relative precision of 5%. (a) How many photons should she count? (b) How many should she count for a relative precision of 0.5%?

9. A star cluster is a collection of gravitationally bound stars. Individual stars in the cluster move about with different velocities but the average of all these should give the velocity of the cluster as a whole. An astronomer measures the radial velocities of four stars in a cluster that contains 1000 stars. They are 74, 41, 61, and 57 km s^{-1}. How any additional stars should he measure if he wishes to achieve a precision of 2 km s^{-1} for the radial velocity of the cluster as a whole?

10. The astronomer in problem 8 discovers that when she points her telescope to the blank sky near the star she is interested in, she measures a background count that is 50% of what she measures when she points to the star. She reasons that the brightness of the star (the interesting quantity) is given by

$$\text{star} = \text{measurement} - \text{background}$$

Revise your earlier estimates. How many measurement photons should she count to achieve a relative precision of 5% in her determination of the star brightness? How many for 0.5%?

11. An astronomer makes three one-second measurements of a star's brightness, counting 4 photons in the first, 81 in the second, and 9 in the third. What is the best estimate of the average photon arrival rate and its uncertainty?

Compute this (a) by using Equation (2.18), then (b) by noting that a total of 94 photons have arrived in 3 seconds.

Explain why these two methods give such different results. Which method is the correct one? Is there anything peculiar about the data that would lead you to question the validity of either method?

12. We repeat problem 14 from Chapter 1, where a single unknown star and standard star are observed in the same field. The data frame is in the figure below. The unknown star is the fainter one. If the magnitude of the standard is 9.000, compute the magnitude of the unknown, as in problem 1.14, but now also compute the uncertainty of your result, in magnitudes. Again data numbers represent the number of photons counted in each pixel.

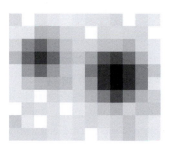

34	16	26	33	37	22	25	25	29	19	28	25
22	20	44	34	22	26	14	30	30	20	19	17
31	70	98	66	37	25	35	36	39	39	23	20
34	99	229	107	38	28	46	102	159	93	37	22
33	67	103	67	36	32	69	240	393	248	69	30
22	33	34	29	36	24	65	241	363	244	68	24
28	22	17	16	32	24	46	85	157	84	42	22
18	25	27	26	17	18	30	29	35	24	30	27
32	23	16	29	25	24	30	28	20	35	22	23
28	28	28	24	26	26	17	19	30	35	30	26

Chapter 3
Place, time, and motion

> Then, just for a minute. . . he turned off the lights. . . . And then while we all still waited I understood that the terror of my dream was not about losing just vision, but the whole of myself, whatever that was. What you lose in blindness is the space around you, the place where you are, and without that you might not exist. You could be nowhere at all.
>
> – Barbara Kingsolver, *Animal Dreams*, 1990

Where is Mars? The center of our Galaxy? The brightest X-ray source? Where, indeed, are we? Astronomers have always needed to locate objects and events in space. As our science evolves, it demands ever more exact locations. Suppose, for example, an astronomer observes with an X-ray telescope and discovers a source that flashes on and off with a curious rhythm. Is this source a planet, a star, or the core of a galaxy? It is possible that the X-ray source will appear to be quite unremarkable at other wavelengths. The exact position for the X-ray source might be the only way to identify its optical or radio counterpart. Astronomers need to know where things are.

Likewise, knowing *when* something happens is often as important as *where* it happens. The rhythms of the spinning and orbiting Earth gave astronomy an early and intimate connection to timekeeping. Because our Universe is always changing, astronomers need to know what time it is.

The "fixed stars" are an old metaphor for the unchanging and eternal, but positions of real celestial objects do change, and the changes tell stories. Planets, stars, gas clouds, and galaxies all trace paths decreed for them. Astronomers who measure these motions, sometimes only through the accumulated labors of many generations, can find in their measurements the outlines of nature's decree. In the most satisfying cases, the measurements uncover fundamental facts, like the distances between stars or galaxies, or the age of the Universe, or the presence of planets orbiting other suns beyond the Sun. Astronomers need to know how things move.

3.1 Astronomical coordinate systems

> Any problem of geometry can easily be reduced to such terms that a knowledge
> of the lengths of certain straight lines is sufficient for its construction.
>
> — Rene Descartes, *La Geometrie, Book I*, 1637

Descartes' brilliant application of coordinate systems to solve geometric problems has direct relevance to astrometry, the business of locating astronomical
objects. Although astrometry has venerably ancient origins,[1] it retains a central
importance in astronomy.

3.1.1 Three-dimensional coordinates

I assume you are familiar with the standard (x, y, z) Cartesian coordinate system
and the related spherical coordinate system (r, ϕ, θ), illustrated in Figure 3.1(a).
Think for a moment how you might set up such a coordinate system in practice.
Many methods could lead to the same result, but consider a process that consists
of four decisions:

1. Locate the origin. In astronomy, this often corresponds to identifying some distinctive
   real or idealized object: the centers of the Earth, Sun, or Galaxy, for example.
2. Locate the x–y plane. We will call this the "fundamental plane." The fundamental
   plane, again, often has physical significance: the plane defined by the Earth's equator –
   or the one that contains Earth's orbit – or the symmetry plane of the Galaxy, for
   example. The z-axis passes through the origin perpendicular to the fundamental plane.
3. Decide on the direction of the positive x-axis. We will call this the "reference direction." Sometimes the reference direction has a physical significance – the direction
   from the Sun to the center of the Galaxy, for example. The y-axis then lies in the
   fundamental plane, perpendicular to the x-axis.
4. Finally, decide on a convention for the signs of the y- and z-axes. These choices
   produce either a left- or right-handed system – see below."

The traditional choice for measuring the *angles* is to measure the first coordinate, ϕ (or λ), within the fundamental plane so that ϕ increases from the
$+x$-axis towards the $+y$-axis (see Figure 3.1). The second angle, θ (or ζ), is
measured in a plane perpendicular to the fundamental plane increasing from
the positive z axis towards the x–y plane. In this scheme, ϕ ranges, in radians,
from 0 to 2π and θ ranges from 0 to π. A common alternative is to measure
the second angle (β in the figure) from the x–y plane, so it ranges between $-\pi/2$
and $+\pi/2$.

---

[1] Systematic Babylonian records go back to about 650 BC but with strong hints that the written
tradition had Sumerian roots in the late third millennium. Ruins of megalithic structures with clear
astronomical alignments date from as early as 4500 BC (Nabta, Egypt).

(a)

Right-hand

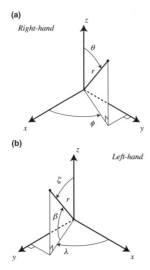

(b)

Left-hand

Fig. 3.1 Three-dimensional coordinate systems. (a) The traditional system is right-handed. (b) This system is left-handed, its axes are a mirror image of those in (a). In either system one can choose to measure the second angle from the fundamental plane (e.g. angle β) instead of from the z axis (angles θ or ζ).

The freedom to choose the signs of the y- and z-axes in step 4 of this procedure implies that there are two (and only two) kinds of coordinate systems. One, illustrated in Figure 3.1a, is **right-handed**: if you wrap the fingers of your right hand around the z axis so the tips point in the $+\phi$ direction (that is, from the $+x$ axis towards the $+y$ axis), then your thumb will point in the $+z$ direction. In a **left-handed** system, like the (r, λ, ζ) system illustrated in Figure 3.1(b), you use your left hand to find the $+z$ direction. The left-handed system is the mirror image of the right-handed system. In either system, Pythagoras gives the radial coordinate as:

$$r = \sqrt{x^2 + y^2 + z^2}$$

3.1.2 Coordinates on a spherical surface

> It is one of the things proper to geography to assume that the Earth as a whole is spherical in shape, as the universe also is. . .
>
> — Strabo, *Geography*, II, 2, 1, *c.* AD 18

If all points of interest are on the surface of a sphere, the r coordinate is superfluous, and we can specify locations with just two angular coordinates like (ϕ, θ) or (λ, β). Many astronomical coordinate systems fit into this category, so it is useful to review some of the characteristics of geometry and trigonometry on a spherical surface.

1. A **great circle** is formed by the intersection of the sphere and a plane that contains the center of the sphere. The shortest distance between two points on the surface of a sphere is an arc of the great circle connecting the points.
2. A **small circle** is formed by the intersection of the sphere and a plane that does not contain the center of the sphere.
3. The **spherical angle** between two great circles is the angle between the planes, or the angle between the straight lines tangent to the two great circle arcs at either of their points of intersection.
4. A **spherical triangle** on the surface of a sphere is one whose sides are all segments of great circles. Since the sides of a spherical triangle are arcs, the sides can be measured in angular measure (i.e. radians or degrees) rather than linear measure. See Figure 3.2.
5. The **law of cosines** for spherical triangles in Figure 3.2 is:

$$\cos a = \cos b \cos c + \sin b \sin c \cos A$$

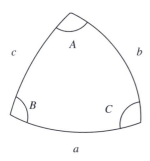

Fig. 3.2 A spherical triangle. You must imagine this figure is drawn on the surface of a sphere. *A, B,* and *C* are spherical angles; *a, b,* and *c* are arcs of great circles.

or

$$\cos A = \cos B \cos C + \sin B \sin C \cos a$$

6. The **law of sines** is

$$\frac{\sin a}{\sin A} = \frac{\sin b}{\sin B} = \frac{\sin c}{\sin C}$$

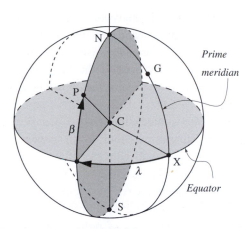

Fig. 3.3 The latitude–longitude system. The center of coordinates is at C. The fundamental direction, line CX, is defined by the intersection of the prime meridian (great circle NGX) and the equator. Latitude, β, and longitude, λ, for some point, P, are measured as shown. Latitude is positive north of the equator, negative south. Astronomical longitude for Solar System bodies is positive in the direction opposite the planet's spin. (i.e. to the west on Earth). On Earth, coordinates traditionally carry no algebraic sign, but are designated as north or south latitude, and west or east longitude. The coordinate, β, is the geocentric latitude. The coordinate actually used in practical systems is the geodetic latitude (see the text).

3.1.3 Terrestrial latitude and longitude

> "I must be getting somewhere near the center of the Earth...yes...but then I wonder what Latitude and Longitude I've got to?" (Alice had not the slightest idea what Latitude was, nor Longitude either, but she thought they were nice grand words to say.)
>
> — Lewis Carroll, *Alice's Adventures in Wonderland*, 1897

Ancient geographers introduced the seine-like latitude–longitude system for specifying locations on Earth well before the time Hipparchus of Rhodes (*c.* 190–120 BC) wrote on geography. Figure 3.3 illustrates the basic features of the system.

In our scheme, the first steps in setting up a coordinate system are to choose an origin and fundamental plane. We can understand why Hipparchus, who believed in a geocentric cosmology, would choose the center of the Earth as the origin. Likewise, choice of the equatorial plane of the Earth as the fundamental plane makes a lot of practical sense. Although the location of the equator may not be obvious to a casual observer like Alice, it is easily determined from simple astronomical observations. Indeed, in his three-volume book on geography, Eratosthenes of Alexandria (*c.* 275 − *c.* 194 BC) is said to have computed the location of the equator relative to the parts of the world known to him. At the time, there was considerable dispute as to the habitability of the (possibly too hot) regions near the equator, but Eratosthenes clearly had little doubt about their location.

Great circles perpendicular to the equator must pass through both poles, and such circles are termed *meridians*. The place where one of these − the *prime meridian* − intersects the equator could constitute a reference direction (*x*-axis). Unfortunately, on Earth, there is no obvious meridian to use for this purpose. Many choices are justifiable, and for a long time geographers simply chose a prime meridian that passed though some locally prominent or worthy place. Thus, the latitude of any point on Earth was unique, but its longitude was not,

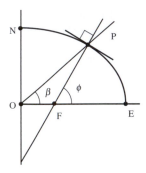

Fig. 3.4. Geocentric (β) and geodetic (ϕ) latitudes. Line PF is perpendicular to the surface of the reference spheroid, and approximately in the direction of the local vertical (local gravitational force).

since it depended on which meridian one chose as prime. This was inconvenient. Eventually, in 1884, the "international" community (in the form of representatives of 25 industrialized countries meeting in Washington, DC, at the First International Meridian Conference) settled the zero point of longitude at the meridian of the Royal Observatory in Greenwich, located just outside London, England.

You should note that the latitude coordinate, β, just discussed, is called the **geocentric latitude**, to distinguish it from ϕ, the **geodetic latitude**. Geodetic latitude is defined in reference to an ellipsoid-of-revolution that approximates the actual shape of the Earth. It is the angle between the equatorial plane and a line perpendicular to the surface of the reference ellipsoid at the point in question.

Figure 3.4 shows the north pole, N, equator, E, and center, O, of the Earth. The geocentric and geodetic latitudes of point P are β and ϕ, respectively. Geodetic latitude is easier to determine and is the one employed in specifying positions on the Earth. The widely used technique of global positioning satellites (GPS), for example, returns geodetic latitude, longitude, and height above a reference ellipsoid. To complicate things a bit more, the most easily determined latitude is the **geographic latitude**, the angle between the local vertical and the equator. Massive objects like mountains affect the geographic but not the geodetic latitude and the two can differ by as much as an arc minute. Further complications on the sub-arc-second scale arise from short- and long-term motion of the geodetic pole itself relative to the Earth's crust due to tides, earthquakes, internal motions, and continental drift.

Planetary scientists establish latitude–longitude systems on other planets, with latitude usually easily defined by the object's rotation, while definition of longitude depends on identifying some feature to mark a prime meridian.

Which of the two poles of a spinning object is the "north" pole? In the Solar System, the preferred (but not universal!) convention is that the **ecliptic** – the plane containing the Earth's orbit – defines a fundamental plane, and a planet's north pole is the one that lies to the (terrestrial) north side of this plane. Longitude should be measured as increasing in the direction opposite the spin direction.

For other objects, practices vary. One system says the north pole is determined by a right-hand rule applied to the direction of spin: wrap the fingers of your right hand around the object's equator so that they point in the direction of its spin. Your thumb then points north (in this case, "north" is in the same direction as the angular momentum vector).

3.1.4 The altitude–azimuth system

Imagine an observer, a shepherd with a well-behaved flock, say, who has some leisure time on the job. Our shepherd is lying in an open field, contemplating the sky. After a little consideration, our observer comes to imagine the sky as a hemisphere – an inverted bowl whose edges rest on the horizon. Astronomical objects, whatever their real distances, can be seen to be stuck onto or projected onto the inside of this hemispherical sky.

This is another situation in which the r-coordinate becomes superfluous. The shepherd will find it difficult or impossible to determine the r coordinate for the objects in the sky. He knows the direction of a star but not its distance from the origin (which he will naturally take to be himself). Astronomers often find themselves in the same situation as the shepherd. A constant theme throughout astronomy is the problem of the third dimension, the r-coordinate: the directions of objects are easily and accurately determined, but their distances are not. This prompts us to use coordinate systems that ignore the r-coordinate and only specify the two direction angles.

In Figure 3.5, we carry the shepherd's fiction of a hemispherical sky a little bit further, and imagine that the hemispherical bowl of the visible sky is matched by a similar hemisphere below the horizon, so that we are able to apply a spherical coordinate scheme like the one illustrated. Here, the origin of the system is at O, the location of the observer. The fundamental plane is that of the "flat" Earth (or, to be precise, a plane tangent to the tiny spherical Earth at point O). This fundamental plane intersects the sphere of the sky at the **celestial horizon** – the great circle passing through the points NES in the figure. **Vertical circles** are great circles on the spherical sky that are perpendicular to the fundamental plane. All vertical circles pass through the overhead point, which is called the **zenith** (point T in the figure), as well as the diametrically opposed point, called the **nadir**. The vertical circle that runs in the north–south direction (circle NTS in the figure) is called the **observer's meridian**.

The fundamental direction in the altitude–azimuth coordinate system runs directly north from the observer to the intersection of the meridian and the celestial horizon (point N in the figure). In this system, a point on the sky, P, has two coordinates:

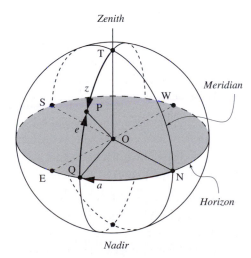

Fig. 3.5. The altitude–azimuth system. The horizon defines the fundamental plane (gray) and the north point on the horizon, N, defines the fundamental direction. Point P has coordinates a (azimuth), which is measured along the horizon circle from north to east, and e (altitude), measured upwards from the horizon. Objects with negative altitudes are below the horizon.

- The **altitude**, or **elevation**, is the angular distance of P above the horizon (∠QOP or e in the figure). Objects below the horizon have negative altitudes.
- The **azimuth** is the angular distance from the reference direction (the north point on the horizon) to the intersection of the horizon and the vertical circle passing through the object (∠NOQ or a in the figure).

Instead of the altitude, astronomers sometimes use its complement, z, the ***zenith distance*** (∠TOP in the figure).

The (a, e) coordinates of an object clearly describe where it is located in an observer's sky. You can readily imagine an instrument that would measure these coordinates: a telescope or other sighting device mounted to rotate on vertical and horizontal circles that are marked with precise graduations.

One of the most elementary astronomical observations, noticed even by the most unobservant shepherd, is that celestial objects don't stay in the same place in the horizon coordinate system. Stars, planets, the Sun, and Moon all execute a ***diurnal motion***: they rise in the east, cross the observer's meridian, and set in the west. This, of course, is a reflection of the spin of our planet on its axis. The altitude and azimuth of celestial objects will change as the Earth executes its daily rotation. Careful measurement will show that stars (but not the Sun and planets, which move relative to the "fixed" stars) will take about 23 hours, 56 minutes and 4.1 seconds between successive meridian crossings. This period of time is known as one ***sidereal day***. Very careful observations would show that the sidereal day is actually getting longer, relative to a stable atomic clock, by about 0.0015 second per century. The spin rate of the Earth is slowing down.

3.1.5 The equatorial system: definition of coordinates

Because the altitude and azimuth of celestial objects change rapidly, we create another reference system, one in which the coordinates of stars remain the same. In this ***equatorial coordinate system***, we carry the fiction of the spherical sky one step further. Imagine that all celestial objects were stuck on a sphere of very large radius, whose center is at the center of the Earth. Furthermore, imagine that the Earth is insignificantly small compared to this ***celestial sphere***. Now adopt a geocentric point of view. You can account for the diurnal motion of celestial objects by presuming that the entire celestial sphere spins east to west on an axis coincident with the Earth's actual spin axis. Relative to one another objects on the sphere never change their positions (not quite true – see below). The star patterns that make up the figures of the constellations stay put, while terrestrials observe the entire sky – the global pattern of constellations – to spin around its north–south axis once each sidereal day. Objects stuck on the celestial sphere thus appear to move east to west across the terrestrial sky, traveling in small circles centered on the nearest celestial pole.

The fictional celestial sphere is an example of a scientific model. Although the model is not the same as the reality, it has features that help one discuss, predict, and understand real behavior. (You might want to think about the meaning of the word "understand" in a situation where model and reality differ so extensively.) The celestial-sphere model allows us to specify the positions of the stars in a coordinate system, the equatorial system, which is independent of time, at least on short scales. Because positions in the equatorial coordinate system are also easy to measure from Earth, it is the system astronomers use most widely to locate objects on the sky.

The equatorial system *nominally* chooses the center of the Earth as the origin and the equatorial plane of the Earth as the fundamental plane. This aligns the z-axis with the Earth's spin axis, and fixes the locations of the two **celestial poles** – the intersection of the z-axis and the celestial sphere. The great circle defined by the intersection of the fundamental plane and the celestial sphere is called the **celestial equator**. One can immediately measure a latitude-like coordinate with respect to the celestial equator. This coordinate is called the **declination** (abbreviated as Dec or δ), whose value is taken to be zero at the equator, and positive in the northern celestial hemisphere; see Figure 3.6.

We choose the fundamental direction in the equatorial system by observing the motion of the Sun relative to the background of "fixed" stars. Because of the Earth's orbital motion, the Sun appears to trace out a great circle on the celestial sphere in the course of a year. This circle is called the **ecliptic** (it is where eclipses happen) and intersects the celestial equator at an angle, ϵ, called the **obliquity of the ecliptic**, equal to about 23.5 degrees. The point where the Sun crosses the equator traveling from south to north is called the **vernal equinox** and this point specifies the reference direction of the equatorial system. The coordinate angle measured in the equatorial plane is called the **right ascension** (abbreviated as RA or α). As shown in Figure 3.6, the equatorial system is right-handed, with RA increasing from west to east.

For reasons that will be apparent shortly, RA is usually measured in hours: minutes:seconds, rather than in degrees (24 hours of RA constitute 360 degrees of arc at the equator, so one hour of RA is 15 degrees of arc long at the equator). To deal with the confusion that arises from both the units of RA and the units of Dec having the names "minutes" and "seconds", one can speak of "minutes (or seconds) *of time*" to distinguish RA measures from the "minutes of arc" used to measure Dec.

3.1.6 The relation between the equatorial and the horizon systems

Figure 3.7 shows the celestial sphere with some of the features of both the horizon and equatorial systems marked. The figure assumes an observer, "O", located at about 60 degrees north latitude on Earth. Note the altitude of

Fig. 3.6 The equatorial coordinate system. In both celestial spheres pictured, the equator is the great circle passing through points V and B, and the ecliptic is the great circle passing through points V and S. The left-hand sphere shows the locations of the north (N) and south (M) celestial poles, the vernal (V) and autumnal (A) equinoxes, the summer (S) solstice, and the hour circles for 0 Hr (arc NVM) and 6 Hr (arc NBM) of right ascension. The right-hand sphere shows the right ascension (∠VOQ, or α) and declination (∠QOP, or δ) of the point P.

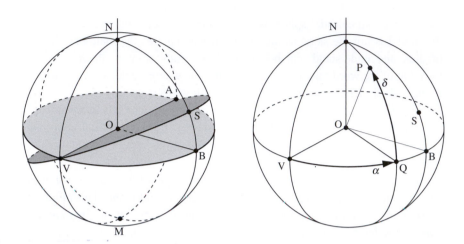

the north celestial pole (angle NOP in Figure 3.7a). You should be able to construct a simple geometric argument to convince yourself that: the altitude angle of the north celestial pole equals the observer's geodetic latitude.

Observer "O," using the horizon system, will watch the celestial sphere turn, and see stars move along the projected circles of constant declination. Figure 3.7a shows the declination circle of a star that just touches the northern horizon. Stars north of this circle never set and are termed *circumpolar*. Figure 3.7a also shows the declination circle that just touches the southern horizon circle, and otherwise lies entirely below it. Unless she changes her latitude, "O" can never see any of the stars south of this declination circle.

Reference to Figure 3.7a also helps define a few other terms. Stars that are neither circumpolar nor permanently below the horizon will rise in the east, cross, or *transit*, the observer's celestial meridian, and set in the west. When a star transits the meridian it has reached its greatest altitude above the horizon, and is said to have reached its *culmination*. Notice that circumpolar stars can be observed to cross the meridian twice each sidereal day (once when they are highest in the sky, and again when they are lowest). To avoid confusion, the observer's celestial meridian is divided into two pieces at the pole. The smaller bit visible between the pole and the horizon (arc NP in the figure) is called the lower meridian, and the remaining piece (arc PTML) is called the *upper meridian*.

Figure 3.7b shows a star, S, which has crossed the upper meridian some time ago and is moving to set in the west. A line of constant right ascension is a great circle called an *hour circle*, and the hour circle for star S is shown in the figure.

You can specify how far an object is from the meridian by giving its *hour angle*. The hour circle of an object and the upper celestial meridian intersect at the pole. The hour angle, HA, is the angle between them. Application of the law of sines to a spherical right triangle shows that the hour angle could also be

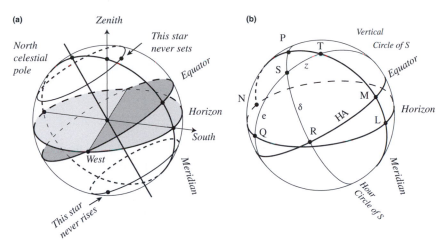

Fig. 3.7 The horizon and equatorial systems. Both spheres show the horizon, equator and observer's meridian, the north celestial pole at P, and the zenith at T. Sphere (a) illustrates the diurnal paths of a circumpolar star and of a star that never rises. Sphere (b) shows the hour circle (PSR) of a star at S, as well as its declination, δ, its hour angle, HA = arc RM = $\angle$MPS, its altitude, e, and its zenith distance, z.

measured along the equator, as the arc that runs from the intersection of the meridian and equator to the intersection of the star's hour circle and the equator (arc RM in Figure 3.7b). Hour angle, like right ascension, is usually measured in time units. Recalling that RA is measured in the plane of the equator, we can state one other definition of the hour angle:

HA of the object = RA on meridian − RA of the object

The hour angle of a star is useful because it tells how long ago (in the case of positive HA) or how long until (negative HA) the star crossed, or will cross, the upper meridian. The best time to observe an object is usually when it is highest in the sky, that is, when the HA is zero and the object is at culmination.

To compute the hour angle from the formula above, you realize that the RA of the object is always known – you can look it up in a catalog or read it from a star chart. How do you know the right ascension of objects on the meridian? You read that from a ***sidereal clock***.

A sidereal clock is based upon the apparent motions of the celestial sphere. A clockmaker creates a clock that ticks off exactly 24 uniform "sidereal" hours between successive upper meridian transits by the vernal equinox (a period of about 23.93 "normal" hours, remember). If one adjusts this clock so that it reads zero hours at precisely the moment the vernal equinox transits, then it gives the correct sidereal time.

Sidereal day = Time between upper meridian transits by the vernal equinox

A sidereal clock mimics the sky, where the hour circle of the vernal equinox can represent the single hand of a 24-hour clock, and the observer's meridian can represent the "zero hour" mark on the clockface. There is a nice correspondence between the reading of any sidereal clock and the right ascension coordinate, namely

sidereal time = right ascension of an object on the upper meridian

It should be clear that we can restate the definition of hour angle as:

HA of object = sidereal time now − sidereal time object culminates

If either the sidereal time or an object's hour angle is known, one can derive the coordinate transformations between equatorial (α, δ) and the horizon (e, a) coordinates for that object. Formulae are given in Appendix D.

3.1.7 Measuring equatorial coordinates.

Astronomers use the equatorial system because RA and Dec are easily determined with great precision from Earth-based observatories. You should have a general idea of how this is done. Consider a specialized instrument, called a **transit telescope** (or **meridian circle**): the transit telescope is constrained to point only at objects on an observer's celestial meridian − it rotates on an axis aligned precisely east–west. The telescope is rigidly attached to a graduated circle centered on this axis. The circle lies in the plane of the meridian and rotates with the telescope. A fixed index, established using a plumb line perhaps, always points to the zenith. By observing where this index falls on the circle, the observer can thus determine the altitude angle (or zenith distance) at which the telescope is pointing. The observer is also equipped with a sidereal clock, which ticks off 24 sidereal hours between upper transits of the vernal equinox.

To use the transit telescope to determine declinations, first locate the celestial pole. Pick out a circumpolar star. Read the graduated circle when you observe the star cross the upper and then again when it crosses the lower meridian. The average of the two readings gives the location of $\pm 90°$ declination (the north or south celestial pole) on your circle. After this calibration you can then read the declination of any other transiting star directly from the circle.

To find the *difference* in the RA of any two objects, subtract the sidereal clock reading when you observe the first object transit from the clock reading when you observe the second object transit. To locate the vernal equinox and the zero point for the RA coordinate, require that the right ascension of the Sun be zero when you observe its declination to be zero in the spring.

Astrometry is the branch of astronomy concerned with measuring the positions, and changes in position, of sources. Chapter 11 of Birney *et al.* (2006) gives a more though introduction to the subject than we will do here, and Monet (1988) gives a more advanced discussion. The Powerpoint presentation on the Gaia website (http://www.rssd.esa.int/Gaia) gives a good introduction to astrometry from space.

Observations with a transit telescope can measure arbitrarily large angles between sources, and the limits to the accuracy of **large-angle astrometry** are different from, and usually much more severe than, the limits to small-angle astrometry. In **small-angle astrometry**, one measures positions of a source relative to a local reference frame (e.g. stars or galaxies) contained on the same detector field. Examples of small-angle astrometry are the measurement of the separation of double stars with a micrometer-equipped eyepiece, the measurement of stellar parallax from a series of photographs, or the measurement of the position of a minor planet in two successive digital images of the same field.

The angular size and regularity of the stellar images formed by the transit telescope limit the precision of large-angle astrometry. The astronomer or her computer must decide when and where the center of the image transits, a task made difficult if the image is faint, diffuse, irregular, or changing shape on a short time scale. In the optical or near infrared, atmospheric seeing usually limits ground-based position measurements to an accuracy of about 0.05 arcsec, or 50 milli-arcsec (mas).

Positional accuracy at radio wavelengths is much greater. The technique of **very long baseline interferometry** (VLBI) can determine coordinates for point-like radio sources (e.g. the centers of active galaxies) with uncertainties less than 1 mas. Unfortunately, most normal stars are not sufficiently powerful radio sources to be detected, and their positions must be determined by optical methods.

There are other sources of error in wide-angle ground-based astrometry. Refraction by the atmosphere (see Figure 3.8 and Appendix D) changes the apparent positions of radio and (especially) optical sources. Variability of the atmosphere can produce inaccuracies in the correction made for refraction. Flexure of telescope and detector parts due to thermal expansion or variations in gravitational loading can cause serious systematic errors. Any change, for example, that moves the vertical index relative to the meridian circle will introduce inconsistencies in declination measurements.

Modern procedures for measuring equatorial coordinates are much more refined than those described at the beginning of this section, but the underlying principles are the same. Most ground-based transit measurements are automated with a variety of electronic image detectors and strategies for determining transit times.

Space-based large-angle astrometry uses principles similar to the ground-based programs. Although ground-based transit telescopes use the spinning Earth as a platform to define both direction and time scale, any uniformly spinning platform and any clock could be equivalently employed. The spin of the artificial satellite HIPPARCOS, for example, allowed it to measure stellar positions by timing transits in two optical telescopes mounted on the satellite. Because images in space are neither blurred by atmospheric seeing or subject to atmospheric refraction, most of the 120,000 stars in the HIPPARCOS catalog

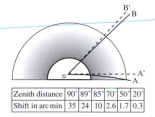

Zenith distance	90°	89°	85°	70°	50°	20°
Shift in arc min	35	24	10	2.6	1.7	0.3

Fig. 3.8 Atmospheric refraction. The observer is on the surface at point O. The actual path of a light ray from object A is curved by the atmosphere, and O receives light from direction A′. Likewise, the image of object B appears at B′ – a smaller shift in position because both the path length and the angle of incidence are smaller. Refraction thus reduces the zenith distance of all objects, affecting those close to the horizon more than those near the zenith. The table below the figure gives approximate shifts in arc minutes for different zenith distances.

have positional accuracies around 0.7 mas in each coordinate. A future mission, Gaia (the European Space Agency expects launch in 2012), will use a similar strategy with vastly improved technology. Gaia anticipates positional accuracies on the order of 0.007 mas (= 7 μas) for bright stars and accuracies better than 0.3 mas for close to a billion objects brighter than $V = 20$.

Catalogs produced with large-angle astrometric methods like transit tele-scope observations or the Gaia and HIPPARCOS missions are usually called *fundamental catalogs*.

It is important to realize that although the relative positions of most "fixed" stars on the celestial sphere normally do not change appreciably on time scales of a year or so, their equatorial coordinates *do* change by as much as 50 arcsec per year due to precession and other effects. Basically, the location of the celestial pole, equator, and equinox are always moving (see Section 3.1.8 below). This is an unfortunate inconvenience. Any measurement of RA and Dec made with a transit circle or other instrument must allow for these changes. What is normally done is to correct measurements to compute the coordinates that the celestial location *would* have at a certain date. Currently, the celestial equator and equinox for the year 2000 (usually written as J2000.0) are likely to be used.

You should also realize that even the *relative* positions of some stars, espe-cially nearby stars, do change very slowly due to their actual motion in space relative to the Sun. This *proper motion*, although small (a large proper motion would be a few arcsec per century), will cause a change in coordinates over time, and an accurate specification of coordinates must give the *epoch* (or date) for which they are valid. See Section 3.4.2 below.

3.1.8 Precession and nutation

Conservation of angular momentum might lead one to expect that the Earth's axis of rotation would maintain a fixed orientation with respect to the stars. However, the Earth has a non-spherical mass distribution, so it does experience gravitational torques from the Moon (primarily) and Sun. In addition to this lunisolar effect, the other planets produce much smaller torques. As a result of all these torques, the spin axis changes its orientation, and the celestial poles and equator change their positions with respect to the stars. This, of course, causes the RA and Dec of the stars to change with time.

This motion is generally separated into two components, a long-term general trend called *precession*, and a short-term oscillatory motion called *nutation*. Figure 3.9 illustrates precession: the north *ecliptic* pole remains fixed with respect to the distant background stars, while the north *celestial* pole (NCP) moves in a small circle whose center is at the ecliptic pole. The precessional circle has a radius equal to the average obliquity (around 23 degrees), with the NCP completing one circuit in about 26,000 years, moving at a very nearly – but

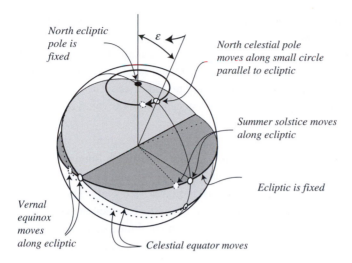

North ecliptic pole is fixed

ε

North celestial pole moves along small circle parallel to ecliptic

Summer solstice moves along ecliptic

Ecliptic is fixed

Vernal equinox moves along ecliptic

Celestial equator moves

Fig. 3.9 Precession of the equinoxes. The location of the ecliptic and the ecliptic poles is fixed on the celestial sphere. The celestial equator moves so that the north celestial pole describes a small circle around the north ecliptic pole of radius equal to the mean obliquity.

not precisely – constant speed. The celestial equator, of course, moves along with the pole, and the vernal equinox, which is the fundamental direction for both the equatorial and ecliptic coordinate systems, moves westward along the ecliptic at the rate (in the year 2000) of 5029.097 arcsec (about 1.4 degrees) per century. Precession will in general cause both the right ascension and declination of every star to change over time, and will also cause the ecliptic longitude (but not the ecliptic latitude) to change as well.

The most influential ancient astronomer, Hipparchus of Rhodes (recorded observations 141–127 BCE) spectacularly combined the rich tradition of Babylonian astronomy, which was concerned with mathematical computation of future planetary positions from extensive historic records, and Greek astronomy, which focused on geometrical physical models that described celestial phenomena. He constructed the first quantitative geocentric models for the motion of the Sun and Moon, developed the trigonometry necessary for his theory, injected the Babylonian sexagesimal numbering system (360° in a circle) into western use, and compiled the first systematic star catalog. Hipparchus discovered lunisolar precessional motion, as a steady regression of the equinoxes, when he compared contemporary observations with the Babylonian records. Unfortunately, almost all his original writings are lost, and we know his work mainly though the admiring Ptolemy, who lived three centuries later.

Since the time of Hipparchus, the vernal equinox has moved about 30° along the ecliptic. In fact, we still refer to the vernal equinox as the "first point of Aries," as did Hipparchus, even though it has moved out of the constellation Aries and through almost the entire length of the constellation Pisces since his time. Precession also means that the star Polaris is only temporarily located near the north celestial pole. About 4500 years ago, at about the time the Egyptians constructed the Great Pyramid, the "North Star" was Thuban, the brightest star in Draco. In 12,000 years, the star Vega will be near the pole, and Polaris will have a declination of 43°.

Unlike lunisolar precession, planetary precession actually changes the angle between the equator and ecliptic. The result is an oscillation in the obliquity so that it ranges from 22° to 24°, with a period of about 41,000 years. At present, the obliquity is decreasing from an accepted J2000 value of 23° 26′ 21.4″ at a rate of about 47 arcsec per century.

Nutation, the short period changes in the location of the NCP, is usually separated into two components. The first, nutation in longitude, is an oscillation of the equinox ahead of and behind the precessional position, with an amplitude of about 9.21 arcsec and a principal period of 18.6 years. The second, nutation in obliquity, is a change in the value of the angle between the equator and ecliptic. This also is a smaller oscillation, with an amplitude of about 6.86 arcsec and an identical principal period. Both components were discovered telescopically by James Bradley (1693–1762), the third British Astronomer Royal.

3.1.9 Barycentric coordinates

Coordinates measured with a transit telescope from the surface of the moving Earth as described in the preceding section are in fact measured in a non-inertial reference frame, since the spin and orbital motions of the Earth accelerate the telescope. These **apparent equatorial coordinates** exhibit variations introduced by this non-inertial frame, and their exact values will depend on the time of observation and the location of the telescope. Catalogs therefore give positions in an equatorial system similar to the one defined as above, but whose origin is at the barycenter (center of mass) of the Solar System. Barycentric coordinates use the mean equinox of the catalog date (a fictitious equinox which moves with precessional motion, but not nutational). The barycentric coordinates are computed from the apparent coordinates by removing several effects. In addition to precession and nutation, we will discuss two others. The first, due to the changing vantage point of the telescope as the Earth executes its orbit, is called **heliocentric stellar parallax**. The small variation in a nearby object's apparent coordinates due to parallax depends on the object's distance and is an important quantity described in Section 3.2.2.

The second effect, caused by the finite velocity of light, is called the **aberration of starlight**, and produces a shift in every object's apparent coordinates. The magnitude of the shift depends only on the angle between the object's direction and the direction of the instantaneous velocity of the Earth. Figure 3.10 shows a telescope in the barycentric coordinate system, drawn so that the velocity of the telescope, at rest on the moving Earth, is in the $+x$ direction. A photon from a distant object enters the telescope at point A, travels at the speed of light, c, and exits at point B. In the barycentric frame, the photon's path makes an angle θ with the x-axis. However, if the photon is to

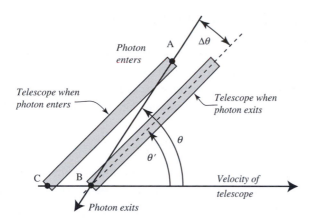

Fig. 3.10 The aberration
of starlight. A telescope
points towards a source.
The diagram shows the
telescope moving to the
right in the barycentric
frame. The apparent
direction of the source, θ',
depends on the direction
and magnitude of the
telescope velocity.

enter and exit the moving telescope successfully, the telescope must make an
angle $\theta' = \theta - \Delta\theta$ with the x-axis in the frame fixed on the Earth. A little
geometry shows that, if V is the speed of the Earth,

$$\Delta\theta = \frac{V}{c}\sin\theta$$

Thus aberration moves the apparent position of the source (the one measured by
a telescope on the moving Earth) towards the x-axis. The magnitude of this
effect is greatest when $\theta = 90°$, where it amounts to about 20.5 arcsec.

3.1.10 The ICRS

The International Astronomical Union (IAU) in 1991 recommended creation of a
special coordinate system whose origin is at the barycenter of the Solar System,
with a fundamental plane approximately coincident with the Earth's equatorial
plane in epoch J2000.0. The x-axis of this ***International Celestial Reference
System* (ICRS)** is taken to be in the direction of the vernal equinox on that date.
However, unlike the equatorial system, or previous barycentric systems, the axes
of the ICRS are defined and fixed in space by the positions of distant galaxies, not
by the apparent motion of the Sun. Unlike Solar System objects or nearby stars,
these distant objects have undetectable angular motions relative to one another.
Their relative positions do not depend on our imperfect knowledge or observations of the Earth's rotation, precession, and nutation. Thus, the ICRS is a very
good approximation of an inertial, non-rotating coordinate system.

In practice, radio-astronomical determinations of the equatorial coordinates of
over 200 compact extragalactic sources (mostly quasars) define this inertial
reference frame in an ongoing observing program coordinated by the International Earth Rotation Service in Paris. Directions of the ICRS axes are now
specified to a precision of about 0.02 mas relative to this frame. The ICRS
positions of optical sources are known primarily through HIPPARCOS and

Hubble Space Telescope (HST) observations near the optical counterparts of the defining radio sources, as well as a larger number of other radio sources. Approximately 100,000 stars measured by HIPPARCOS thus have ICRS coordinates known with uncertainties typical of that satellite's measurements, around 1 mas. Through the HIPPARCOS measurements, ICRS positions can be linked to the Earth-based fundamental catalog positions like FK5 (see Chapter 4).

3.1.11 The ecliptic coordinate system

The ecliptic, the apparent path of the Sun on the celestial sphere, can also be defined as the intersection of the Earth's orbital plane with the celestial sphere. The orbital angular momentum of the Earth is much greater than its spin angular momentum, and the nature of the torques acting on each system suggests that the orbital plane is far more likely to remain invariant in space than is the equatorial plane. Moreover, the ecliptic plane is virtually coincident with the plane of symmetry of the Solar System as well as lying nearly perpendicular to the Solar System's total angular momentum vector. As such, it is an important reference plane for observations and dynamical studies of Solar System objects.

Astronomers define a geocentric coordinate system in which the ecliptic is the fundamental plane and the vernal equinox is the fundamental direction. Measure ecliptic longitude, λ, from west to east in the fundamental plane. Measure the ecliptic latitude, β, positive northward from the ecliptic. Since the vernal equinox is also the fundamental direction of the equatorial system, the north ecliptic pole is located at RA = 18 hours and Dec = $90° - \varepsilon$, where ε is the obliquity of the ecliptic.

The ecliptic is so nearly an invariant plane in an inertial system that, unlike the equatorial coordinates, the ecliptic latitudes of distant stars or galaxies will *not* change with time because of precession and nutation. Ecliptic longitudes on the other hand, are tied to the location of the equinox, which is in turn defined by the spin of the Earth, so longitudes will have a precessional change of about 50″ per year.

3.1.12 The Galactic coordinate system

> Whoever turns his eye to the starry heavens on a clear night will perceive that band of light... designated by the name Milky Way... it is seen to occupy the direction of a great circle, and to pass in uninterrupted connection round the whole heavens:... so perceptibly different from the indefiniteness of chance, that attentive astronomers ought to have been thereby led, as a matter of course, to seek carefully for the explanation of such a phenomenon.
> — Immanuel Kant, *Universal Natural History and a Theory of the Heavens*, 1755

Kant's explanation for the Milky Way envisions our own Galaxy as a flattened system with approximately cylindrical symmetry composed of a large number of stars, each similar to the Sun. Astronomers are still adding detail to Kant's essentially correct vision: we know the Sun is offset from the center by a large fraction of the radius of the system, although the precise distance is uncertain by at least 5%. We know the Milky Way, if viewed from above the plane, would show spiral structure, but are uncertain of its precise form. Astronomers are currently investigating extensive evidence of remarkable activity in the central regions.

It is clear that the central plane of the disk-shaped Milky Way Galaxy is another reference plane of physical significance. Astronomers have specified a great circle (the **Galactic plane**) that approximates the center-line of the Milky Way on the celestial sphere to constitute the fundamental plane of the Galactic coordinate system. We take the fundamental direction to be the direction of the center of the galaxy. Galactic latitude (b or b^{II}) is then measured positive north (the Galactic hemisphere contains the north celestial pole) of the plane, and Galactic longitude (l or l^{II}) is measured from Galactic center so as to constitute a right-handed system.

Since neither precession nor nutation affects the Galactic latitude and longitude, these coordinates would seem to constitute a superior system. However, it is difficult to measure l and b directly, so the Galactic coordinates of any object are in practice derived from its equatorial coordinates. The important parameters are that the north Galactic pole ($b = +90°$) is defined to be at

$$\alpha = 12\!:\!49\!:\!00, \delta = +27.4° \text{ (equator and equinox of 1950)}$$

and the Galactic center ($l = b = 0$) at

$$\alpha = 17\!:\!42\!:\!24, \delta = -28°55' \text{ (equator and equinox of 1950)}$$

3.1.13 Transformation of coordinates

Transformation of coordinates involves a combination of rotations and (sometimes) translations. Note that for very precise work, (the transformation of geocentric to ICRS coordinates, for example) some general-relativistic modeling may be needed.

Some of the more common transformations are addressed in the various national almanacs, and for systems related just by rotation (equatorial and Galactic, for example), you can work transformations out by using spherical trigonometry (see Section 3.1.2). Some important transformations are given in Appendix D, and calculators for most can be found on the Internet.

3.2 The third dimension

Determining the distance of almost any object in astronomy is notoriously difficult, and uncertainties in the coordinate r are usually enormous compared to uncertainties in direction. For example, the position of Alpha Centauri, the nearest star after the Sun, is uncertain in the ICRS by about 0.4 mas (three parts in 10^9 of a full circle), yet its distance, one of the best known, is uncertain by about one part in 2500. A more extreme example would be one of the quasars that define the ICRS, with a typical positional uncertainty of 0.02 mas (six parts in 10^{10}). Estimates of the distances to these objects depend on our understanding of the expansion and deceleration of the Universe, and are probably uncertain by at least 10%. This section deals with the first two rungs in what has been called the "cosmic distance ladder," the sequence of methods and calibrations that ultimately allow us to measure distances (perhaps "estimate distances" would be a better phrase) of the most remote objects.

3.2.1 The astronomical unit

We begin in our own Solar System. Kepler's third law gives the scale of planetary orbits:

$$a = P^{2/3}$$

where a is the average distance between the planet and the Sun measured in **astronomical units** (AU, or, preferably, au) and P is the orbital period in years. This law sets the *relative* sizes of planetary orbits. One au is defined to be the mean distance between the Earth and Sun, but the length of the au in meters, and the absolute scale of the Solar System, must be measured empirically.

Figure 3.11 illustrates one method for calibrating the au. The figure shows the Earth and the planet Venus when they are in a position such that apparent angular separation between Venus and the Sun, as seen from Earth, (the *elongation* of Venus) is at a maximum. At this moment, a radio (radar) pulse is sent from the Earth towards Venus, and a reflected pulse returns after elapsed time Δt. The Earth-to-Venus distance is just

$$\frac{1}{2} c \Delta t$$

Thus, from the right triangle in the figure, the length of the line ES is one au or

$$1 \text{ au} = \frac{c \Delta t}{2 \cos \theta}$$

Clearly, some corrections need to be made because the orbit of neither planet is a perfect circle, but the geometry is known rather precisely. Spacecraft in orbit around Venus and other planets (Mars, Jupiter, and Saturn) also provide the opportunity to measure light-travel times, and similar geometric analyses yield absolute orbit sizes. The presently accepted value for the length of the au is

$$1 \text{ au} = 1.49\,5978 \times 10^{11} \text{ m}$$

with an uncertainty of 1 part in 10^6.

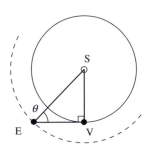

Fig. 3.11 Radar ranging to Venus. The astronomical unit is the length of the line ES, which scales with EV, the Earth-to-Venus distance.

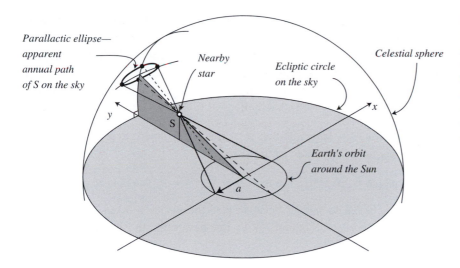

Parallactic ellipse— apparent annual path of S on the sky

Nearby star

Celestial sphere

Ecliptic circle on the sky

y

S

x

Earth's orbit around the Sun

a

Fig. 3.12 The parallactic ellipse. The apparent position of the nearby star, S, as seen from Earth, traces out an elliptical path on the very distant celestial sphere as a result of the Earth's orbital motion.

3.2.2 Stellar parallax

Once the length of the au has been established, we can determine the distances to nearby stars through observations of **heliocentric stellar parallax**. Figure 3.12 depicts the orbit of the Earth around the Sun. The plane of the orbit is the ecliptic plane, and we set up a Sun-centered coordinate system with the ecliptic as the fundamental plane, the z-axis pointing towards the ecliptic pole, and the y-axis chosen so that a nearby star, S, is in the y–z plane. The distance from the Sun to S is r. As the Earth travels in its orbit, the apparent position of the nearby star shifts in relation to very distant objects. Compared to the background objects, the nearby star appears to move around the perimeter of the **parallactic ellipse**, reflecting the Earth's orbital motion.

Figure 3.13 shows the plane that contains the x-axis and the star. The parallax angle, p, is half the total angular shift in the star's position (the semi-major axis of the parallactic ellipse in angular units). From the right triangle formed by the Sun–star–Earth:

$$\tan p = \frac{a}{r}$$

where a is one au. Since p is in every case going to be very small, we make the small angle approximation: for $p \ll 1$:

$$\tan p \cong \sin p \cong p$$

So that for any right triangle where p is small:

$$p = \frac{a}{r}$$

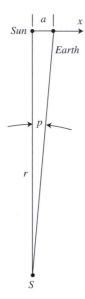

Fig. 3.13 The parallax angle.

In this equation, it is understood that a and r are measured in the same units (aus, for example) and p is measured in radians. Radian measure is somewhat inconvenient for small angles, so, noting that there are about 206 265 arcsec per radian, we can rewrite the small-angle formula as

$$p[\text{arcsec}] = 206\,265\,\frac{a}{r}\;[a, r \text{ in same units}]$$

Finally, to avoid very large numbers for r, it is both convenient and traditional to define a new unit, the **parsec**, with the length:

$$1 \text{ parsec} = 206\,265 \text{ au} = 3.085\,678 \times 10^{16}\,\text{m} = 3.261\,633 \text{ light years}$$

The parsec (pc) is so named because it is the distance of an object whose *par*allax is one *sec*ond of arc. With the new unit, the parallax equation becomes:

$$p[\text{arcsec}] = \frac{a[\text{au}]}{r[\text{pc}]} \tag{3.1}$$

This equation represents a fundamental relationship between the small angle and the sides of the ***astronomical triangle*** (any right triangle with one very short side). For example, suppose a supergiant star is 20 pc away, and we measure its angular diameter with the technique of speckle interferometry as 0.023 arcsec. Then the physical diameter of of the star, which is the short side of the relevant astronomical triangle (the quantity a in Equation (3.1)), must be 20×0.023 pc arcsec = 0.46 au.

In the case of stellar parallax, the short side of the triangle is always 1 au. If $a = 1$ in Equation (3.1), we have:

$$p[\text{arcsec}] = \frac{1}{r[\text{pc}]} \tag{3.2}$$

In the literature, the parallax angle is often symbolized as π instead of p. Note that the parallactic ellipse will have a semi-major axis equal to p, and a semi-minor axis equal to $p \sin \lambda$, where λ is the ecliptic latitude of the star. The axes of an ellipse fit to multiple observations of the position of a nearby star will therefore estimate its parallax.

There are, of course, uncertainties in the measurement of small angles like the parallax angle. Images of stars formed by Earth-based telescopes are typically blurred by the atmosphere, and are seldom smaller than a half arc second in diameter, and are often much larger. In the early days of telescopic astronomy, a great visual observer, James Bradley (1693–1762), like many astronomers before him, undertook the task of measuring stellar parallax. Bradley could measure stellar positions with a precision of about 0.5 arcsec (500 milli-arcseconds or mas). This precision was sufficient to discover the phenomena of nutation and aberration, but not to detect a stellar parallax (the largest parallax of a bright star visible at Bradley's latitude is that of Sirius, i.e. 379 mas).

A few generations later, Friedrich Wilhelm Bessel (1784–1846), a young clerk in an importer's office in Bremen, began to study navigation in order to move ahead in the business world. Instead of mercantile success, Bessel discovered his love of astronomical calculation. He revised his career plans and secured a post as assistant to the astronomer Johann Hieronymous Schroeter, and began an analysis of Bradley's observations. Bessel deduced the systematic errors in Bradley's instruments (about 4 arcsec in declination, and 1 second of time in RA – much worse than Bradley's random errors), and demonstrated that major improvements in positional accuracy should be possible. Convinced that reducing systematic and random errors would eventually lead to a parallax measurement, Bessel mounted a near-heroic campaign to monitor the double star 61 Cygni along with two "background" stars. In 1838, after a 25-year effort, he succeeded in measuring the parallax with a precision that satisfied him. (His value for the parallax, 320 mas, is close to the modern value of 286 mas.) Bessel's labor was typical of his ambition[2] and meticulous attention to error reduction. The 61 Cygni parallax project was Herculean, and parallaxes for any but the very nearest stars emerged only after the introduction of photography.[3]

Beginning in the late 1880s, photography gradually transformed the practice of astronomy by providing an objective record of images and by allowing the accumulation of many photons in a long exposure. With photography, human eyesight no longer limited human ability to detect faint objects with a telescope. Photography vastly augmented the power of small-angle astrometry. Astronomical photographs (negatives) were usually recorded on emulsion-covered glass plates at the telescope, then developed in a darkroom.

Away from the telescope, astronomers could then measure the positions of objects on the plate, usually with *a plate-measuring machine* (Figure 3.14). During the twentieth century, such machines became increasingly automated,

---

[2] Bessel pioneered mathematical analysis using the functions that now bear his name. He spent 30 years measuring the "Prussian degree" – the length, in meters, of a degree of arc of geodetic latitude. This was part of an international effort to determine the shape of the Earth from astronomical measurements. After his publication of the corrected positions of the 3222 stars in Bradley's catalog, Bessel went on to measure his own positions for 62,000 other stars, and inspired his student, F.W. Argelander, to organize a project to determine the transit-circle positions for all stars brighter than ninth magnitude in the northern hemisphere – about a third of a million objects.

[3] Wilhelm Struve published his measurement of the parallax of Vega immediately after Bessel's publication (he had actually completed his analysis before Bessel had) and Thomas Henderson published the parallax of Alpha Centauri in the next year. Bessel's measurement was the most accurate.

Fig. 3.14 Schematic of a
measuring engine. A
fixed microscope views
an object on a
photographic plate. The
(x, y) position of the plate
is controlled by two
precision screws, and its
value can be read from
scales on the moving
stages.

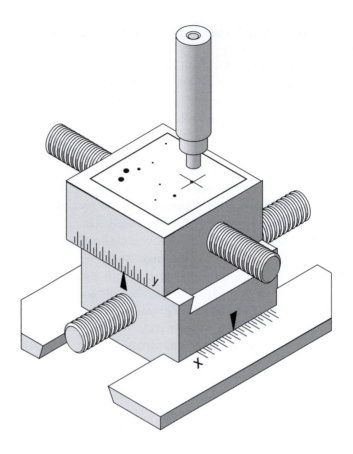

Fig. 3.14 Schematic of a measuring engine. A fixed microscope views an object on a photographic plate. The (x, y) position of the plate is controlled by two precision screws, and its value can be read from scales on the moving stages.

precise, and expensive. Computer-controlled measuring machines called ***micro-densitometers***, which record the darkness of the image at each position on a plate, became important astronomical facilities. Direct digital recording of images with electronic arrays, beginning in the 1970s, gradually made the measuring machine unnecessary for many observations. Nevertheless, photography is still employed in some vital areas of astronomy, and measuring machines still continue to produce important data.

Conventional modern small-angle astrometry, using photographic or electronic array detectors, can measure the relative position of a nearby star with a precision of the something like 50 mas in a single measurement. This uncertainty can be reduced by special equipment and techniques, and by repeated measurements as a star shifts around its parallactic ellipse. One's ability to measure a parallax depends on the presence of suitable background objects and the stability of the observing system over several years. Uncertainty in p from conventional ground-based small-angle astrometry can be routinely reduced to around 5 mas with repeated measurements (50 observations of a single star are not unusual). Even so, this means that only parallaxes larger than 50 mas will have uncertainties smaller than 10%, so only those stars nearer than $1/0.05 = 20$ pc can be considered to have distances precisely known by the ground-based parallax method. There are approximately 1000 stars detected closer than 20 pc, a rather small number compared to the 10^{11} or so stars in the Milky Way. Appendix D1 lists the nearest stars, based upon the best current parallaxes.

Ground-based parallax measurements can approach 0.5 mas precision with suitable technique and special instruments, at least for a small number of the best-studied stars. Nevertheless, spaced-based methods have produced the greatest volume of precision measurements. The HIPPARCOS space mission measured the positions, parallaxes, and proper motions of 117,955 stars in the four years between 1989 and 1993. The median precision of the parallaxes is 0.97 mas for stars brighter than $V = 8.0$. The Hubble Space Telescope has made a much smaller number of measurements of similar accuracy. The planned Gaia mission anticipates precisions of 7 mas for $V = 7$, 10–25 µas for $V = 15$, and around 500 µas at $V = 20$. The Space Interferometry Mission (SIM lite) space-craft, with a planned launch sometime after 2015, would use a new observing method (a Michaelson interferometer) to achieve precisions of 4 µas for stars brighter than $V = 15$, and of 12 µas at $V = 20$.

3.3 Time

> Alice sighed wearily. "I think you might do something better with the time," she said, "than wasting it in asking riddles that have no answers."
> "If you knew Time as well as I do," said the Hatter," you wouldn't talk about wasting IT. It's HIM. . . . I dare say you never even spoke to Time!"
> "Perhaps not," Alice cautiously replied; "but I know I have to beat time when I learn music."
>
> — Lewis Carroll, *Alice's Adventures in Wonderland*, 1897

Time is a physical quantity of which we have never enough, save for when we have too much and it gets on our hands. Ambition to understand its nature has consumed the time of many. It is unclear how much of it has thereby been wasted in asking riddles with no answers. Perhaps time will tell.

3.3.1 Atomic time

Measuring time is a lot easier than understanding it. The way to measure time is to "beat" it, like Alice. In grammar school, I learned to count seconds by pronouncing syllables: "Mississippi one, Mississippi two, Mississippi three. . . ." A second of time is thus, roughly, the duration required to enunciate five syllables. A similar definition, this one set by international agreement, invokes a more objective counting operation:

> **1 second** (Système International, or SI second) = the duration of 9,192,631,770 periods of the radiation corresponding to the transition between the two hyper-fine levels of the ground state of the cesium-133 atom.

A device that counts the crests of a light wave and keeps a continuous total of the elapsed SI seconds is an *atomic clock*. An atomic clock located at rest on the surface of the Earth keeps **TAI** or *international atomic time* (TAI = *Temps Atomique International*). Practical atomic clocks have a precision of about 2 parts in 10^{13}. International atomic time is the basis for dynamical computations involving time as a physical parameter and for recording observations made on the surface of the Earth. Things get a little complicated if you compare an atomic clock on the surface of the Earth with one located elsewhere (like the barycenter of the Solar System). Relativity theory and observation show that such clocks will not necessarily run at the same rate, but, when compared, will differ according to their relative velocities (special relativity) and according to their accelerations or local gravitational fields (general relativity). Precise timekeeping accounts for relativity effects, but the starting time scale in these computations is generally TAI.

The *astronomical day* is defined as 86,400 SI seconds. There are, however, other kinds of days.

3.3.2 Solar time

Early timekeepers found it most useful and natural to count days, months, and years, and to subdivide these units. The day is the most obvious and practical of these units since it correlates with the light–dark cycle on the surface of the Earth. Much of early timekeeping was a matter of counting, grouping, and subdividing days. Since the rotation of the Earth establishes the length of the day, counting days is equivalent to counting rotations.

Figure 3.15 illustrates an imaginary scheme for counting and subdividing days. The view is of the Solar System, looking down from above the Earth's north pole, which is point P. The plane of the page is the Earth's equatorial plane, and the large circle represents the equator itself. The small circle represents the position of the Sun projected onto the equatorial plane. In the figure, we assume the Sun is motionless, and we attach hour markers just outside the equator as if we were painting the face on an Earth-sized 24-hour clock. These markers are motionless as well, and are labeled so that they increase counterclockwise, and so that the marker in the direction of the Sun is labeled 12, the one opposite the Sun, 0. The choice of 24 hours around the circle, as well as the subdivision into 60 minutes per hour and 60 seconds per minute, originated with the ancient Babylonian sexagesimal (base 60) number system.

Point O in the figure is the location of a terrestrial observer projected onto the equatorial plane. This observer's meridian projects as a straight line passing through O and P. The figure extends the projected meridian as an arrow, like the hand of a clock, which will sweep around the face with the painted numbers as the Earth rotates relative to the Sun. Since we are using the Sun as the reference marker, this turning of the Earth is actually a combination of spin and orbital

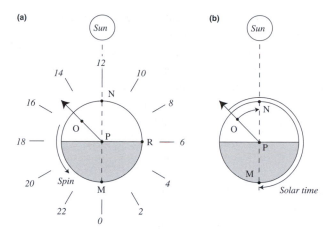

Fig. 3.15 A view of the equatorial plane from above the Earth's north pole, P. (a) The meridian of the observer points to the apparent solar time. (b) The apparent solar time, ∠OPM, equals the hour angle of the Sun, ∠OPN, plus 12 hours.

motion. The meridian points to the number we will call the ***local apparent solar time***. Each cycle of the meridian past the zero mark (midnight) starts a new day for the observer. Every longitude has a different meridian and thus a different solar time. The local solar time, for example, is 12 hours for an observer at point N in the figure, and 6 hours for an observer at R.

A little consideration of Figure 3.15b should convince you of the following definition:

> ***local apparent solar time*** = the hour angle of the Sun as it appears on the sky
> (∠OPN), plus 12 hours

Simple observations (for example, with a sundial) will yield the local apparent solar time, but this method of timekeeping has a serious deficiency. Compared to TAI, local apparent solar time is non-uniform, mainly because of the Earth's orbital motion. Because of the obliquity of the ecliptic, motion in the orbit has a greater east–west component at the solstices than at the equinoxes. In addition, because Earth's orbit is elliptical, Earth's orbital speed varies; it is greatest when it is closest to the Sun (at perihelion, around January 4) and slowest when furthest away (aphelion). As a result, apparent solar days throughout the year have different lengths compared to the defined astronomical day of 86,400 SI seconds.

This non-uniformity is troublesome for precise timekeeping. To remove it, one strategy is to average out the variations by introducing the idea of the ***mean Sun***: a fictitious body that moves along the celestial equator at uniform angular speed, completing one circuit in one tropical year (i.e. equinox to equinox). If we redefine the "Sun" in Figure 3.15 as the mean Sun, we can define a more uniform time scale:

> ***local mean solar time*** = the hour angle of the fictitious mean Sun, plus 12 hours

The difference between the apparent and the mean solar times is called the *equation of time*:

equation of time = local apparent solar time − local mean solar time

The equation of time takes on values in the range ±15 minutes in the course of a year. See Appendix D for more information.

To circumvent the difficulty arising from the fact that every longitude on Earth will have a different mean solar time, one often records or predicts the time of an event using the reading from a mean solar clock located at the zero of longitude. This is called the universal time (UT):

Universal time (UT or UT1) = mean solar time at Greenwich

The UT clock, of course, is actually located in your laboratory – it is simply set to agree with the mean solar time at the longitude of Greenwich. Thus, if the Moon were to explode, everyone on Earth would agree about the UT of the mishap, but only people at the same longitude would agree about the mean solar time at which it occurs.

Although a big improvement on apparent solar time, the UT1 is still not completely uniform. For one thing, the precession rate (needed to compute the mean Sun) is imperfectly known and changes over long time scales. The major difficulty, however, is that the spin of the Earth is not quite uniform. The largest variations are due to varying tidal effects that have monthly and half-monthly periods, as well as seasonal (yearly) variations probably due to thermal and meteorological effects. A smaller, random variation, with a time scale of decades, is probably due to poorly understood core–mantle interactions. Finally, over the very long term, tidal friction causes a steady slowing of the spin of the Earth. As result of this long-term trend, the mean solar day is getting longer (as measured in SI seconds) at the rate of about 0.0015 seconds per century. Thus, on the time scale of centuries, a day on the UT1 clock, (the mean solar day) is increasing in duration compared to the constant astronomical day of 86,400 SI seconds, and is fluctuating in length by small amounts on shorter time scales.

The International Earth Rotation Service (IERS), in Paris, has taken the monitoring of the variations in the terrestrial spin rate as one of its missions. In order to coordinate the Earth's rotation with TAI, the US Naval Observatory, working for the IERS, maintains the *coordinated universal time* (UTC) clock. Coordinated universal time approximates UT1, but uses SI seconds as its basic unit. To keep pace with UT1 to within a second, the UTC clock introduces an integral number of "leap" seconds as needed. Because of the random component in the acceleration of the Earth's spin, it is not possible to know in advance when it will be necessary to add (or remove) a leap second. A total of 22 leap seconds were counted by TAI (but not by UTC) between 1972 and the end of 1998. Coordinated universal time is the basis for most legal time systems (zone time).

Unlike UT or UTC, local solar time at least has the practical advantage of approximate coordination with local daylight: at 12 noon on the local mean solar clock, you can assume the Sun is near the meridian. However, every longitude will have a different meridian and a different local solar time. Even nearby points will use different clocks. To deal in a practical fashion with the change in mean solar time with longitude, most legal clocks keep zone time:

$$\text{zone time} = \text{UTC} + \text{longitude correction for the zone}$$

This strategy ensures that the legal time is the same everywhere inside the zone. Zones are usually about 15° wide in longitude, so the longitude correction is usually an integral number of hours. (Remember the Earth spins at a rate of 15 degrees per hour.) For example, Eastern Standard Time (longitude 75°) = UTC − 5 hours, Pacific Standard Time (longitude 120°) = UTC − 8 hours.

Time services provide signals for setting clocks to the current UTC value. In the USA, the National Institute of Standards and Technology broadcasts a radio signal "(stations WWV and WWVH) at 2.5, 5, 10, 15 and 20 MHz that contain time announcements and related information, and at 60 kHz (station WWVB) for synchronization of clocks. Computer networks can synchronize to UTC using standard protocols (ITS and ACTS). A convenient one-time check on UTC is at the US Naval Observatory website (http://www.usno.navy.mil/ USNO), which is also a good source for details about various times scales and terrestrial coordinate systems.

Sidereal time is also defined by the rotation of the Earth and its precessional variations, and therefore does not flow uniformly, but follows the variations manifest in UT1:

$$\text{sidereal time} = \text{the hour angle of the mean vernal equinox of date}$$

Having defined the day, astronomers find it useful to maintain a continuous count of them:

$$\text{Julian date} = \text{number of elapsed UT or UTC days since}$$
$$\text{4713 BC January 1.5 (12 hrs UT on January 1.)}$$

It is also common to use a Julian date (JD), rather than a UT date, to specify the date. The date of the equator and equinox in a catalog of equatorial coordinates might be specified as

$$\text{J2000.0} = \text{"Julian epoch 2000.0"} = \text{2000 Jan 1.5 UT} = \text{JD 2451545.0}$$

Appendix A summarizes some other time units.

Fig. 3.16 Displacement in space and space velocity: (a) illustrates the relation between proper motion, μ, and the displacement in a unit time; (b) shows the two components of the space velocity.

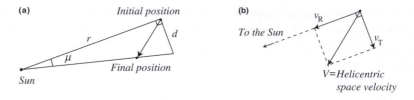

3.4 Motion

3.4.1 Space motion

Consider an object that moves relative to the Sun. Figure 3.16, which is drawn in a three-dimensional coordinate system centered at the barycenter of the Solar System, shows the motion, that is, the displacement, of such an object over a suitably long time. The plane of the figure contains both the origin of coordinates and the displacement vector. Part (a) of the figure shows the actual displacement, while part (b) shows the displacement divided by the time interval, that is, the velocity. Both displacement and velocity vectors can be decomposed into radial and tangential components. The total velocity, usually called the **space velocity**, is the vector sum of the **tangential velocity** and **the radial velocity**:

$$\vec{V} = \vec{v}_T + \vec{v}_R$$

$$V = \sqrt{v_T^2 + v_R^2}$$

Measuring the two components requires two very different observing strategies. Astronomers can measure radial velocity directly with a spectrograph, and can measure tangential velocities indirectly by observing changes in position.

3.4.2 Proper motion

Suppose you have some quality observing time tonight, and you measure the position of a certain star in ICRS coordinates. Suppose, also, that 10 years in the future you observe the same star a second time. If this star were truly motionless with respect to the center of the Solar System and the distant galaxies that define the axes of the ICRS, then the coordinates you measure 10 years from now will be the same as those you measure tonight. Remember, effects like precession and parallax are not present in the ICRS.

On the other hand, most stars *do* move with respect to the ICRS axes. Especially if the star is nearby, its coordinates may very well change after only 10 years. The **rate of change** in coordinates is called the **proper motion** of the object. As the name suggests, proper motion reflects the "true" motion of the star with respect to the barycenter of the Solar System, and does not include those coordinate changes like aberration, precession, nutation, or heliocentric parallax that result from terrestrial motions. Proper motion, of course, is relative, and it may not be possible (or meaningful) to decide if it is due to motion of the barycenter or of the star.

Think about the objects that will *not* exhibit a proper motion over your 10-year observing interval. Certainly, these will include very distant objects like quasars, since they define the coordinate system. Also, any nearby star that has no tangential velocity will have no proper motion. Finally, you will not detect a proper motion for any object that is so far away that it does not change its angular position by an amount detectable by your instruments, even though its tangential velocity might be substantial.

The basic methods for measuring proper motion are fairly easy to understand. First some terminology: in astrometry, the *epoch* of an observation means the time of observation. The equatorial coordinate system used to record an observation also has a time associated with it, which is the date(s) of the *equator and equinox* employed. The two dates need not be the same. Unfortunately, even astronomers are not careful with this terminology, and will occasionally say "epoch" when they really mean "equator and equinox".

Keeping the above distinction in mind, you could determine the proper motion of a star by comparing its positions in fundamental catalogs for two different epochs (dates of observation), being careful to express the coordinates using the same barycentric equator and equinox. For example, look up the position of your star in a catalog for epoch 1934, which lists coordinates using the 1950 equator and equinox. Then find the same star in a second catalog, which gives its epoch 1994 position. The second catalog uses the equator and equinox of J2000. Now you must transform the epoch 1934 coordinates so that they are given in equator and equinox 2000 coordinates. Now that both positions are expressed in the same coordinate system (J2000), compute the difference between the 1994 position and the 1934 position. The difference, divided by the time interval (60 years, in this case) is the proper motion. Proper motions determined in this fashion are often called *fundamental proper motions*. The method depends on astronomers doing the hard work of assembling at least two fundamental catalogs.

You can also measure proper motions using small-angle astrometry. Compare a photograph of a star field taken in 1994 with one taken with the same instrument in 1934. Align the photographs so that most of the images coincide, especially the faint background stars and galaxies. Any object that has shifted its position with respect to these "background objects" is exhibiting *relative proper motion*. The possibility that there might be some net proper motion in the background objects limits the accuracy of this sort of measurement, as does the likelihood of changes in the instrument over a 60-year span. Nevertheless, relative proper motions are more easily determined than fundamental motions, and are therefore very valuable because they are available for many more stars.

You can, of course, use observations from different instruments (an old photograph and a recent CCD frame for example) to measure relative proper motions, but the analysis becomes a bit more complex and prone to systematic error.

Proper motion, represented by the symbol μ, is usually expressed in units of seconds of arc per year, or sometimes in seconds of arc per century. Since μ is a vector quantity, proper motion is generally tabulated as its RA and Dec components, μ_α and μ_δ.

The tangential component of the space velocity is responsible for proper motion. For the same tangential speed, nearer objects have larger proper motions. Refer to Figure 3.16. If an object at distance r has a tangential displacement $d = v_T t$ in time t, then, for small μ,

$$\mu = \frac{d/t}{r} = \frac{v_T}{r} \quad (3.3)$$

The statistical implications of Equation (3.3) are so important they are expressed in an astronomical "proverb": **_swiftness means nearness_**. That is, given a group of objects with some distribution of tangential velocities, the objects with the largest values for μ (swiftness) will tend, statistically, to have the smallest values for r (nearness). Putting the quantities in Equation (3.3) in their usual units (km s^{-1} for velocity, parsecs for distance, seconds of arc per year for μ), it becomes

$$\mu = \frac{v_T}{4.74r}$$

This means, of course, that you can compute the tangential velocity if you observe both the proper motion and the parallax (p):

$$v_T = 4.74\frac{\mu}{p}$$

3.4.3 Radial velocity

On May 25, 1842, Christian Doppler (1803–1853) delivered a lecture to the Royal Bohemian Scientific Society in Prague. Doppler considered the situation in which an observer and a wave source are in motion relative to one another. He made the analogy between the behavior of both water and sound waves on the one hand, and of light waves on the other. Doppler correctly suggested that, in all three cases, the observer would measure a frequency or wavelength change that depended on the radial velocity of the source. The formula that expresses his argument is exact for the case of light waves from sources with small velocities:

$$\frac{\lambda - \lambda_0}{\lambda_0} = \frac{\Delta\lambda}{\lambda_0} = \frac{v_R}{c} \quad (3.4)$$

Here λ_0 is the wavelength observed when the source is motionless, λ is the wavelength observed when the source has radial velocity v_R, and c is the speed of light. In his lecture, Doppler speculated that the differing radial velocities of stars were largely responsible for their different colors. To reach this conclusion, he assumed that many stars move at a considerable fraction of the speed of light relative to the Sun. This is wrong. But even though he was incorrect about the colors of the stars, the **Doppler effect**, as expressed in Equation (3.4), was soon verified experimentally, and is the basis for all astronomical direct measurements of radial velocity. It is interesting to note that first Arnand Fizeau, in Paris in 1848, and then Ernst Mach, in Vienna in 1860, each independently worked out the theory of the Doppler effect without knowledge of the 1842 lecture.

Fizeau and Mach made it clear to astronomers how to *measure* a radial velocity. The idea is to observe a known absorption or emission *line* in the spectrum of a moving astronomical source, and compare its wavelength with some zero-velocity reference. The first references were simply the wavelength scales in visual spectrographs. Angelo Secci, in Paris, and William Huggins, in London, both attempted visual measurements for the brighter stars during the period 1868–1876, with disappointing results. Probable errors for visual measurements were on the order of 30 km s^{-1}, a value similar to the actual velocities of most of the bright stars. James Keeler, at Lick Observatory in California, eventually was able to make precision visual measurements (errors of about 2–4 km s^{-1}), at about the same time (1888–1891) that astronomers at Potsdam first began photographing spectra. **Spectrographs** (with photographic recording) immediately proved vastly superior to **spectroscopes**. Observers soon began recording **comparison spectra**, usually from electrically activated iron arcs or hydrogen gas discharges, to provide a recorded wavelength scale. Figure 3.17 shows a photographic spectrum and comparison. A measuring engine (see Figure 3.14), a microscope whose stages are moved by screws equipped with micrometer read-outs, soon became essential for determining positions of the lines in the source spectrum relative to the lines in the comparison. In current practice, astronomers record spectra and comparisons digitally and compute shifts and velocities directly from the data.

Precise radial velocities. What limits the precision of a radial velocity measurement? We consider spectrometry in detail in Chapter 11. For now, just

Fig. 3.17 A conventional photographic spectrum. A stellar spectrum, with absorption lines, lies between two emission-line comparisons.

note that, since the important measurement is the physical location of spectral lines on the detector, an astronomer certainly would want to use a detector/ spectrometer capable of showing as much detail as possible. The **resolving power** of a spectrograph is the ratio

$$R = \frac{\lambda}{\delta\lambda}$$

where $\delta\lambda$ is wavelength resolution (i.e. two narrow spectral lines that are closer than $\delta\lambda$ in wavelength will appear as a single line in the spectrogram). Limits to resolving power will be set by the design of the spectrograph, but also by the brightness of the object being investigated, and the size and efficiency of the telescope feeding the spectrograph. As is usual in astronomy, the most precise measurements can be made on the brightest objects.

Early spectroscopists soon discovered other limits to precision. They found that errors arose if a spectrograph had poor mechanical or thermal stability, or if the path taken by light from the source was not equivalent to the path taken by light from the comparison. New spectrograph designs improved resolving power, efficiency, stability, and the reliability of wavelength calibration. At the present time, random errors of less than 100 m s^{-1} in absolute stellar radial velocities are possible with the best optical spectrographs. At radio wavelengths, even greater precision is routine.

Greater precision is possible in differential measurements. Here the astronomer is concerned only with *changes* in the velocity of the object, not the actual value. Very precise optical work, for example, has been done in connection with searches for planets orbiting solar-type stars. The presence of a planet will cause the radial velocity of its star to vary as they both orbit the barycenter of the system. Precisions at a number of observatories now approach 3 m s^{-1} or better for differential measurements of brighter stars.

Large redshifts. When the radial velocity of the source is a considerable fraction of the speed of light, special relativity replaces Equation (3.4) with the correct version:

$$z = \frac{\lambda - \lambda_0}{\lambda_0} = \frac{\Delta\lambda}{\lambda_0} = \frac{\left(1 - \beta^2\right)^{1/2}}{(1 - \beta)} - 1 \qquad (3.5)$$

where

$$\beta = \frac{v_R}{c} = \frac{1(z+1)^2 - 1}{1(z+1)^2 + 1}$$

Here z is called the **redshift parameter**, or just the redshift. If the source moves away from the observer, both v_R and z are positive, and a spectral feature in the visual (yellow-green) will be shifted to longer wavelengths (i.e. towards the red). The spectrum is then said to be **redshifted** (even if the observed feature were a microwave line that was shifted to longer wavelengths and thus *away* from the red). Likewise, if the source moves towards the observer, v_R and z are negative, and the spectrum is said to be **blueshifted.**

In an early result from the spectroscopy of non-stellar objects, Vesto Melvin Slipher, in 1914, noticed that the vast majority of the spiral nebulae (galaxies) had redshifted spectra. By 1931, Milton Humason and Edwin Hubble had recorded galaxy radial velocities up to 20,000 km s^{-1}, and were able to demonstrate that the redshift of a galaxy was directly proportional to its distance. Most astronomers interpret **Hubble's law,**

$$v_R = H_0 d \qquad (3.6)$$

as indicating that our Universe is expanding (the distances between galaxies are increasing). In Equation (3.6), it is customary to measure v in km s^{-1} and d in megaparsecs, so H_0, which is called the **Hubble constant**, has units of km s^{-1} Mpc^{-1}. In these units, recent measurements of the Hubble constant fall in the range 67–77. Actually, the redshifts are not interpreted as due to the Doppler effect, but as the result of the expansion of space itself.

The object with the largest spectroscopic redshift (as of early 2007) is a galaxy, IOK-1, which has $z = 6.96$. You can expect additional detections in this range. Doppler's 1842 assumption that major components of the Universe have significant shifts in their spectra was quite correct after all.

Summary

- Coordinate systems can be characterized by a particular origin, reference plane, reference direction, and sign convention.

- Astronomical coordinates are treated as coordinates on the surface of a sphere. The laws of **spherical trigonometry** apply. Concepts:

  great circle   *law of cosines*   *law of sines*

- The geocentric terrestrial **latitude and longitude** system uses the equatorial plane and **prime meridian** as references. Concepts:

  geocentric latitude   *geodetic latitude*   *geographic latitude*
  Greenwich   *polar motion*

- The altitude–azimuth system has its origin at the observer and uses the horizontal plane and geographic north as references. Concepts:

  vertical circle   *zenith*   *nadir*
  zenith distance   *meridian*   *diurnal motion*
  sidereal day

- The equatorial system of right ascension and declination locates objects on the celestial sphere. The Earth's equatorial plane and the vernal equinox are the references. This system rotates with respect to the altitude–azimuth system. Concepts:

  (continued)

Summary (*cont.*)

celestial pole	*ecliptic*	*obliquity*
altitude of pole=	*upper meridian*	*circumpolar star*
observer's latitude	*sidereal time*	*hour circle*
transit		
hour angle		

- Astrometry establishes the positions of celestial objects. Positions are best transformed into the International Celestial Reference Frame (**ICRS**) which is independent of motions of the Earth. Concepts:

transit telescope	*meridian circle*	*VLBI*
HIPPARCOS	*Hipparchus*	*atmospheric refraction*
fundamental catalog	*Gaia*	*epoch*
precession	*nutation*	*apparent coordinates*
aberration of starlight	*ecliptic coordinates*	*J2000*
Galactic coordinates		

- Heliocentric stellar parallax is an effect that permits measurement of distances to nearby stars. Concepts:

astronomical unit (au)	*astronomical triangle*
parallax angle	*parsec (pc)*

$$p[\text{arcsec}] = \frac{a[\text{au}]}{r[\text{pc}]}$$

- Physicists define time in terms of the behavior of light, but practical time measurements have been historically tied to the rotation of the Earth. Concepts:

atomic clock	*local apparent solar time*
TAI second	*local mean solar time*
universal time	*coordinated universal time*
zone time	*Julian date*

- The tangential component of an object's velocity in the ICRS system gives rise to a change in angular position whose rate of change is called the proper motion.

$$v_T[\text{km s}^{-1}] = 4.74 \frac{\mu[\text{arcsec yr}^{-1}]}{p[\text{arcsec}]}$$

- The radial component of an object's velocity can be measured by a shift in its spectrum due to the Doppler effect. Similar shifts are caused by the expansion of the universe. Concepts:

redshift parameter: $z = \Delta\lambda/\lambda \approx v_R/c$

spectroscopic resolving power (R)   *relativistic Doppler effect*

Hubble's law: $v_R = H_0 d$

Exercises

> Each problem that I solved became a rule which served afterwards to solve other problems.
>
> — Rene Descartes, *Discours de la Méthode* ..., 1637

1. Two objects differ in RA by an amount $\Delta\alpha$, and have declinations δ_1 and δ_2. Show that their angular separation, θ, is given by

$$\cos\theta = \sin\delta_1 \sin\delta_2 + \cos\delta_1 \cos\delta_2 \cos\Delta\alpha$$

2. Which city is closer to New York (74° W, 41° N): Los Angeles (118° W, 34° N) or Mexico City (99° W, 19° N)? By how much? (The radius of the Earth is 6300 km).

3. A. Kustner conducted one of the first systematic radial velocity studies. In 1905, he found that the velocity of stars in the ecliptic plane varied with an amplitude of $29.617 \pm .057$ km s^{-1} in the course of a sidereal year. Assume that the Earth's orbit is circular and use this information to derive the length (and uncertainty) of the au in kilometers.

4. Position angles are measured from north through east on the sky. For example, the figure at right shows a double star system in which component B is located in position angle θ with respect to component A. The two have an angular separation of r seconds of arc. If component A has equatorial coordinates (α, δ), and B has coordinates ($\alpha + \Delta\alpha$, $\delta + \Delta\delta$), derive expressions for $\Delta\alpha$ and $\Delta\delta$.

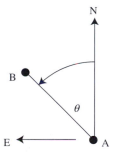

5. The field of view of the Vassar 32-inch CCD camera is a square 1000 seconds of arc on each side. Fill in the width of the field in the RA coordinate (i.e. in H:M:S units) when the telescope is pointed at declinations listed in the table:

Declination (degrees)	Width of field (minutes:seconds of RA)
0	1:06.7
20	
40	
60	
70	
80	
85	

6. The winter solstice (December 22) is the date of the longest night of the year in the northern hemisphere. However, the date of the earliest sunset in the northern hemisphere occurs much earlier in the month (at about 16:35 zone time on December 8 for longitude 0 degrees and latitude 40 degrees N). Examine the curve for the equation of time and suggest why this might be the case. Explain how this observation would depend upon one's exact longitude within a time zone.

7. On the date of the winter solstice, what are the approximate local sidereal times and local apparent solar times at sunset? (Assume 40° N latitude and use a celestial sphere.)

8. A certain supernova remnant in our galaxy is an expanding spherical shell of glowing gas. The angular diameter of the remnant, as seen from Earth, is 22.0 arcsec. The parallax of the remnant is known to be 4.17 mas from space telescope measurements. Compute its distance in parsecs and radius in astronomical units.

9. An astronomer obtains a spectrum of the central part of the above remnant, which shows emission lines. Close examination of the line due to hydrogen near wavelength 656 nm reveals that it is actually double. The components, presumably from the front and back of the shell, are separated by 0.160 nanometers. (a) With what velocity is the nebula expanding? (b) Assuming this has remained constant, estimate the age of the remnant. (c) The astronomer compares images of the remnant taken 60 years apart, and finds that the nebula has grown in diameter from 18.4 to 22.0 arcsec. Use this data to make a new computation for the distance of the remnant independent of the parallax.

10. In 1840, the estimated value of the au, 1.535×10^8 km, was based upon Encke's 1824 analysis of the observations of the transits of Venus in 1761 and 1769. Encke's result should have been accorded a relative uncertainty of around 5%. If Bessel's (1838) parallax for 61 Cygni was 0.32 ± 0.04 arcsec, compute the distance and the total relative uncertainty in the distance to this star, in kilometers, from the data available in 1840. If the presently accepted value for the parallax is 287.1 ± 0.5 mas, compute the modern estimate of the distance, again in kilometers, and its uncertainty.

11. The angular diameter of the Sun is 32 arc minutes when it is at the zenith. Using the table below (you will need to interpolate), plot a curve showing the apparent shape of the Sun as it sets. You should plot the ellipticity of the apparent solar disk as a function of the elevation angle of the lower limb, for elevations between 0 and 10 degrees. (If a and b are the semi-major and semi-minor axes of an ellipse, its ellipticity, ε, is $(a - b)/a$. The ellipticity varies between 0 and 1.) Is your result consistent with your visual impression of the setting Sun?

Apparent zenith distance (degrees)	75	80	83	85	86	87	88	89	89.5	90
Atmospheric refraction (arcsec)	215	320	445	590	700	860	1103	1480	1760	2123

12. The Foggy Bottom Observatory has discovered an unusual object near the ecliptic, an object some students suggest is a very nearby sub-luminous star, and others think is a trans-Neptunian asteroid. The object was near opposition on the date of discovery.

Below are sketches of four CCD images of this object, taken 0, 3, 9 and 12 months after discovery. Sketches are oriented so that ecliptic longitude is in the horizontal direction. The small squares in the grid surrounding each frame measure 250 mas $\times$ 250 mas. Note that the alignment of the grid and stars varies from frame to frame.

(a) Why is there no frame 6 months after discovery?

(b) Compute the proper motion, parallax, and distance to this object.

(c) Is it a star or an asteroid? Explain your reasoning.

(d) Compute its tangential velocity.

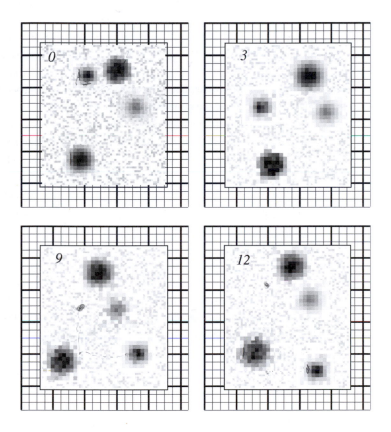

13. Current telescopes can detect stellar objects with apparent magnitudes as faint as $V = 24$ with good precision. What is the greatest distance that a supernova of magnitude -20 can be detected? Compute the expected redshift parameter of such an object.

14. The distances (there are actually several definitions of "distance" in an expanding universe) corresponding to redshift parameters larger than 1 actually depend on several cosmological parameters, not the simple relations in Equations (3.5) and (3.6). For example, the time light has taken to travel from the galaxy IOK-1 ($z = 6.96$) to us, using current values of these parameters, is about 12.88 Gyr. Compare this with the light travel time computed naively from Equations (3.5) and (3.6).

Chapter 4
Names, catalogs, and databases

> ... the descriptions which we have applied to the individual stars as parts of the
> constellation are not in every case the same as those of our predecessors (just as
> their descriptions differ from their predecessors') ... However, one has a ready
> means of identifying those stars which are described differently; this can be done
> simply by comparing the recorded positions.
>
> Claudius Ptolemy[1], *c.* AD 150, *The Almagest*, Book VII, H37

You can discover traces of the history of astronomy scattered in the names of
the objects astronomers discuss – a history that starts with the mythological
interpretation of the sky echoed in constellation names, and that continues to
an era when comets are named after spacecraft and quasars after radio telescopes. As discoveries accumulate, so too do the names. As the number of
objects of interest has risen to the hundreds of millions, tracking their identities and aliases has grown to a daunting enterprise, made tractable only by
the use of worldwide computer networks and meta-database software. In this
chapter we introduce the methods for identifying a particular celestial
object, but more importantly, the methods for discovering what is known
about it.

Very early in the history of astronomy, as Ptolemy tells us, astronomers
realized the obvious. The identities of most objects in the sky, like the identities
of mountains or cities, could be safely tied to their locations. However, a difficult problem arises in our Solar System (the subject of most of *The Almagest*),
where objects move around the sky quickly. For 1400 years, Ptolemy was the
supreme astronomical authority. He provided his students with laborious and
ultimately inaccurate methods for predicting the positions of Solar System

---

[1] Claudius Ptolemy (*c.* AD 100–165) spent most of his life in Alexandria, Egypt, and worked on
many branches of applied mathematics. His major work on astronomy, written in Greek, was one
of the few classical works on the subject to survive intact. This book, μεγαλη συνταξιξ (*Megale
Syntaxis* – the "great composition") became *Al Magisti* ("the Greatest") in the Arabic translation.
When the Arabic version reached Spain in the twelfth century, translators rendered this title as
"*Almagest*" in Latin. The *Almagest* remained the unchallenged authority on astronomy until
Copernicus published *De Revolutionibus* in 1543.

objects. You will find that even though you will have to deal with many more objects, modern computer technology means your experience will be more pleasant than theirs. It is, of course, up to you whether or not your experience is more worthwhile.

4.1 Star names

> You can know the name of a bird in all the languages of the world, but when you're finished, you'll know absolutely nothing whatever about the bird.... I learned very early the difference between knowing the name of something and knowing something.
>
> – Richard Feynman, "What is Science?", 1966

We are about to spend several pages discussing the names of celestial objects. Bear in mind Feynman's point: a name is only a link to what others actually know about an object. It is satisfying to know all the names of the red star at Orion's shoulder, but more significant to know it is larger than the orbit of Earth, varies in brightness and size, has spots on its surface, and has nearly run out of nuclear fuel.

4.1.1. Proper names

Abd al Rahman Abu al Husain (AD 903–986), known as Al-Sufi (The Mystic), named the stars. In his *Book of the Fixed Stars* he combined traditional Arabic nomenclature with Ptolemy's catalog of 1022 stars in 48 constellations. Al-Sufi (probably a Persian) wrote in Baghdad, but his book reached Europe and became a primary source for our present names of the brightest stars. Names like Rigel, Sirius, and Altair are medieval Latin transliterations or corruptions of Al-Sufi's Arabic names, which he in turn often composed from the Greek designations collected or suggested by Ptolemy. For example, in his list for the constellation Orion, Ptolemy describes star number 2 as "the bright reddish star on the right shoulder." Al-Sufi gives the Arabic name, *Ibt al Jauzah* (Armpit of the Central One), which medieval Latin transliterates to something like *Bed Elgueze*, and hence the modern *Betelgeuse*. Independent of Al-Sufi, some modern names derive with little change from the classical Latin (Arcturus, Spica), Greek (Procyon), or Arabic (Vega) designations. See Appendix E2 for a list of the most commonly recognized star names, and Allen (1899) for an extended discussion. In addition to the bright stars, some fainter stars have acquired enough fame to deserve "proper" names, like *Barnard's Star* (largest proper motion), *Plaskett's Star* (very high mass binary), and *Proxima Centauri* (nearest star to the Sun).

Professional astronomers generally avoid using such traditional names, except for the most universally recognized.

4.1.2 Bayer designations

In AD 1603, Johann Bayer produced *Uranometria*, a star atlas based in part upon the superb positional data for 777 stars from Tycho's[2] catalog of 1598. Bayer used star names and constellation designations derived in large part from Al-Sufi. Many of the fainter stars had no traditional names, or had very obscure names, so Bayer invented a naming convention. In each constellation he assigned a Greek letter to every star. Letters went in order of brightness, with alpha allotted to the brightest star, beta to second brightest, and so on. A star's two-part name then consisted of a Greek letter followed by the Latin genitive form of its constellation name. So, for example:

$$\text{Betelgeuse, the brightest star in Orion} = \text{Alpha Orionis} = \alpha\,\text{Ori}$$

$$\text{Sirius, the brightest star in Canis Major} = \text{Alpha Canis Majoris} = \alpha\,\text{CMa}$$

The International Astronomical Union in 1930 defined the boundaries of the 88 modern constellations and established a standard three-letter abbreviation for each (see Appendix E1 for a list). The pictorial figures these constellations are imagined to represent have various origins. Many are the 48 transmitted by Ptolemy from classical sources, and many of these are probably of very ancient origin indeed (e.g. Leo as a lion, Taurus as a bull — see Schaefer, 2006), but several are relatively new, especially in the south. Eighteenth-century astronomers proposed many of these to represent the tools of science and art (e.g. Telescopium, Microscopium, Antila, Pictor) — figures that some find lacking mythic resonance.

4.1.3 Flamsteed designations

> But beyond the stars of sixth magnitude you will behold through the telescope a host of other stars, which escape the unassisted sight, so numerous as to be almost beyond belief . . .
>
> — Galileo Galilei, *The Sidereal Messenger*, 1610

---

[2] Tycho Brahe (1546–1601) may well have been the greatest observational astronomer who ever lived, and introduced several revolutionary practices that today characterize astronomy. (His name should be pronounced "Tee-ko Bra-hee". The first name is the Latinized version of the Danish "Tyge", and is usually mispronounced "Tye-ko".) Tycho used technology to push the precision of measurement to its fundamental limits. (In Tycho's case the limits were set by the resolving power of the human eye, about one arc minute. The best positional measures prior to Tycho have precisions of 10–15 arc minutes.) For twenty years, he built and then directed the great observatory complex, Uraniborg, on the island of Hveen in the Baltic. At Uraniborg, Tycho tested the reliability of new instruments by examining reproducibility of measurements and agreement with other instruments. Tycho's brand of astronomy was an expensive enterprise, and he marshaled the financial support of his own noble family as well as the state (in the person of King of Denmark, and briefly, Emperor Rudolph II of Bohemia). He practiced astronomy in an international arena, and relied on a network of scientists and technologists throughout northern Europe for advice and collaboration. See Thoren (1990) for a modern biography of Tycho.

Tycho's Catalog recorded the position and brightness of each star investigated by the greatest naked-eye observer in history. His positions are accurate to about 60 seconds of arc, a phenomenal achievement. In 1609, however, a mere six years after the publication of *Uranometria*, Galileo turned a primitive telescope to the sky. This momentous act completely transformed astronomy and coincidentally unleashed a nightmare for stellar nomenclature. Telescopes unveiled many more stars than were visible to Tycho, a number "almost beyond belief" and certainly beyond naming. Obviously, Bayer's Greek letter scheme could not gracefully be extended to include these telescopic stars. Fortunately, the telescope held out the promise of great positional accuracy, and this capability has been the key to subsequent naming practices.

The first Astronomer Royal, John Flamsteed, introduced an important convention in his *British Catalog,* published in 1722. This, the first reliable *telescopic* catalog, gave equatorial positions for 2935 stars (12 are duplicates), with names constructed within each constellation by assigning numbers in order of increasing RA, e.g.

$$\text{Betelgeuse} = 58 \text{ Orionis}$$

4.1.4 Variable stars

One well-established modern naming convention concerns stars that change brightness. The brightest variables are simply known by their popular or Bayer designations. Polaris, Delta Cephei, and Betelgeuse are all variables. Fainter variables are named *in order of discovery* by concatenating capital roman letters with the constellation name, beginning with the letters R through Z, then continuing with RR, RS, through RZ. The sequence then continues with SS to SZ, TT to TZ, and so on until ZZ. Then comes AA to AZ, BB to BZ, etc. until QZ. This provides for a total of 334 stars per constellation (the letter J is not used). If more names are required, the lettering continues with V 335, then V 336 etc. For example, all the following are variable star names:

S And

RR Lyr

V341 Ori

Supernovae are stars that flare to spectacular luminosities in a single explosive episode, and then fade out forever. Almost all the supernovae discovered occur in other galaxies. Although technically variable stars, supernovae have a unique naming convention: They are assigned roman letters in the order in which they are discovered in a particular year. Thus, SN 1987A was the first supernova discovered in 1987. After reaching the letter Z, the sequence continues with aa, ab...az, ba, bb etc. (Note the change to lower case.)

4.1.5 Durchmusterung Numbers

> ...if one seriously wants to aim at discovering all the principal planets that
> belong to the Solar System, the complete cataloging of stars must be carried out.
>
> – Freidrich William Bessel, Letter to the Royal Academy of Sciences in Berlin, 1824

Bessel (1784–1846) initiated the visionary project of a complete catalog of all stars brighter than ninth magnitude, but it was Freidrich Wilhelm August Argelander (1799–1875), his student, who carried out the work. He oversaw the production and publication, in 1859–1962, of the *Bonner Durchmusterung*, or BD, a catalog and atlas that remains a source of star names. The BD gives the positions and visually estimated magnitudes for 324,198 stars brighter than magnitude 9.5 and north of Dec = −2°. Three other catalogs begun in the nineteenth century completed the stellar mapping on the remaining parts of the celestial sphere and employed the same naming scheme as the BD. Eduard Schoenfeld, a student of Argelander, published the *Bonner Durchmusterung Extended* (1886) (extended south to Dec = −23°). It is also abbreviated BD, or sometimes BDE, and contains 133,659 stars. Juan Macon Thome and Charles Dillon Perrine, in Argentina, published the *Cordoba Durchmusterung* (1892–1914). Abbreviated CD or CoD, it contains 613,953 stars with $m < 10.5$, Dec $< -23°$). Finally, taking advantage of a new technology, David Gill and Jacobus Cornelius Kapteyn published the *Cape Photographic Durchmusterung* (1896) (CPD, 450,000 stars). The naming scheme for the *Durchmusterungs* is as follows: within each 1-degree wide band of declination, stars are numbered consecutively in order of right ascension around the sky. The three-part name consists of the catalog abbreviation, the declination band, and the right ascension ordinal. For example:

$$\text{Betelgeuse} = \text{BD} + 07 \ 1055$$

The BD, CD, and CPD designations are still in widespread use in astronomical literature and their practice of designating objects by specifying a catalog name plus numerical identifier has become the preferred method in astronomy. There are thousands of catalogs and discovery lists, some of them quite specialized, and nomenclature can get obscure and confusing, even for professional astronomers. Except for bright objects, it is good astronomical practice to recall Ptolemy's advice, and give both name and position when identifying an object.

4.1.6 The Henry Draper Catalog

At the close of the nineteenth century, European astronomers largely accepted Bessel's heritage, and committed themselves to several huge programs in positional astronomy (most notably, a very high precision fundamental catalog of transit telescope positions, the first *Astronomische Gesellschaft Katalog* (*AGK1*) and the astrographic *Carte du Ciel*). The inheritance proved to be a burden that prevented many Europeans from stepping nimbly into new areas of science. In the United States, astronomers had just started to grow in number. Less burdened by tradition, they were light-footed enough to step into some productive research areas. In 1886, for example, Edward Charles Pickering, the Director of the Harvard College Observatory, began a long-term project aimed at classifying the photographic spectra of stars brighter than $m = 10$. Spectroscopy proved to be the foundation of modern astrophysics.

Harvard astronomer Henry Draper (1837–82) was the first ever to photograph the spectrum of a star (Vega, in 1872). His widow, Anna Palmer Draper (1839–1914), established a memorial fund of several hundred thousand dollars to support Pickering's project. The primary workers in this monumental work were Williamina Fleming, Antonia Maury, Annie Jump Cannon, and over a dozen other women. It was Maury, Draper's niece, who first recognized the proper sequence of spectral types in 1897. Cannon (1863–1941) joined the project in 1896 and developed an uncanny skill at rapid classification. In the four years (1911–1914) prior to the completion of the Henry Draper Catalog (HD), Cannon classified an average of 200 stars per day. Her work on the Henry Draper Extension (HDE), from 1923 onward, saw her classify an additional 400,000 stars. She expanded the nomenclature that Fleming (in 1890) had initially proposed into a system resembling the modern one. The nine volumes of the *Henry Draper Catalog,* published between 1918 and 1924, contain the Cannon spectral types for 225,300 stars. The stars in the HD are numbered consecutively, and star names are constructed from the HD sequence number:

$$\text{Betelgeuse} = \text{HD } 39801$$

The use of HD numbers is widespread in the astronomical literature.

4.1.7 The Hubble Guide Star Catalog

This catalog was first prepared in advance of the launch of the Hubble Space Telescope, partly to provide precise positions and magnitudes for the faint stars that could serve as guide objects for maintaining telescope pointing. Two editions of the catalog were produced from microdensitometer scans of wide-field photographic plates. The GSC I (final version 1.2) contains roughly 15 million stars and 4 million non-stellar objects brighter than $m = 16$. For purposes of naming objects, the GSC I divides the sky into 9537 regions, and constructs object names by giving the region number followed by a sequence number within the region. So, for example,

$$\text{Betelgeuse} = \text{GSC } 129 - 1873 = \text{GSC } 00129 - 01873$$

The astrometric and photometric quality of version 1.2 rests on HIPPARCOS/ Tycho data (see next section) for brighter stars, as well as ground-based photometry. The GSC I is distributed as CD-ROM sets, and many planetarium/sky atlas computer programs access it in this form to generate star maps and identify objects.

The GSC II (currently version 2.3.2 in 2008) uses plate material from two epochs in multiple band passes and will support space missions like the James Webb Space Telescope (JWST) and Gaia. It includes proper motions and colors, as well as extended magnitude coverage to $V < 18$, for about *one billion* objects. Object names are ten-digit alphanumeric codes similar to the GSC I system. (Betelgeuse = GSC2.3 N9I5-000041) Positions are accurate to better than 0.28 arcsec in the HIPPARCOS (optical ICRS) system. The GSC II Catalog (Lasker *et al.*, 2008), is not generally available in physical media, but only as a web-based resource (see *VizieR*, below, or http://www-gsss.stsci.edu/Catalogs/GSC/ GSC2/GSC2.htm).

4.1.8 The HIPPARCOS and Tycho catalogs

Two catalogs published in 1997 summarize data from the HIPPARCOS space astrometry satellite (see Chapter 3), whose goal was to determine the parallaxes of over 100,000 stars *in the solar neighborhood* that are brighter than $V \sim 12.4$. The resulting **HIPPARCOS Catalog** is a very important reference because of the unprecedented precision and homogeneity of its observational results. The 118,218 stars in the catalog are simply designated by a sequence number:

$$\text{Betelgeuse} = \text{HIP } 27989$$

A second goal of the mission was to determine positions and magnitudes in a system (B_T, V_T) similar to the Johnson B and V for the 2.5 million brightest stars – a magnitude limit of about $V_T = 11.5$. Entries in the **Tycho 2 Catalog** are identified by a scheme nearly identical to the one employed by the GSC I. In general, the TYC designation is the same as the GSC I designation, except for objects not in the GSC, and for GCS stars that Tycho showed to have multiple components (a third number is added to the Tycho designation to cover such cases). Thus,

$$\text{Betelgeuse} = \text{TYC } 129 - 1873 - 1$$

The observations for Tycho come from a single instrument outside the atmosphere and their reduction closely ties positions to the astrometric system of the HIPPARCOS stars. The catalog has excellent positional accuracy (about 7 mas for $V < 9.0$, 60 mas at worst) and photometric uniformity (standard errors in V_T of 0.012 magnitudes for $V < 9$, 0.10 magnitudes at worst). The catalog uses ground-based positions from different epochs for all but 4% of the entries to derive and list proper motions with precisions in the 1–3 mas yr^{-1} range.

4.1.9 The US Naval Observatory catalogs

At about the time Cannon joined Pickering at Harvard to work on the HD, an international conference in Paris initiated a program to photograph the entire sky, using new photographic refractors ("astrographs") installed in twenty observatories, from Helingsfors, Finland, in the north, to Melbourne in the south. Each observatory was assigned a declination zone. The intent was to measure the positions of all stars brighter than 11th magnitude to at least 0.5 arcsec, and also to print photographs of the entire sky as an atlas showing stars to magnitude 13 or 14. This was a spectacularly ambitious – and ill-fated – undertaking, at a time when astronomical photography was in its infancy.

The photographic atlas, called the ***Carte du Ciel***, was only partially completed due to the great expense of photographic reproduction. The positional measurements were completed and eventually published (the last in 1950) by the individual observatories, in a heterogeneous and inconvenient collection known as the ***Astrographic Catalog (AC)***.

The United States Naval Observatory emerged in the late twentieth century as the major US center for ground-based astrometric studies. In cooperation with several other observatories, the USNO re-issued the AC data as a modern catalog. The current version, the ***AC 2000***, published electronically in 2001, is a re-analysis of the AC measurements, deriving the equatorial positions of 4.6 million stars in the Tycho-2/ICRS system. The AC 2000 also re-computes the magnitudes given by AC using the Tycho B-band photometry as a calibration. The mean epoch of the AC observations is about 1905, so the AC 2000 is extremely valuable as a source of accurate positions and magnitudes of an enormous number of faint stars at a very early epoch.

The ***US Naval Observatory CCD Astrographic Catalog (UCAC)***, in its current version (UCAC2, published in 2003) provides ground-based positions from new CCD observations for about 48 million stars south of declination +40°. The UCAC2 basically extends the Tycho-2 catalog, but also observes the ICRS-defining galaxies, so that positional accuracy ranges from 20 mas for stars brighter than R magnitude 14 to around 70 mas at the faint limit ($R = 16$). Proper motions, with errors in the range 1–7 mas yr^{-1} are derived for every star from earlier epoch data, including the AC 2000, but photometric accuracy is poor. The most recent version of this catalog, UCAC3, released in 2009, extends coverage to the entire sky (100 million objects) and has improved positions and photometry.

Another catalog, ***USNO-B1.0***, is very similar to GSC II. It contains entries for slightly over one billion stars/galaxies detected in the digitized images of several photographic sky surveys. The catalog can only be accessed over the Internet, and presents right ascension and declination, proper motion, and magnitude estimates. The positional error is near 200–300 mas.

Names in AC 2000 and UCAC are catalog-sequential, while USNO is zone + sequence number:

Betelgeuse $= $ AC2000.2 241908 $=$ UCAC2 50060561 $=$ USNOB1.0 $-$ 0974 $-$ 0088855

4.1.10 The nomenclature problem

Most of the catalogs mentioned so far aim to provide complete listings (and names) of stars brighter than a certain magnitude limit, while others have more restrictive criteria for the objects they include (HIPPARCOS, for example). There are a very large number (thousands) of other catalogs, most of which have specialized purposes (e.g. colors, characteristics of variable stars, radial velocities, spectra, etc.) and most of which result in additional names for the stars they include.

The International Astronomical Union (IAU) has recognized that progress in astronomy will generate many more catalogs and names, and has tried to enforce some guidelines for generating a new name for any object (star, galaxy, nebula...) outside the Solar System. The IAU guidelines are lengthy, but approach the problem by regularizing and restricting some of the more widespread conventions already common in astronomy. They propose that new object names consist of two parts: a unique three (or more) character acronym (like TYC, UCAC2, or GSC) followed by a sequence number. The sequence number can be the order in a list (as in the HD), a combination of field number and sequence (as in the BD or GSC) or, preferably, some specification of coordinate position.

The IAU also instigated the creation of the *Reference Dictionary of Nomenclature of Celestial Objects*. The dictionary is an important resource for identifying object references as well as for avoiding ambiguity or duplication in new designations. It currently (2010) lists over 19,000 acronyms and can you access it through the *VizieR* site (vizier.inasan.ru/viz-bin/Dic).

4.1.11 Other star catalogs

The previous sections omitted discussion of many important star catalogs, and it would be impossible in a book like this to even list them by title. Fortunately, the power of computer networks has proved a good match to the explosive growth in the number, length, and variety of catalogs. A number of national and international data centers now hold most important catalogs online, and each site provides methods for interrogating its holdings via the Internet (see Section 4.5). Table 4.1 is a small sample of the catalogs of stars (and therefore star names) that await the curious.

Table 4.1. A very limited selection of star catalogs, some of mainly historical interest

Example star name	Reference	Comment
HR 2061	Yale Bright Star Catalog, 5th edition, D. Hoffleit and W. H. Warren, Jr. (1991), On-line only, e.g. http://adc.astro.umd.edu	Widely used compilation of basic astronomical data for the 9096 stars brighter than magnitude 6.5.
SAO 113271	Smithsonian Astrophysical Observatory Star Catalog, SAO Staff (1966,1984), USNO (1990)	Compilation of precision data for 258,997 stars from previous catalogs.
Lalande 21185	A Catalog of ... De Lalande*F. Baily, British Association of Advanced Science, London (1847)	Positions of 47,390 stars from the observations of J. de Lalande, first published in Histoire Celeste Francaise in 1795.
FK5 224	Fifth Fundamental Catalog, W. Fricke, et al., Veröffentlichungen Astronomisches Rechen-Institut Heidelberg, vol. 32 (1988)	Fundamental positions for 1500 bright stars. Defined the preferred barycentric system prior to ICRS. Update of FK4.
Giclas 123–109	Lowell Proper Motion Survey, H. L. Giclas et al., Lowell Observatory Bulletin (1971–1978)	12,000 stars with large proper motions
GC 7451	General Catalogue of 33342 stars for the Epoch 1950, Dudley Observatory Albany, NY, B. Boss, Carnegie Institution of Washington, Publ. No 468, vol. 1–5 (1937)	Survey of all stars brighter than $V = 7$: positions and proper motions
CCDM J01016–1014AB	Un catalogue des composantes d'etoiles doubles et multiples (C.C.D.M.), J. Dommanget, Bull. Inf. Centre Donnees Stellaires, 24, 83–90 (1983)	Components for 34,031 double or multiple systems.
GCRV 10221	General Catalog of Stellar Radial Velocities, R. E. Wilson, Carnegie Institution of Washington DC, Publ. 601 (1953)	Radial velocities of 15,000 stars
AG+07 681	Astronomishe Gesellschaft Katalog, 3rd edition, Astronomical Journal, 103, 1698 (1992)	Positions and proper motions in FK4 for a selected set of 170,000 stars, based on photographic data 1928–1985
V* EW Lac	Combined General Catalog of Variable Stars, 4th edition, P. N. Kholopov, 1998, Moscow State University	40,000 variable stars, Galactic and extragalactic

* A catalog of those stars in the "Histoire Celeste Francaise" of J. De Lalande for which tables of reduction to the epoch 1800 have been published by Professor Schumacher

4.2 Names and catalogs of non-stellar objects outside the Solar System

> Wonderful are certain luminous Spots or Patches, which discover themselves only by the Telescope, . . . in reality are nothing else but light coming from an extra-ordinary great Space in the Aether; through which a lucid Medium is diffused, that shines with its own proper Lustre . . . . These are Six in Number . . . .There are undoubtedly more of these which have not yet come to our Knowledge . . . .
>
> – Edmund Halley, *Philosophical Transactions of the Royal Society*, vol. 4, 1721

4.2.1 Bright objects

Simon Marius (1570–1624) probably used a telescope to examine astronomical objects before Galileo, and almost certainly discovered the first of Halley's "luminous spots" – the spiral galaxy in Andromeda – late in 1612. Marius described the object as "like a candle flame seen through the horn window of a lantern". One hundred years later, Halley called attention to the six extended, cloud-like objects – *nebulae* – so far discovered. Astronomers would puzzle over these fuzzy objects for two more centuries before they fully realized the great physical variety lurking behind a superficially similar appearance in the telescope. Not all lanterns hold candles.

Charles Messier (1730–1817) published the first important catalog (of 40 nebulae) in 1774. The final (1783) version lists 103 nebulae. These include relatively nearby clouds of incandescent gas, illuminated dust, star clusters in our own galaxy, and many galaxies outside our own. (Modern authors sometimes add numbers 104–110 to Messier's list). Messier was a comet-hunter, and had made his lists partly to help avoid the embarrassment of mistaken "discoveries". His list, and not his comet discoveries, has perpetuated his name, or at least his initial. The ***Messier objects*** include the most prominent non-stellar objects in the northern sky, many of which are commonly referred to by their Messier numbers. The great nebula in Andromeda that Marius noted is probably called "M 31" more often than anything else. Table 4.2 lists some famous Messier objects.

In 1758, Messier observed a comet and coincidently discovered M 1. At about the same time, a young musician in the Hanoverian army decided to emigrate to England rather than see further action in the Seven Years' War. In England, William Herschel (1738–1822) made his living as a popular music teacher in the city of Bath, where he was joined by his sister Caroline (1750–1848) in 1772. Shortly after Caroline's arrival, William developed an interest in telescope-making, and after each day's work as a musician, would labor obsessively to grind or polish his latest mirror. The story goes that Caroline would place food in William's mouth and read to him while he worked on his optics. With their telescopes, Caroline discovered several comets, and William, on 13 March 1781, happened to discover the planet Uranus. This feat brought instant fame and eventual financial support from King George III. William retired from the music business in 1782 to become a full-time astronomer, assisted, as ever, by Caroline.

Table 4.2. *Messier objects*

Messier	NGC	Type	Individual name
1	1952	Supernova remnant	Crab Nebula
8	6523	Emission nebula	Lagoon Nebula
13	6205	Globular star cluster	Hercules Cluster
16	6611	Emission nebula	Eagle Nebula
17	6618	Emission nebula	Omega Nebula, Horseshoe Nebula
20	6514	Emission nebula	Trifid Nebula
27	6853	Planetary nebula	Dumbbell Nebula
31	224	Spiral galaxy	Andromeda Galaxy
32	221	Elliptical galaxy	Satellite of Andromeda
33	598	Spiral galaxy	Local group galaxy
42	1976	Emission nebula	Great Nebula in Orion
44	2632	Open star cluster	Praesepe, The Beehive
45		Open star cluster	Pleiades, The Seven Sisters
51	5194	Spiral galaxy	Whirlpool Galaxy
57	6720	Planetary nebula	Ring Nebula
101	5457	Spiral galaxy	Pinwheel Galaxy

Familiar with Messier's lists and equipped with a superior telescope, Herschel set about discovering nebulae, and by the time of his death, he and Caroline had compiled a list of around 2000 objects. William's son, John Herschel (1792–1871), continued the search (primarily with William's 18-inch telescope) and extended it to the southern hemisphere. In 1864, John presented a catalog of 5079 objects, which was further expanded by John L. E. Dreyer, a Danish–Irish astronomer, and published in 1887 as the **New General Catalog** of 7840 nebulae, listed in order of RA. Many nebulae are today known and loved by their numbers in this catalog: NGC 6822 is a faint nearby irregular galaxy, NGC 7027 is a planetary nebula, NGC 6960 is a supernova remnant (the Veil), NGC 2264 is a very young star cluster. In 1895 and 1908, Dreyer published two additional lists (6900 nebulae, total) as supplements to the NGC. Together, these lists constitute the **Index Catalog**. Only a few of these fainter objects are famous: IC 434 is the Horsehead Nebula; IC 1613 is an irregular galaxy in the local group.

4.2.2 Faint non-stellar objects

Twentieth-century astronomers soon recognized that the NGC and IC contain non-stellar objects of vastly differing physical natures, and subsequent catalogs of nebulae tended to be more specialized. Table 4.3 gives a very incomplete sampling of some of these.

Table 4.3. *Catalogs of non-stellar objects. Code: HHhh = hours, decimal hours of RA, MMSS = minutes, seconds of time or arc, DDdd degrees and decimal degrees of arc, LLlI, BBbb = degrees and decimal degrees of Galactic longitude and latitude; FF = field number*

Type of object	Sample designation	Reference
89 Stellar associations	Assoc 34	*Un catalogue des composantes d'etoiles doubles et multiples (C.C.D.M.),* J. Dommanget, *Bull. Inf. Centre Donnees Stellaires,* **24,** 83–90 (1983) (plus supplements)
137 Globular star clusters	GCl 101	*Catalogue of Star Clusters and Associations,* G. Alter, J. Ruprecht, and V. Vanysek, Akad. Kiado, Budapest, Hungary (1970) (plus supplements)
1112 Open star clusters	OCl 925	*Catalogue of Star Clusters and Associations,* G. Alter, J. Ruprecht, and V. Vanysek, Akad. Kiado, Budapest, Hungary (1970) (plus supplements)
1154 Open star clusters	Lund 877, Lynga 877 or C HHMM+DDd	*Catalogue of Open Cluster Data,* 5th edition, Lynga (Lund Obsrevatory) (1987). Revised: 1996AJ.112.2013S (electronic)
37 Open star clusters	Trumpler 21	*Preliminary Results on the Distances, Dimensions and Space Distribution of Open Star Clusters,* R. J. Trumpler, *Lick Obs. Bull.,* **14,** 154–188 (1930)
1125 Emission nebulae	LBN 1090 or LBN LLL.lI±BB.bb	*Catalogue of Bright Nebulae,* B.T. Lynds, , *Astrophys. J., Suppl. Ser.,* **12,** 163 (1965)

Object	Designation	Reference
1036 Planetary nebulae	PK LLL+BB	*Catalogue of Galactic Planetary Nebulae*, L. Perek and L. Kohoutek, *Acad. Publ. Czech. Acad. Sci.*, 1–276 (1967)
1802 Dark nebulae	LDN 1234	*Catalogue of Dark Nebulae*, B.T. Lynds, *Astrophys. J., Suppl. Ser.*, **7**, 1–52 (1962)
287 Million objects, mostly, galaxies and stars	SDSS JHHMMSS.ss+ DDMMSS.s	Current data release of the Sloan Digital Sky survey (images, photometry, and spectra). On-line access at http://www.sdss.org/
12,921 Galaxies	UGC 12345	*Uppsala General Catalogue of Galaxies*, P. Nilson, *Nova Acta Regiae Soc. Sci. Upsaliensis, Ser.* V (1973). Data for 12,921 galaxies north of delta = −23
1200 Galaxies	MCG + FF-FF-1234	*Morphological Catalogue of Galaxies. Part V* (Parts 1-IV earlier), B. A. Vorontsov-Vel'yaminov, V.P. Arkhipova, *Trudy Gosud. Astron. Inst. Shternberga*, **46**, 1–67 (1974)
338 Peculiar galaxies	Arp 123= APG 123	*Atlas of Peculiar Galaxies*, H. Arp, *Astrophys. J., Suppl. Ser.*, **14**, 1–20 (1966)
31350 Galaxies and 970 clusters of galaxies	Z FFF-123 = ZHHMM.m± DDMM ZwCl FFF-123 = ZwCl HHMM±DDMM	*Catalogue of Galaxies and of Clusters of Galaxies*, F. Zwicky, E. Herzog, P. Wild, M. Karpowicz, and C.T. Kowal, *California Inst. Techn.* vol. I to vol. VIIIi (1961–68)
5200 Clusters of galaxies	ACO 1234 *or* ACO S 1234	*A Catalog of Rich Clusters of Galaxies*, G.O. Abell, H.G. Corwin, Jr., and R.P. Olowin, *Astrophys. J., Suppl. Ser.*, **70**, 1–138 (1989)

The ***Sloan Digital Sky Survey*** is emblematic of modern projects. Concerned mainly with extragalactic objects and their redshifts, SDSS has surveyed 25% of the entire sky (mainly avoiding the Galactic plane) with an imaging five-color CCD camera, and has followed up with spectra of objects of interest. Efficient instruments and automated data reduction have so far identified about 300 million objects, and taken spectra of about 1.2 million of these (mostly galaxies brighter than $V = 17.5$, but also about 100,000 quasars and 150,000 unusual stars). In terms of data volume, SDSS can currently claim to be the largest single astronomical project (about 15 terabytes, roughly the same information content as the Library of Congress).

4.3 Objects at non-optical wavelengths

Optical observers had a head start of several thousand years in the task of cataloging celestial objects. The opening of the electromagnetic spectrum to observation, first with radio detections from the surface of the Earth in the 1940s, then with observations at all wavelengths from space beginning in the 1960s, added a huge number of catalogs to our libraries as well as complexity and variety to astronomical names in use. Astronomers making the very first detections of objects in new bands tended to mimic the early Bayer-like naming conventions in optical astronomy: for example, the radio sources Centaurus A, Sagittarius B, etc., and X-ray sources Cygnus X-1 and Cygnus X-2.

However, as the numbers of non-visible sources accumulated, they tended to follow the IAU recommendations on nomenclature. Thus, all except the brightest sources have useful (but perhaps unromantic) names like the examples listed in Table 4.4. As you can see, if objects detected in the non-visual bands coincide with more "famous" optical sources, astronomers will tend to use the optical name even when discussing X-ray or radio properties.

4.4 Atlases and finding charts

The previous sections emphasize two facts. First, the number of interesting objects as well as the number of aliases an object might have will increase as more observations are published. Second, although the multiple names for an object may be obscure, its *position* at a particular epoch does not change. If the position is expressed with sufficient precision in a well-defined coordinate system, there can be no ambiguity about the identity of the object.

It might seem that the solution to the nomenclature problem is simple: refer to every object not only by name, but also by position. This is certainly done. However, establishing precise coordinates is not a trivial task, nor is it trivial to point a telescope with the same precision. Often, an astronomer may wonder which of the two faint stars near the center of the field of view is the one she

Table 4.4. *Examples of some source designations at non-optical wavelengths*

Modern designation	Source	Other designations
4U0900-40	Entry in the fourth catalog from the Uhuru X-ray satellite at position RA 9:00 Dec −40	Vela X-1, GP Vel, a binary star; one component is a neutron star
CXO JHHMMSS.s +DDMMSS	Chandra X-ray satellite catalog. First release expected 2009	Source at the specified J2000 coordinates
4C 02.32	Entry in the 4th Cambridge Catalog of Radio Sources	3C 273, first quasar discovered = QSO J1229 +0203 = ICRF J122906.6 +020308
GBS 0526–661	Gamma-ray burst source at this location	Possibly N 49, a supernova remnant in the Large Magellanic Cloud
2MASS J05551028 +0724255	2-Micron All-Sky Survey (near-IR JHK photometry)	Betelgeuse
FIRST J022107.4− 020230	Catalog of Faint Images of the Radio Sky at Twenty centimeters	Very faint anonymous galaxy
GeV J0534+ 2159	Compton Gamma-ray Observatory, EGRET instrument, J2000 position	Crab Nebula: M1, Taurus A, Tau X-1, and many other designations
IRAS 05314+ 2200	Infrared Astronomical Satellite (12, 25 60 and 100 micron photometry) 1950 position	Crab Nebula

wants to spend the next hour observing. **Atlases** give a pictorial representation of the sky, and can be of tremendous help in identification. In practice, astronomers will routinely make use of ***finding charts***, images of a small area of the sky near the object of interest. In many cases, astronomers will publish finding charts along with their results as aides to the identification of the objects of interest.

4.5 Websites and other computer resources

Proliferation of the number and expansion of the depth of catalogs, atlases, and other research results challenges astronomers' ability to maintain access to available information. The sheer volume makes storage on paper unthinkable, and the difficulties in searching (e.g. what is known about object A?) by traditional methods become distressing.

Compressing data on computer media has helped. For example, paper and film prints of the **Palomar Sky Survey**, a photographic atlas, have occupied a 2-meter-long cabinet in my college's library for the past 50 years. This same data, stored on CD-ROMs, now occupies 40 cm of a single shelf in my office, and I can carry a compressed 2-CD version in my pocket (but I never do – see the next paragraph).

The real power to tame the beastly growth of astronomical information lies in computer networks. Several sites implement this great idea to some degree: organize and store all relevant catalog, atlas, or other results online, and provide access to any computer on the Internet for searching this data. The most important of these are the **CDS** site, maintained by the University of Strasbourg for non-Solar System objects, which contains the special **Simbad** (object database), **VizieR** (catalog information) and **Aladin** (sky atlas) sites, and the **MPC** site, maintained by the Smithsonian Astrophysical Observatory, for Solar System objects.

It is important to realize that any search of the astronomical professional research literature is best undertaken with a specialized search engine (Google and its competitors can miss a great deal). The **Astronomical Data System**, **ADS**, is probably the preferred site. Table 4.5 gives the URLs for some important resources.

4.6 Solar System objects

Unlike stars and nebulae, Solar System objects move around the sky quickly, a behavior which sometimes produces difficulty in identification and naming. There are many examples of minor planets, comets, and small moons that were discovered and subsequently lost because their orbits were not known with sufficient accuracy.[3] There are even more examples of "new" objects that turn out to have been previously observed. This potential for confusion has led astronomers to a system of provisional designations for newly discovered small bodies, so that the provisional name can be superceded by a permanent name

---

[3] To determine the orbit with accuracy, one usually needs to observe the object's motion over a significant arc of the complete orbit. For a newly discovered asteroid in the belt, for example, this means following the object for about 3 or 4 weeks (about 2% of a complete orbit) in order to establish an ephemeris (prediction of future positions) for the next opposition. Only after four observed oppositions (almost a complete sidereal orbit), however, can the orbit be regarded as precisely known from ground-based observations. Opposition occurs when the RA of the Sun and the object differ by 12 hours. For periodic comets, the requirement is usually two perihelion passages, which may be separated by as long as 200 years.

once the orbit is accurately known, or the object is identified as one previously cataloged.

For example, the **Minor Planet Center** of the IAU (Table 4.5) manages the identification process for minor planets. If an observer reports the positions of a possible discovery from at least two different nights, the Center assigns a provisional designation based on the *date* of the report, and attempts to identify the candidate as an object already provisionally or permanently named. If the candidate is observed on four successive oppositions without identification with a

Table 4.5. *Some important astronomical reference sites. Since websites can disappear, a search engine may be needed to find current URLs*

Title	URL (prefix http:// to locator given)	Description
CDS	cdsweb.u-strasbg.fr	International centre for astronomical data at Strasbourg. Links to most important databases and bibliographic services
SIMBAD	simbad.u-strasbg.fr or simbad.harvard.edu	CDS reference for identification of astronomical objects
VizieR	webviz.u-strasbg.fr/ viz-bin/VizieR	Access software for about 7500 astronomical catalogs
Aladin	aladin.u-strasbg.fr/ aladin.gml	Interactive software sky atlas with catalog identifications
NED	www.ipac.caltech.edu	Extragalactic database
ADS	adsabs.harvard.edu/ ads_abstracts.html	Searchable database of abstracts and texts of astronomical research papers
Sky View	skyview.gsfc.nasa.gov	Sky survey images at multiple wavelengths
DSS	stdatu.stsci.edu/dss/	Digitized Palomar sky survey
IRSA	irsa.ipac.caltech.edu	Infrared data archive from multiple sources
MPC	www.cfa.harvard.edu/ iau/mpc.html	Minor Planet Center
NSSDC	nssdc.gsfc.nasa.gov/	Space sciences data center for spacecraft investigations of Solar System objects
Horizons	ssd.jpl.nasa.gov/ horizons.cgi#top	Ephemeris generator

previously designated object, MPC assigns a permanent designation consisting of a catalog number and a name suggested by the discoverer.

Similar identification schemes are used for comets, small moons, and rings. Appendix E3 gives some of the rules for provisional and permanent designations.

The IAU Working Group on Planetary System Nomenclature coordinates the naming of regions and features like craters, basins, fissures, and mountains on the surfaces of large and small bodies. The group attempts to enforce an international approach in theme selection, to prevent duplication of the same name for features on different objects, to maintain a thematic structure in the names, and to avoid political or religious references.

If you know the name of an object outside the Solar System, it is relatively easy to use the SIMBAD resource to find all its aliases and produce a finding chart for it (you can also easily generate a bibliography of most of the astronomical works that mention it). To generate a finding chart for a Solar System object, however, an additional step is necessary – you need to compute the position the object will occupy at the time you wish to observe it. A table of object positions as a function of time is called an *ephemeris*, and the ephemerides for bright objects are traditionally listed in the annual edition of the *Astronomical Almanac*. However, for these, and especially for fainter objects, online resources are the best option (an alternative is to compute the position yourself). The ***Horizons ephemeris generator*** at the Jet Propulsion Laboratory is a good example, as is the generator at the Minor Planet Center (see Table 4.5 for links).

Summary

- Many systems exist for naming stars. All useful names reference a catalog entry that provides information about the star, and many names themselves contain some positional information.
- Bayer, Flamsteed, and variable star designations use constellation locations in generating a star's name.
- Other schemes simply use a sequential number, some combination of zone and sequence, or actual equatorial coordinates to produce a name. Some historically important or currently useful catalogs:

  Henry Draper (HD)   *Bonner Durchmunsterung (BD, CD, CPD)*
  Tycho Catalog (HIP, TYC)   *Hubble Guide Star Catalog (GSC)*
  US Naval Observatory Catalogs (AC 2000, UCAC, USNO-B)

- Astronomers attempt to follow some IAU-sanctioned conventions in assigning new names to objects.
- We usually know bright non-stellar objects by their entry numbers in either the Messier (M) catalog, the New General Catalog (NGC) or Index Catalog (IC). Fainter objects tend to occupy catalogs devoted to particular types of objects: galaxies, clusters, planetary nebulae, etc.

- Observations at non-optical wavelengths commonly generate new catalogs and new names.
- Atlases and finding charts greatly aid the identification process, as do accurate positions of sufficient precision.
- Motion of even bright Solar System objects complicates their location and therefore identification. Astronomers use a system of provisional names for newly discovered objects until accurate orbits can be computed.
- Internet-accessible resources are essential for coping with problems of identification, nomenclature, and links to bibliographic information for a particular object. Very important sites are:
  - CDS – with subsets Simbad, VizieR and Aladin, for the identification of objects outside the Solar System, links to basic catalog information, and a sky atlas
  - ADS – for searches of the astronomical research literature

Exercises

1. Take the full tutorial at the CDS website. Use the SIMBAD *criteria query* feature to produce a list of all stars with a parallax greater than 250 mas. Include the parallax value, V magnitude, and spectral type in the list you produce.

2. As you know, α Cen is the nearest star. In the southern sky it forms a striking pair with β Cen. How far away is β Cen? Cite the source of this information (reference to the actual measurement).

3. Use the SIMBAD site to investigate the star cluster NGC 7790. Produce a finding chart for the cluster with the Aladin application. Then:
   (a) Use VizieR to produce a list of all known and suspected variable stars within 8 arc minutes of the cluster center. Identify the three stars (one is a close double) that are Cepheid variables.
   (b) Find the name and apparent magnitude at maximum brightness of the non-double Cepheid.
   (c) Use SIMBAD/VizieR to look up the catalog entry in the Combined General Catalog of Variable Stars for this star and record its period and epoch of maximum light.
   (d) Use a spreadsheet to predict the universal dates and times of mdaximum light for this star over the next month. On the same sheet, produce a calendar for the next month that shows the phase (fraction of advance over a full period) the star will have at 9 pm local time on each day.
   (e) Find the reference to a recent paper (use ADS) that discusses the distance to NGC 7790.

4. Use the Horizons site to find the position of the dwarf planets Eris and Makemake at 0 hours UT on the first day of next month. One of these should be visible at night. Generate a finding chart that would enable you to identify the dwarf planet on that date and compute the approximate local standard time at which it crosses the meridian.

Fig. 5.1 (a) Dotted lines represent rays from the source and the smaller solid curve is a wavefront that has traveled a distance s in time t. The larger solid curve is a wavefront at distance s_1. (b) Rays and wavefronts in a medium where the index of refraction is not homogeneous. Each point on a wavefront has the same optical path length from the source.

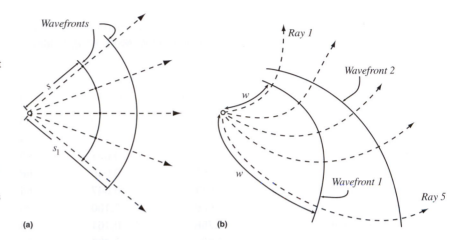

(a)   (b)

picture light as energy sliding along rays, or you can imagine light propagating as a series of wavefronts that expand away from the source at the phase velocity.

Figure 5.1b shows a more complicated situation, where the medium is inhomogeneous (here n increases towards the upper right). Light rays are no longer straight, and wavefronts no longer spherical. For example, because it travels through a higher index, light moving along ray 1 moves a shorter physical distance than light moving for the same time along ray 5. Wavefront 1 locates photons that have left the source together. We say that wavefront 1 locates the ends of rays of equal **optical path length**. If ds is an infinitesimal element of length along a path, the light travel time along a ray is just

$$t = \int \frac{ds}{v} = \frac{1}{c} \int nds = \frac{w}{c}$$

where the quantity

$$w = \int nds$$

is the optical path length. Everywhere on a wavefront, then, the optical path length and travel time to the source are constants. The wavefront concept is a very useful *geometrical* concept, and does not depend on the actual behavior of light as a wave phenomenon. Nevertheless, in some situations with a coherent source (where all waves are emitted in phase) the geometrical wavefronts also correspond to surfaces of constant phase (the **phase fronts**).

5.1.2. Fermat's principle and Snell's law

In Figure 5.2, a plane perpendicular to the plane of the diagram separates two different materials. The index of refraction is larger in the material on the right. In the plane of the diagram (the **plane of incidence**) a light ray travels upwards and to the right, striking the normal to the interface at the **angle of incidence**, θ_1.

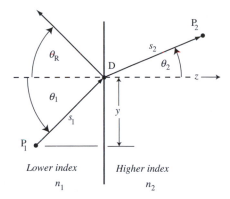

Fig. 5.2 Reflection and refraction at a plane boundary. The dashed line is the normal to the interface, and the arrows show the reflected and refracted rays for a ray incident from the lower left.

At the interface, the ray splits into two components — a reflected ray, which stays in the original material, and a refracted ray, which crosses the interface. These two rays respectively make angles θ_R and θ_2 with the normal. By convention, we measure positive angles counterclockwise from the normal.

In 1652, Pierre de Fermat[1] formulated a curious statement that can be used to determine the exact paths of the reflected and refracted rays. ***Fermat's principle*** asserts that the path of a ray between two points will always be the one that constitutes an extremum (i.e. a local minimum or, occasionally, maximum) in the total travel time, or, equivalently, in the total optical path length. Fermat's principle, in the form of a simple geometric argument (see Problem 1), implies the familiar law of reflection. That is,

$$\theta_1 = -\theta_R \qquad (5.1)$$

With regard to refraction an example may convince you that Fermat's principle should lead to a change in direction upon crossing the interface. In Figure 5.3, Tarzan, who is lounging on the sand at point A, observes that Jane is about to be devoured by crocodiles in the water at point B. Tarzan knows his running speed on smooth sand is much higher than his swimming speed in crocodile-infested water. He reasons that the straight-line path ACB will actually take longer to traverse than path ADB, since ADB involves considerably less swimming and fewer vexatious reptiles. The "ray" Tarzan actually traverses is thus "refracted" at the sand–water interface. The angle of

---

[1] Fermat (1601–1665) lived quietly and published little, although he was well respected as a superb mathematician and corresponded with the leading scientists and mathematicians of his day, including Huygens and Pascal. He was moved to publish his principle in optics by a dispute with Descartes. (Fermat was correct, but Descartes had great influence, and managed to damage Fermat's reputation and delay acceptance of his principle.) Many of Fermat's most important mathematical results — many without proof — were discovered only after his death in his private papers and in marginal notes in texts in his library. These included his famous "last theorem," which withstood proof until 1994.

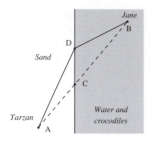

Fig. 5.3 Fermat's principle applied by Tarzan.

refraction (Tarzan's choice of point D) will depend on his relative speeds in sand and water.

Returning to Figure 5.2, we can apply Fermat's principle to deduce the path of the refracted ray by requiring the optical path between the fixed points P_1 and P_2 to be an extremum – a minimum in this case. Treat the distance y as the variable that locates point D. Then Fermat's principle demands

$$\frac{dw}{dy} = \frac{d}{dy}(s_1 n_1 + s_2 n_2) = 0 \tag{5.2}$$

Substitution for the distances s_1 and s_2 leads (see Problem 2) to **_Snell's law of refraction_**:

$$n_1 \sin(\theta_1) = n_2 \sin(\theta_2) \tag{5.3}$$

The sense of Snell's[2] law is that rays traveling from a lower index medium to a higher index medium (the situation in Figure 5.3) will bend towards the perpendicular to the interface. Note that Equation (5.3) reduces to the law of reflection if we take $n_1 = -n_2$.

An equivalent conception of refraction describes the turning of a wavefront, as in Figure 5.4. This view suggests that when one part of a wavefront is slowed down by the medium on the right, it turns so that it can keep pace with its faster-moving portion in the left-hand region.

Snell's law applies equally well if the incident ray travels from right to left in Figure 5.3. In this case, moving from a higher to a lower index medium, the refracted ray bends away from the perpendicular to the interface. In fact, there must be a certain angle of incidence, called the **_critical angle_**, which produces a refracted ray that bends so far from the perpendicular that it never leaves the higher index medium. From Equation (5.3), you can see that the critical angle is given by

$$\theta_C = \sin^{-1}\left(\frac{n_1}{n_2}\right)$$

What actually happens is called **_total internal reflection_** – for angles of incidence greater than critical, there is no refracted ray, and all light that reaches the interface is reflected back into the higher index medium.

Snell's law is a general result that applies to interfaces of any shape, and (with reflection as a special case) can be used as the foundation of almost all of geometrical optics.

---

[2] Ibn Sahl (Abu Sa`d al-`Ala' ibn Sahl, _c_. 940–1000 CE) apparently published the law of refraction (in the form of a diagram with the correct ratios of the sides of right triangles) considerably before its rediscovery by the Dutch mathematician Willebrord Snellius (1580–1626) and its subsequent popularization by Descartes.

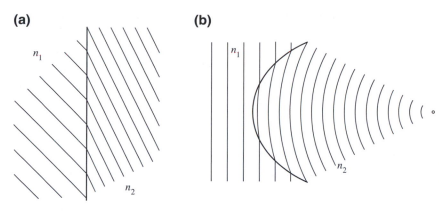

Fig. 5.4 (a) Plane wavefronts traversing a plane interface between a low index medium on the left and a higher index medium on the right. (b) Plane wavefronts traversing a curved interface producing a focusing effect – wavefronts become spherical after refraction.

5.1.3 Reflection and transmission coefficients

The laws governing the relative *intensities* of the incident beam that are reflected and refracted fall outside the realm of geometrical optics, and are deduced by rather messy applications of the theory of electromagnetic waves. **Fresnel's formulas**, for the reflection and transmission coefficients, give the amplitudes of the reflected and refracted waves as a function of angle of incidence, polarization, and indices of refraction. You should be aware of a few results:

(a) Polarization makes a difference. Waves polarized with the electric field vectors perpendicular to the plane of incidence (the transverse electric, or TE, case) in general are reflected differently from waves polarized with the magnetic field perpendicular to the plane of incidence (the transverse magnetic, or TM, case).

(b) The **reflectance**, R, is the fraction of the power of the incident wave that is reflected. At normal incidence ($\theta_1 = 0$) for all cases (TE, TM, external, or internal):

$$R = \left(\frac{n_1 - n_2}{n_1 + n_2}\right)^2$$

(c) For both the TE and TM polarizations, the reflectance becomes large at large angles of incidence. In the external case, $R \rightarrow 1.0$ as $\theta_1 \rightarrow 90°$, and light rays that strike a surface at **grazing incidence** θ_1 close to $90°$ will be mostly reflected. For the internal case, $R = 1.0$ (a perfect mirror) for all angles greater than the critical angle.

(d) For all values of θ_1 other than those described above, R is smaller for the TM polarization than for the TE polarization. Thus, initially unpolarized light will become partially polarized after reflection from a dielectric surface. At one particular angle (Brewster's angle, $\theta_p = \tan^{-1}(n_1/n_2)$), in fact, $R = 0$ for the TM polarization, and only one polarization is reflected.

5.1.4 Reflection from a spherical surface

An important result in geometrical optics describes reflection and refraction at a spherical interface. We begin with reflection, since it is somewhat simpler, and

Fig. 5.5 Reflection from a
spherical surface.

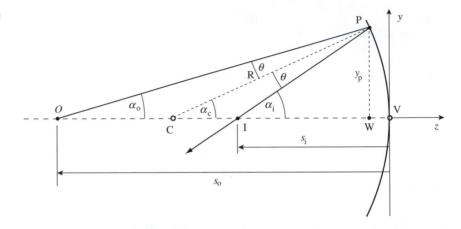

since most modern astronomical telescopes use one or more curved mirrors as
their primary light-gathering and image-forming elements. In Figure 5.5, parallel rays of light are traveling from left to right, and reflect from a concave
spherical surface of radius R whose center is at C. This figure illustrates a
number of conventions that apply later in this section. First, the horizontal dotted
line that is coincident with the axis of symmetry of the system is called
the ***optical axis***. Second, rays entering the system are assumed incident
from the left. Finally, we set up a conventional Cartesian coordinate system, where
the z-axis is coincident with the optical axis, and the origin is at the ***vertex*** of
the mirror: the intersection of the optical axis and the mirror surface – point
V in the figure. The y-axis goes upwards on the page, the x-axis inwards.

Now consider the special case of the ***paraxial approximation*** – the assumption
that all incident rays are nearly parallel to the optical axis, and that all angles
of reflection are small. This latter assumption means that the diameter of the
mirror (e.g. in the y direction) is small compared to its radius of curvature. Some
simple and very useful relationships apply in this approximation. In Figure 5.5,
consider the ray that originates at the ***object*** at point O on the optical axis, and is
reflected to reach the ***image*** point I. In the paraxial approximation, then, all
angles labeled α and θ must be very small, and the distance WV is also very
small.

In this case, triangles with side y_p in common yield

$$\alpha_o \approx \tan(\alpha_o) \approx \frac{y_p}{s_o}$$

$$\alpha_c \approx \tan(\alpha_c) \approx \frac{y_p}{R}$$

$$\alpha_i \approx \tan(\alpha_i) \approx \frac{y_p}{s_i}$$

If we also consider triangles OPC and CPI, we have

$$\theta = \alpha_c - \alpha_o = \alpha_i - \alpha_c$$

or

$$2\alpha_c = \alpha_o + \alpha_i$$

Finally, substituting into the last equation from the first three approximations, we have the paraxial equation for mirrors:

$$\frac{2}{R} = \frac{1}{s_o} + \frac{1}{s_i} \tag{5.4}$$

The distance $R/2$ is termed the **focal length** of the mirror, and is often symbolized by f, so the above equation is usually written as:

$$\frac{1}{f} = \frac{1}{s_o} + \frac{1}{s_i} = -P \tag{5.5}$$

The sign convention (z value increase to the right) means that R, f, s_0 and s_i are all negative numbers. The quantity P on the right-hand side of this equation is called the **power** of the surface. (The units of P are m^{-1}, or **diopters**.) Note that if the **object distance** s_o approaches infinity, the **image distance** s_i, approaches f. In other words, the mirror reflects every ray arriving parallel to the optical axis to a common point.

Figure 5.6 illustrates this, where every ray parallel to the axis passes through point F, and where the distance FV is the focal length, f, of the mirror We often say that the concave mirror in Figure 5.6 **gathers** electromagnetic radiation and concentrates it at point F. Gathering light is a simple matter of orienting the optical axis to point at a source, and allowing the mirror to bring all rays it intercepts to a common focus. Clearly, the **light-gathering power** of a mirror will be directly proportional to its surface area, which is in turn proportional to the square of its **aperture** (diameter). Bigger mirrors or dishes (radio astronomers use the word "dish" instead of "mirror") gather more light, and, since most astronomical sources are faint, bigger is, in this one regard, better.

A convex mirror, illustrated in Figure 5.7, has a positive focal length, meaning that the focal point lies to the right of the vertex. Clearly, convex mirrors disperse, rather than gather, a bundle of parallel rays. For convex mirrors, the paraxial approximation still results in the same expression (Equation (5.5)) that applied to concave mirrors. The sign convention means that the power, P, of a convex mirror is negative, and its focal length, f, is positive.

5.1.5 Refraction at a spherical surface

Figure 5.8 illustrates a ray refracted at a spherical interface between media of differing indices of refraction. You can derive the paraxial equation for refraction in a way analogous to the derivation for spherical mirrors. Begin with

Fig. 5.6 Rays parallel to
the optical axis all pass
though point F, which is
one focal length from the
vertex.

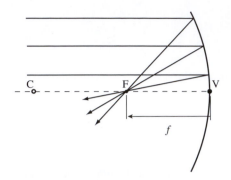

Fig. 5.7 The focal point of
a convex spherical
mirror.

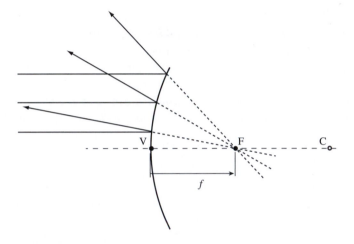

Snell's law and assume all angles are small. The result, which we simply state
without proof, is

$$\frac{n_2}{s_2} - \frac{n_1}{s_1} = \frac{(n_2 - n_1)}{R_{12}}$$

As in the case for mirrors, take the focal length, f, to be the value of S_2 when S_1
approaches infinity:

$$f_2 = \frac{n_2 R_{12}}{(n_2 - n_1)}$$

So, for refraction, the paraxial equation for image and object distances is

$$\frac{n_2}{s_2} - \frac{n_1}{s_1} = \frac{n_2}{f_2} = -\frac{n_1}{f_1} = P_{12} \tag{5.6}$$

Again, the quantity P_{12} is called the **power** of the surface. The power, like the
focal length, measures how strongly converging (or diverging, if P is negative)
an interface is. For both mirrors and refracting surfaces, a plane has zero
power.

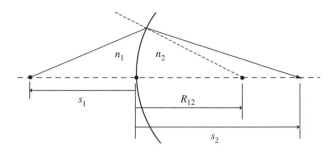

Fig. 5.8 Refraction at a spherical interface. R and s_2 are positive; s_1 is negative.

5.2 Lenses, mirrors, and simple optical elements

5.2.1 Thick lenses

Figure 5.9 shows the cross-section of a ***thick lens*** in air ($n_1 = n_3 = 1, n_2 = n$). If you apply Equation (5.6) to each surface in succession, the power of the combined surfaces turns out to be

$$P = \frac{1}{f} = P_{12} + P_{23} - \frac{d}{n}P_{12}P_{23} \qquad (5.7)$$

Here the focal length, image, and object distances are measured from the effective center of the lens as shown in the figure, and $P_{ij} = (n_i - n_j)/R_{ij}$.

5.2.2 Thin lenses

If you further simplify to the limiting case of a ***thin lens***, the assumption that d is negligibly small reduces Equation (5.7) to

$$\frac{1}{s_2} - \frac{1}{s_1} = (n-1)\left(\frac{1}{R_{12}} - \frac{1}{R_{23}}\right) = \frac{1}{f} = P_{12} + P_{23} = P \qquad (5.8)$$

Note that the focal length (and power) of a thick or thin lens (unlike a single refracting surface) is the same for rays moving in the $+z$ direction as for rays the $-z$ direction. Likewise, note that with the substitution $n_2 = -n_1 = -1$, the equation for a thin lens (5.8) is identical to the equation for a mirror (5.5). In the discussion that follows, then, it is permissible to replace a mirror with a thin lens of the same aperture and focal length, or vice versa.

5.2.3 Graphical ray tracing

An important method for evaluating an optical design is to trace the paths of several rays from an object through all the optical elements until they form a final image. This is best done with a computer program that follows each ray through the system and applies Snell's law and/or the law of reflection at every interface encountered, usually employing more exact formulations than the

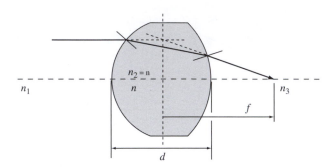

paraxial approximation. Such ***ray-tracing programs*** are essential tools for the
design of astronomical telescopes and instruments.

The paraxial approximation itself permits a very simple method requiring
just ruler and pencil, suitable for rough estimates. Figure 5.10 illustrates this
graphical method for ray tracing through a thin lens. Here we have an object (an
arrow) located to the left of the diagram and we are interested in tracing the
paths of light rays given off by the arrow and passing through the thin lens.
Equation (5.8) predicts the trajectories of rays that originate on-axis at the tail of
the arrow. For rays that originate off-axis, it is easy to predict the paths of three
rays, which is one more than needed to locate the image of the arrowhead.

The specific rules for ***graphical ray tracing*** for a thin lens are:

1. Rays incident parallel to the axis emerge through the right focal point.
2. Rays incident through the left focal point emerge parallel to the axis.
3. Rays through the vertex do not change direction.

A similar set of rules applies for a spherical mirror, illustrated in Figure 5.11:

1. Incident rays parallel to the optical axis are reflected through the focal point, F.
2. Incident rays through the focal point are reflected parallel to the axis.
3. Incident rays that reach the vertex are reflected back at an equal and opposite angle.
4. Incident rays through the center of curvature, C, are reflected back on themselves.

5.2.4 Multiple lenses

Most practical optical systems have multiple elements, and many of these are
designed by ray-tracing computer programs. However, it is sometimes useful to
estimate the properties of such systems through manual graphical tracing techniques, or through algebraic formulae. For example, in the thin-lens limit, the
formula for the combined power of two aligned lenses separated by distance d is
given by Equation (5.7), with the index set to 1:

$$P = \frac{1}{f} = P_1 + P_2 - dP_1P_2 \tag{5.9}$$

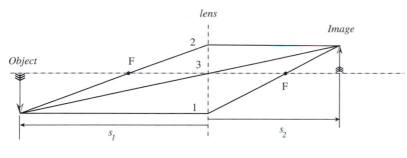

Fig. 5.10 Graphical ray tracing for a thin lens.

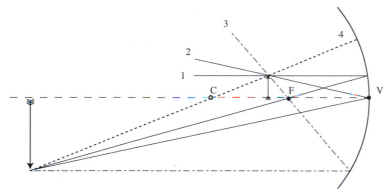

Fig. 5.11 Graphical ray tracing for a spherical mirror in the paraxial approximation.

5.2.5 The thick plane-parallel plate

A thick plane-parallel plate has a power of zero, so does not cause rays to change direction (see Figure 5.12). However, for a converging or diverging beam, the plate will displace the location of the focus by an amount

$$D = d\left(1 - \frac{1}{n_2}\right)$$

Parallel plates frequently appear in astronomical instruments as filters, windows, and specialized elements.

5.2.6 Optical fibers

The optical fiber is an important application of total internal reflection. As illustrated in Figure 5.13, a long cylinder of high index material can serve as a guide for rays that enter one face of the cylinder and strike its side at angles greater that the critical angle. Such rays will travel the length of the cylinder, reflecting from the walls, to emerge from the other face.

Typical fibers are manufactured with a core of high-index glass or plastic enclosed in a cladding of a lower-index material. Internal reflections occur at the core–cladding boundary. Although the cladding serves to protect the core

Fig 5.12 Refraction by a thick plane-parallel plate.

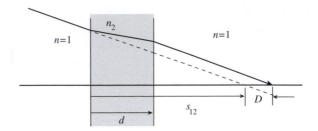

Fig. 5.13 Structure of an optical fiber. Total internal reflection occurs if q_W is greater that the critical angle.

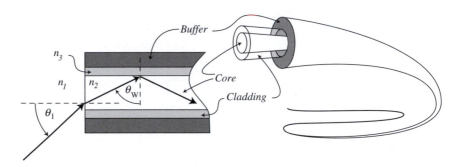

from scratches and dirt (either would generate light loss), many fibers are also coated with an additional layer called the buffer, which provides further protection from mechanical and chemical hazards. The core diameters of fibers used in astronomy are typically in the 50–200 micron range, are made of special materials, are quite flexible, and have multiple applications. A single fiber, for example, can conduct light from the focus of a (moving) telescope to a large or delicate stationary instrument like a spectrograph, or many fibers can simultaneously feed light from multiple images (one fiber per image in a star cluster, for example) to the input of a multi-channel detector.

5.2.7 Optical materials

Reflecting materials
An ideal mirror should have a reflectivity of 1.0 for all wavelengths of interest. The substrate should be easy to shape to an accuracy of a fraction of the shortest of those wavelengths, and once shaped, the substrate and its coating, if any, should be mechanically and chemically stable. Mirrors in astronomical telescopes are often both large and mobile, and may even need to be placed into Earth orbit; so low mass is a virtue. Since telescopes are normally in environments where the temperature can change rapidly, high thermal conductivity and low coefficient of thermal expansion are also essential.

No materials match this ideal, but some are better than others. For the reflecting telescope's first two centuries, telescope makers fashioned mirrors out of **speculum metal**, an alloy that is difficult to prepare, primarily of copper and tin. Although speculum mirrors produced some historic discoveries (e.g. Herschel's finding of Uranus), speculum is heavy, only 45% reflective at best, and tarnishes easily. Astronomers quickly switched to silvered glass mirrors once the technology became available in the 1880s. Compared to speculum, glass is much more easily worked, has a lower density, and better mechanical and thermal stability. The silver coating is highly reflective in the visible and infrared.

Most modern optical telescope mirrors generally utilize substrates made with special glasses (e.g. Pyrex) or ceramics (Cervit or Zerodur) that have very low coefficients of thermal expansion. Often large mirrors are ribbed or honeycombed on the back face to minimize mass while retaining rigidity. Choice of surface coating depends on the application. A coating of metallic aluminum, over-coated with a protective layer of silicon monoxide, is the usual choice in the near ultraviolet and optical because of durability and low cost. Silver, which is poor in the ultraviolet, is superior to aluminum longward of 450 nm, and gold is a superb infrared reflector longward of 650 nm. Solid metal mirrors still have limited application in situations where their high thermal conductivity is especially useful. Beryllium, although toxic, is the lowest density workable metal, with excellent rigidity. The Spitzer Space Telescope has a bare polished beryllium mirror, and the James Webb Space Telescope mirror will be gold-coated beryllium segments. Extremely large ground-based telescopes now in the planning stages will probably utilize low-density materials like beryllium and silicon carbide, which, although expensive, are superior to glass for mirror substrates.

Very short wavelengths (extreme ultraviolet (EUV) and shorter) present two difficulties: First, energetic photons tend to be absorbed, scattered, or transmitted by most materials, rather than reflected. Second, curved mirrors in general need to be shaped with an accuracy of at least $\lambda/4$, which, for a 1 nm X-ray, amounts to one atomic diameter. X-Ray and EUV focusing telescopes have usually been designed with metal mirrors operating in "grazing incidence" mode.

Transmitting materials

Transmitting materials form lenses, windows, correctors, prisms, filters, fibers, and many other more specialized elements. Of primary concern are index of refraction, dispersion, and absorption. Other properties of relevance to astronomical applications include: homogeneity, thermal expansion, frequency of bubbles and inclusions, and dependence of refractive index on temperature. Thermal properties are especially important for instruments (e.g. in spacecraft) that cannot be easily adjusted.

At visible wavelengths, a variety of optical glasses exhibit indices of refraction ranging from 1.5 to 1.9, and dispersions (at 588 nm, the Fraunhofer D line)

Fig. 5.14 Refractive index as a function of wavelength for some optical glasses in the Schott Catalog. Curves end where the glass becomes nearly opaque, except we show the curve for borosilicate crown glass (N-BK7) as a dotted line in the region where it is highly absorbing. SF4 is one of the most dispersive of the flints. N-LAK10 is an unusual crown with low dispersion and high index. SiO$_2$ is fused quartz, and CaF$_2$ is calcium fluoride.

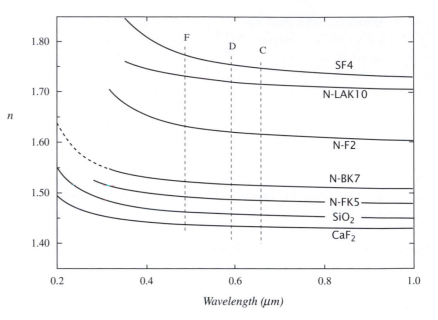

in the range -0.03 to -0.18 μm^{-1}. Such a variety is important in designing systems free from chromatic aberration (see below). Generally, glasses with a high index will tend to have a high dispersion and are termed "flints," while those with dispersions above -0.06 μm^{-1} (closer to zero) are called "crowns." Figure 5.14 shows the actual run of index with wavelength for several optical glasses. Note that as the index and dispersion rise at short wavelengths, glasses become highly absorbing.

In the ultraviolet (from about 150 to 400 nm) ordinary glasses become opaque. Fused quartz (SiO$_2$) is the exception. It transmits well over all but the shortest wavelengths in this range, has a low thermal expansion coefficient, and can be shaped to high accuracy. All other ultraviolet-transmitting materials are crystalline, rather than glass, and are more difficult to shape and more likely to chip and scratch. The most useful of these is probably calcium fluoride CaF$_2$, which transmits from 160 nm to 7 μm. Other fluoride crystals (BaF$_2$, LiF, MgF$_2$) have similar properties. Fused quartz and the fluorides do not transmit well below 180 nm, and some birefringent crystals (most notably sapphire, Al$_2$O$_3$) find limited use as windows in the very far ultraviolet. Optics for wavelengths shorter than 150 nm must be reflecting, and for wavelengths below about 15 nm, only grazing-incidence reflections are practical.

In the infrared, ordinary optical glasses transmit to about 2.2 μm, and some special glasses transmit to 2.7 μm. Infrared-grade fused quartz transmits to about 3 μm. A large selection of crystalline materials, many identical to those useful in the ultraviolet, transmit to much longer wavelengths, but most are soft, or fragile, or sensitive to humidity, so can only be used in protected environments. Germanium (transmits 1.8 to 12 μm) has a high index of refraction (4.0) and low

Table 5.2. *Some materials transparent in the infrared*

Material	Wavelength range (μm)	n	Comments
Sapphire	0.14–5	1.7	Slightly birefringent
LiF	0.18–6	1.4	Slowly degrades with humidity
BaF_2	0.2–11	1.4	Slowly degrades with humidity
ZnS	0.5–12	2.4	Strong
ZnSe	0.6–18	2.4	Soft
NaCl	0.25–16	1.5	Water soluble
CsI	0.4–45	1.7	Water soluble

dispersion, so is especially useful for making lenses. A few infrared- and ultra-violet-transmitting materials are listed in Table 5.2

Surface coatings

Coating the surface of an optical element with a thin film can exploit the wave properties of light to either increase or decrease its reflectance. A thin film exactly 1/4 wavelength thick applied to a glass surface, for example, will intro-duce two reflected beams, one from the front surface, and the other from the film-glass interface. The second beam will emerge one-half wavelength out of phase from the first, and the two reflected beams will destructively interfere. If their amplitudes are equal, then the interference will be total, and the reflectance reduced to zero for that wavelength. If the index of refraction of the glass is n_s, the condition for equal amplitudes is that the index of the film be

$$n_F = \sqrt{n_s}$$

For glass with an index of 1.5, this implies a coating with index 1.22 – amorphous MgF_2 ($n = 1.38$) is the common practical choice for single-layer coatings. An antireflection coating works best at only one wavelength. It will work to some extent for a broad band centered near the design wavelength, and multiple coatings of varied indices can improve the width of that band. A similar treatment can *enhance* the reflectivity of a surface – multiple layers of alternat-ing high and low index materials can improve the reflectivity of a mirror over a broad range of wavelengths.

Cleaning

Because poor transmission or reflection means lost photons, anti- and enhanced-reflection coatings are of great value in astronomy, not only on the surfaces of transmitting or reflecting elements like mirrors, lenses, and windows, but also on absorbing surfaces like the front side of a detector. For

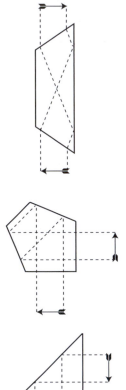

similar reasons, any dirt or dust that accumulates on optical surfaces should be removed if possible.

Cleaning is potentially quite important, since the accumulation of dust on a single surface can reduce transmission or reflection by several percent per year even in relatively clean environments. However, cleaning always presents the opportunity for surface damage, so requires great care. Gentle washing with detergent, followed by thorough rinsing, is effective, economical, but not always practical. Pressurized streams of gas or carbon dioxide snow can remove looser particles. Certain gels applied to the surface can solidify and remove dirt particles when peeled off, a process that is quite effective, although expensive.

5.2.8 Prisms

Prisms are geometric solids with two parallel polygonal faces joined by a number of faces shaped like parallelograms. Figure 5.15 illustrates a few prisms in cross-section that are useful because their internal reflections invert an image or change the path of a beam.

Prisms are especially useful because of their dispersing properties. Figure 5.16 shows the path of a ray through a triangular prism with apex angle A. After two refractions, a ray entering the left face at angle of incidence, α, emerges from the second face at angle θ from the original direction (the angular deviation). Application of Snell's law at both surfaces of this prism gives

$$\sin(\theta + A - \alpha) = \left(n^2 - \sin^2 \alpha\right)^{1/2} \sin A - \sin \alpha \cos A$$

You can show that the angular deviation is a minimum when α and the final angle of refraction are equal (Figure 5.16b), in which case

$$\sin\left(\frac{\theta_0 + A}{2}\right) = n \sin\left(\frac{A}{2}\right) \tag{5.10}$$

Fig. 5.15 Some reflecting prisms, from top to bottom: a right-angle prism bends the beam 90° and inverts the image, a pentaprism bends the beam but leaves the image unchanged, and a dove prism inverts the beam without changing its direction.

This relationship provides a simple method for measuring the index of refraction of transparent material: in monochromatic light, measure the minimum angular deviation of a prism of known apex angle. These equations make it clear that, because the index of refraction is a function of wavelength, then so are θ and θ_0. We define the ***angular dispersion*** as $\partial\theta/\partial\lambda$, and note that since only n and θ are functions of wavelength

$$\frac{\partial\theta}{\partial\lambda} = \frac{\partial n}{\partial\lambda} \frac{\partial\theta}{\partial n}$$

The first factor on the right-hand side depends on the dispersion of the prism material, which is a strong function of wavelength (note the variation in the *slope* of any curve in Figure 15.14), while the second factor is only a very weak function of wavelength (see Problem 5). A good approximation of the curves in Figure 15.14 is given by the Cauchy formula:

$$n(\lambda) = K_0 + \frac{K_2}{\lambda^2} + \frac{K_4}{\lambda^4} + \dots$$

where K_0, K_2, and K_4 are constants that depend on the material. Ignoring all but the first two terms, and substituting for $\partial n/\partial \lambda$ gives

$$\frac{\partial \theta}{\partial \lambda} \cong \frac{g}{\lambda^3} \qquad (5.11)$$

where g is a constant that depends primarily on the prism geometry. In the case of minimum deviation,

$$g = -4K_2 \frac{\sin(A/2)}{\cos(\alpha)} \qquad (5.12)$$

Thus, the absolute value of the angular dispersion of a glass prism will be much higher in the blue than in the red, which can be a disadvantage for some astronomical spectroscopic applications.

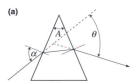
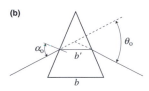

Fig. 5.16 Dispersion by an isosceles triangular prism: (a) the angular deviation, θ, depends on the wavelength because $n = n(\lambda)$. Longer wavelengths have smaller deviations. (b) The minimum angular deviation occurs when the path through the prism is parallel to its base.

5.3 Simple telescopes

> The next care to be taken, in respect of the Senses, is a supplying of their infirmities with Instruments, and, as it were, the adding of artificial Organs to the natural; this in one of them has been of late years accomplist with prodigious benefit to all sorts of useful knowledge, by the invention of Optical Glasses. By the means of Telescopes, there is nothing so far distant but may be represented to our view. . . . By this means the Heavens are open'd, and a vast number of new Stars, and new Motions, and new Productions appear in them, to which all the ancient Astronomers were utterly Strangers.
>
> – Robert Hooke, *Micrographia*, 1665

5.3.1 Telescopes as cameras

Most astronomical telescopes are *cameras* — they form images of objects both on and off the optical axis in the *focal plane*. Although a telescope's optics can be complex, we can profitably represent them with a single "equivalent thin lens" of matching aperture and a focal length that reproduces the image-forming properties of the telescope. For example, the upper part of Figure 5.17 shows a two-mirror reflecting telescope in the Cassegrain configuration, the most common arrangement for modern research telescopes. The lower part of the figure shows the equivalent thin-lens diagram for this telescope.

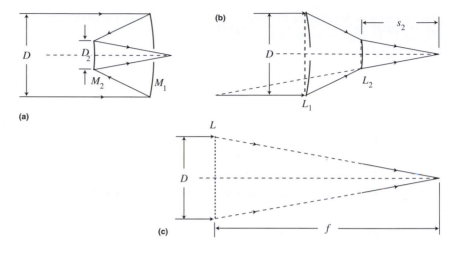

Fig. 5.17 (a) A Cassegrain reflecting telescope, with primary mirror M_1 and secondary mirror M_2. The secondary has diameter D_2 and object distance s_2. (b) The "unfolded" equivalent telescope with mirrors replaced by lenses of the same size, focal length and spacing. (c) The equivalent telescope with a single lens that has the same aperture, D, and reproduces the geometry of the converging beam with effective focal length $f = D(s_2/D_2)$.

5.3.2 Image scale and image size

The **image scale**, s, describes the mapping of the sky by any camera. The image scale is the angular distance on the sky that corresponds to a unit linear distance in the focal plane of the camera. Figure 5.18 shows the equivalent lens diagram of a camera of focal length f. We draw the paths followed by two rays, one from a star on the optical axis, the other from a star separated from the first by a small angle θ on the sky. Rays pass through the vertex of the lens without deviation, so assuming the paraxial approximation, $\theta \approx \tan \theta$, it should be clear from the diagram that

$$s = \frac{\theta}{y} = \frac{1}{f} \ [\theta \text{ in radians}]$$

Since it is usually convenient to express image scales in arcsec per mm, then

$$s = \frac{206,265}{f} \ [\text{arcsec per unit length}]$$

Typical modern focal-plane detectors are composed of many identical light-sensitive pixels. If the center of each (usually square) pixel is separated from its nearest neighbors by distance, d, then the **pixel scale** of a telescope (that is, the angular size on the sky imaged by one pixel) is just $s_p = sd$.

The size and shape of the detector often determine the **field of view** of a camera (the angular diameter or dimensions of a single image). A rectangular detector, for example, with physical length, l, and width, w, will have a field of view of sl by sw. More rarely, the detector may be oversized and the field of view set by obstructions in the telescope's optical system or by the limit of acceptable image quality at large distances from the optical axis.

5.3.3 Focal ratio and image rightness

Focal ratio is defined as

$$\Re = \frac{f}{D} = \frac{\{\text{focal length}\}}{\{\text{diameter of entrance aperture}\}}$$

For example the 20-inch telescope at Vassar College Observatory has a focal length of 200 inches, so $\Re = 10$. This is usually expressed as "$f/10$".

You can show that the brightness (energy per unit area in the focal plane) of an *extended* source in the focal plane is proportional to $\Re^{-2}$, so that images in an $f/5$ system, for example, will be four times as bright as images in an $f/10$ system.

5.3.4 Telescopes with oculars

For the first three centuries, telescopes augmented direct human vision, and an astronomer's eyesight, persistence, skill, and sometimes even physical bravery were important factors in telescopic work. To use a telescope visually, an ocular, or eyepiece, is needed to examine the image plane of the objective (the light-gathering element). Figure 5.19 shows the arrangement where the objective, ocular, and human eye lens are represented as thin lenses. The two stars in the diagram are separated by angle θ on the sky. Viewed through the telescope, they appear to be separated by the angle θ'. The angular magnification is the ratio

$$M = \frac{\theta'}{\theta}$$

From the diagram, making the paraxial approximation, this is just

$$M \approx \frac{\tan \theta'}{\tan \theta} = \frac{y/f'}{y/f} = \frac{f}{f'}$$

So the magnification is the ratio of the focal lengths of the objective and eyepiece. Oculars are subject to chromatic and other aberrations (see below), and benefit from careful design and matching with telescope optics. Most contain several lenses. Rutten and van Venrooij (1988) discuss oculars at length.

5.4 Image quality: telescopic resolution

> That telescope was christened the Star-splitter,
> Because it didn't do a thing but split
> A star in two or three the way you split
> A globule of quicksilver in your hand
>   — Robert Frost, The Star-Splitter, 1923

Fig. 5.18 Definition of image scale. Stars separated by angle θ on the sky form images that are separated by distance y in the focal plane.

Fig. 5.19 A Keplerian telescope. A positive ocular lens located one focal length from the image plane of the objective forms a virtual image at infinity that is seen by the eye. Galileo's original telescope used a negative lens for the ocular, a design with serious disadvantages.

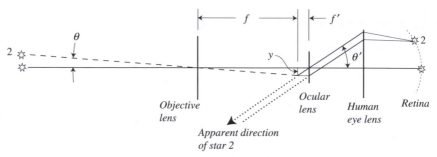

5.4.1. The diffraction limit

The wave properties of light set a fundamental limit on the quality of a telescopic image. Figure 5.20 illustrates the formation of an image by a telescope outside the atmosphere. Wavefronts from a point source arrive at the telescope as perfectly planar and parallel surfaces, provided the source is very distant. The *entrance aperture* is the light-gathering element of the telescope, and its diameter, D, is usually that of the mirror or lens that the incoming wave first encounters. Despite the fact that the source is a point, its image – created by a perfect telescope operating in empty space – will have a finite size because of diffraction of the wave. This size, the *diffraction limit* of the telescope, depends on both the wavelength of light and on D. The diffraction of a plane wavefront by a circular aperture is a messy problem in wave theory, and its solution, first worked out in detail by the English astronomer George Airy in 1831, says that the image is a bulls-eye-like pattern, with the majority (84%) of the light focused into a spot or "disk."[3] Concentric bright rings, whose brightness decreases with distance from the center, surround the very bright central spot, the *Airy disk*. The angular radius of the dark ring that borders the Airy disk is

$$\alpha_A = \frac{1.22\lambda}{D} \text{ [radians]} = \frac{0.252\lambda}{D} \text{ [arcsec m } \mu\text{m}^{-1}] \qquad (5.13)$$

The full-width at half-maximum (FWHM) of the disk is $0.9\alpha_A$. If two point sources lie close together, their blended Airy patterns may not be distinguishable from that of a single source. If we can say for sure that a particular pattern is due to two sources, not one, the sources are *resolved*. The **Rayleigh criterion** for

---

[3] The shape of the Airy pattern for a clear, circular aperture is proportional to the function $[J_1(\theta)/\theta]^2$, where J_1 is the first Bessel function. A central obstruction (e.g. secondary mirror) will alter the pattern by moving intensity from the central disk to the bright rings. Other obstructions (e.g. secondary supports) will produce "diffraction spikes" on images.

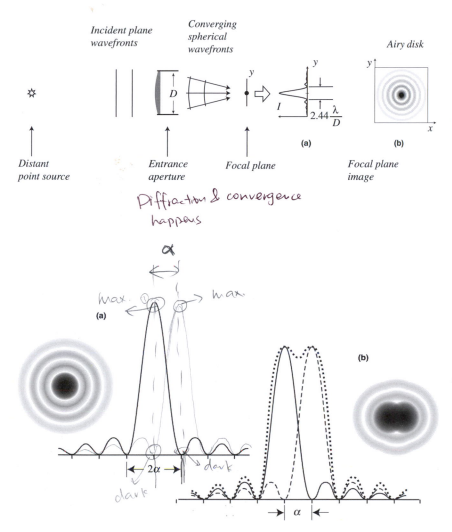

Incident plane wavefronts

Converging spherical wavefronts

Airy disk

$2.44\dfrac{\lambda}{D}$

(a)

(b)

Distant point source

Entrance aperture

Focal plane

Focal plane image

Diffraction & convergence happens

α

max.   max.

(a)

2α

dark

dark

(b)

α

Fig. 5.20 Telescopic images in the absence of an atmosphere. Plane wavefronts diffract upon encountering the circular aperture, and focus as an Airy disk: a bright central spot surrounded by rings of decreasing brightness. (a) The intensity of the resulting image vs. distance on the *y*-axis. The central peak has a full width of twice the Airy radius: $2\alpha = 2.44\lambda/D$ (b) A negative of the two-dimensional diffraction pattern.

Fig. 5.21 (a) The Airy pattern as a negative image and plotted as intensity vs. radius. (b) The negative image of two identical monochromatic point sources separated by an angle equal to Rayleigh's limit. The plot shows intensity along the line joining the two images. The unblended images are plotted as the solid and dashed curves, their sum as the dotted curve.

resolving power requires that to resolve two sources, the centers of their Airy discs must be no closer than α_A, the angular radius of either central spot (both radii are the same). At this limiting resolution, the maximum intensity of one pattern coincides with the first dark ring of the other; see Figure 5.21. At a wavelength of 510 nm, according to Equation (5.13), a 1-meter telescope should have a resolution of 0.128 arcsec. Details smaller than this size will be lost.

5.4.2 Seeing

Rayleigh's criterion is a good predictor of the performance of space telescopes. On the surface of the Earth, however, turbulence, which causes dynamic density

Fig. 5.22 Astronomical seeing: image formation by a telescope in a turbulent atmosphere. (a) In a short exposure, wavefront distortions caused by variations in refractive index in the atmosphere produce interfering Airy patterns. In (b) turbulent motion in the atmosphere during a long exposure moves the individual maxima around the image plane to produce a large seeing disk.

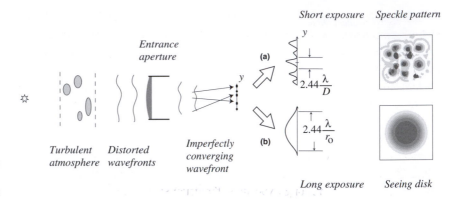

variations in the Earth's atmosphere, limits the resolving power of all but the smallest telescopes. This loss of resolution is termed **seeing**. Seeing (measured as the angular FWHM of the image of a point source) may be as small as several tenths of a second of arc on the very best nights at the very best sites on Earth, but it can reach several seconds of arc at other sites.

Figure 5.22 shows how seeing affects an Earth-bound telescope's image. Temperature differences create compressions and rarefactions – "lumps" of air – with indices of refraction higher or lower than their surroundings. Near Earth's surface, a temperature change of $1\,°C$ will cause an index change of about 10^{-6}. These lumps retard some sections of a plane wavefront while allowing other sections to race ahead. When the distorted front reaches the entrance aperture, different segments are traveling in slightly different directions, and each segment is imaged in a slightly different spot in the focal plane. Even parallel segments that focus on the same spot will in general arrive with different phases. In Figure 5.22, for example, we can approximate the distorted wavefront as 9 or 10 different plane wavefronts, each producing an Airy disk of radius α_A at a different location in the focal plane. The combination of these Airy patterns produces a multi-spotted image called a **speckle pattern**, in which each local spot is the diffraction-limited image of the source as seen through a coherent lump of air. Turbulence, moreover, moves the lumps of air around at high velocity so a particular speckle pattern has only a momentary existence, to be quickly replaced by a different pattern.

In practice, you can "freeze" the speckle pattern on very short exposures (below about 1/20 second). The number of speckles in a single exposure depends on the typical diameter of an atmospheric lump, r_0, and is approximately equal to $(D/r_0)^2$. A technique called **speckle interferometry** can remove the interference effects, and recover the diffraction-limited image. The limitation of the technique is the short exposure required, so it is useful only with large-aperture telescopes and bright objects. On long exposures, the speckle pattern is hopelessly blurred by turbulence, producing a "seeing disk" with full radius $\alpha_s \simeq \lambda/r_0$.

Seeing effects blur all objects observed through the atmosphere. Advanced techniques called **adaptive optics** (Chapter 6) can correct wavefront distortions and sharpen blurred images on long exposures. Adaptive optics systems are expensive, but can be quite successful.

Atmospheric turbulence produces somewhat different phenomena for small apertures. **Scintillation** is the intensity variation produced by a concave (bright) or convex (faint) wavefront distortion. Scintillation is very apparent as "twinkling" to the unaided eye (aperture about 7 mm), and as photometric variability in telescopes. Telescopes with apertures smaller than r_0 produce a single Airy pattern, not a speckle image. However, the pattern moves around the focal plane, and produces, on long exposures, the usual seeing disk.

5.4.3 Atmospheric refraction

We can approximate the Earth's atmosphere as a series of plane-parallel plates, and the surface as an infinite plane. In Figure 5.23a, for example, we imagine an atmosphere of just two layers that have indices, $n_2 > n_1$. A ray incident at angle α refracts at each of the two interfaces, and ultimately makes a new angle, $\alpha + \Delta\alpha$, with the surface: thus, refraction shifts the apparent position of a source towards the zenith. In Figure 5.23b, we imagine that the atmosphere consists of a very large number of thin layers, so in the limit, the effect of refraction is to curve the path of the incident ray. In this limit, the plane-parallel model gives

$$-\Delta\alpha = R_0 \tan\alpha = \frac{(n^2 - 1)}{2n^2}\tan\alpha \approx (n-1)\tan\alpha \qquad (5.14)$$

where n is the index of refraction at the surface. The quantity $(n-1) \times 10^6$ is called the **refractivity**.

Since the index is a function of wavelength (see Figure 5.23c and Table 5.3), rays of different colors are refracted at slightly different angles, and images observed through the atmosphere at large zenith distances are actually very low resolution spectra – with the blue image shifted more towards the zenith than the red. Atmospheric dispersion is quite small in the near infrared. As discussed in Chapter 3, very precise correction for refraction is site-dependent, and Equation (5.14) fails near the horizon. What is relevant here is that the size and shape of a multi-wavelength image will be distorted at large zenith distances.

5.4.4 Chromatic aberration

> ...it is not the spherical Figures of Glasses, but the different Refrangibility of the Rays which hinders the perfection of Telescopes.... Improvement of Telescopes of a given length by Refractions is desparate.
>
> – Isaac Newton, *Opticks*, 1718

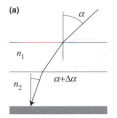

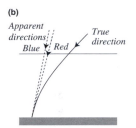

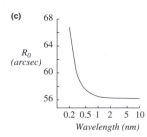

Fig. 5.23 The plane-parallel model for atmospheric refraction: (a) Refraction in a model atmosphere that has only two different uniform layers; (b) shows the limit of this model as the number of layers becomes large. In (c) R_0 is in seconds of arc as a function of wavelength (the refractivity of air is R_0 in radians times 10^6).

Table 5.3. *Atmospheric refraction as a function of wavelength. Data are for dry air at 0°C and standard pressure. Actual refractivity will depend on local temperature, humidity and pressure. Data here are from the formulation by Owens (1967). More detailed treatment is given by Young (2006) gives other data and formulae*

λ (nm)	$(n-1) \times 10^6$	R_0 (arcsec)
200	324	66.8
300	291	60.1
400	283	58.3
550	278	57.3
700	275.7	56.9
1000	274.0	56.5
4000	272.6	56.2
10,000	272.5	56.2
10 cm	355	73.2

Fig. 5.24 Chromatic aberration: (a) shows a thin lens, which, because of dispersion, focuses blue rays nearer the vertex than red rays. The image at blue focus will be a blue spot surrounded by a rainbow-hued blur, with red outermost. The image at the red focus will be a red spot surrounded by a large rainbow blur with blue outermost. A cemented achromat (b) consists of a positive and negative lens of differing powers and dispersions. Focal length differences and surface curvatures are greatly exaggerated.

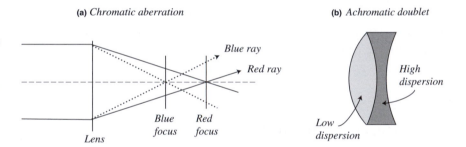

(a) *Chromatic aberration*   **(b)** *Achromatic doublet*

Since $n(\lambda)$ for optical glasses decreases with wavelength in the visible, then the focal length of a convex lens will be longer for red wavelengths than for blue. Different colors in an image will focus at different spots. This inability to obtain perfect focus is called ***chromatic aberration***, and limits the resolving power of a telescope with lenses; see Figure 5.24a.

Chromatic aberration will not occur in all-reflecting optics. To correct it in a lens, the usual strategy is to cement together two lenses made of different glasses, a positive (convex) lens with low chromatic dispersion, and a negative lens with lower absolute power, but higher chromatic dispersion. If powers are chosen to be inversely proportional to dispersions, then the combination will have a finite focal length but reduced dispersion over a significant part of the spectrum; see Figure 5.24b.

An ***achromat*** is a lens whose focal length is the same at two different wavelengths. The first useful achromats, ***doublets*** of crown (convex lens) and flint (concave lens) glass, began to appear in the 1760s in France and England, an appearance certainly delayed by Newton's declaration of its impossibility. Achromats were initially of only small aperture owing to the difficulty in producing large flint glass blanks free from flaws. With Fraunhofer's perfection of achromat production and design, large refractors became the instruments of choice at most observatories after the 1820s. Opticians usually designed these refractors to have equal focal lengths at the wavelengths of the red Fraunhofer C-line at 656.3 nm (Hydrogen α) and the blue F-line at 486.2 nm (Hydrogen β), producing nearly zero chromatic aberration over the most sensitive range (green and yellow) of the eye.

Magnificently suited to human vision, these telescopes poorly matched the new technology of photography introduced in the 1880s, since they had large residual chromatic aberration in the violet and ultraviolet, where the emulsions had their sensitivity. Astronomers soon constructed refracting "astrographs" optimized to short wavelengths – for example, the large number of "standard astrographs" of 33 cm aperture built for the Carte du Ceil project; see Chapter 4.

Modern glasses now provide the achromat designer with a much wider choice of indices and dispersions than were available to Fraunhofer. A properly designed doublet, called an ***apochromat***, can bring three different wavelengths to a common focus, but only through use of expensive materials. (The most useful material is "fluorite," made from manufactured crystalline calcium fluoride. It has a very low index (1.43) and dispersion; see Figure 5.14.) A properly designed triplet, called a ***superapochromat***, can bring four different wavelengths to a common focus.

5.5 Aberrations

5.5.1 Monochromatic wavefront aberrations

An aberration is an imperfection in telescope design that degrades an image. Chromatic aberration, which we discussed above, is present when rays of differing wavelength fail to reach a common focus. Unfortunately, some aberrations are present in monochromatic light.

Consider a "perfect" telescope. It should transform an incident plane wavefront into a converging spherical wavefront whose center is at the focus

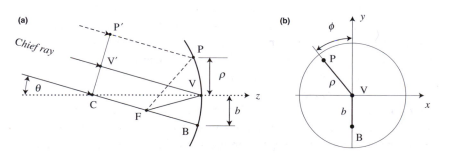

Fig. 5.25 Specification of the location of an arbitrary test ray from an off-axis source. Diagram (a) shows rays through the center of curvature (CFB) and vertex (V'V, the chief ray), which, along with the optical axis, define the plane of the diagram, known as the meridional plane. Point P is outside the plane of the diagram. Diagram (b) locates points P, V, and B in the plane of the aperture when looking down the optical axis.

predicted by the paraxial approximation – the *Gaussian focus*. For point sources off-axis, the perfect telescope should produce spherical wavefronts converging somewhere on the *Gaussian image plane*, also as predicted by paraxial theory. Moreover, straight lines in the sky should produce straight lines in the image plane. This mapping of the points and lines on the sky to points and lines in the image plane is called a *collinear transformation*. If an optical system fails to produce this collinear transformation it is said to exhibit *aberrations*.

There are a number of ways to discuss aberrations. One concentrates on differences between the actual wavefront and the perfect wavefront. Figure 5.25 defines a coordinate system that will specify a "test" ray anywhere on a wavefront. The idea is to measure the wavefront errors at a large number of such test locations so that the aberrated wavefront can be constructed. Figure 5.25a shows two reference rays and a test ray from the same source. After passing though an optical system represented by a curved mirror, all three are intended to focus at F. The first ray, the *chief ray*, V'VF, is the one passing from the source though the center of the entrance aperture (here it is the mirror vertex). The plane of the left-hand diagram, called the *meridional*, or *tangential*, *plane*, contains the chief ray and the optical axis. The plane perpendicular to the meridional plane that contains the chief ray is called the *sagittal plane*. Tangential rays and sagittal rays are confined to their respective planes.

The second ray, CFBFC, passes through the center of curvature of the mirror and reflects back over its original path. This meridional ray will help specify the direction of the source relative to the optical axis.

The third ray, P'PF, is the arbitrary "test" ray, which may be outside the plane of the diagram. In a perfect optical system, the optical path lengths V'VF, P'PF and CBF will all be equal. Figure 5.25b views the ray geometry in the plane of the entrance aperture, looking down the optical axis, and locates points P, V, and B. In this diagram, the distance b is related to the angle θ ($b = R \sin \theta$), that is, it measures how far the source is from the axis. The circular coordinates ϕ and ρ locate the intersection of the test ray and the aperture. If point P is on the

Fig. 5.26 The difference in optical path length between a ray through point P and one through the vertex. The perfect wavefront is a sphere centered on the Gaussian image point at F, and a perfect ray is perpendicular to this front.

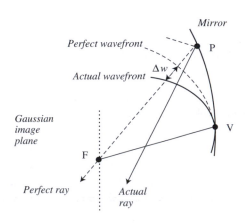

y-axis of this diagram, the test ray is tangential; if P is on the *x*-axis, the ray is sagittal.

We are concerned with any possible difference between the optical path lengths of this test ray (i.e. ray P′PF in Figure 5.25a) and the ray through the vertex. If you analyze the angles of refraction or reflection for a curved surface using the approximation

$$\sin \theta \approx \theta - \frac{\theta^3}{3!}$$

you obtain the results of ***third-order aberration theory*** – a much more accurate computation of where rays travel than the one given by paraxial theory (which assumes $\sin \theta = \tan \theta = \theta$.)

In the third-order treatment, you can show that the optical path difference between the test ray and the chief ray (Figure 5.26) takes the form:

$$\Delta w(\rho, \phi, b) = C_1 \rho^4 + C_2 \rho^3 b \cos \phi$$
$$+ C_3 \rho^2 b^2 \cos^2 \phi + C_4 \rho^2 b^2 + C_5 \rho b^3 \cos \phi \tag{5.15}$$

where the C_i values depend on the shapes of the optical surfaces and (if refractions are involved) indices of refraction. Since each of the terms in Equation (5.15) has a different functional dependence, we distinguish five monochromatic third-order aberrations, also known as the ***Seidel aberrations***.

Table 5.4 lists the aberrations in order of importance for large telescopes, where the exponent on ρ is crucial. This is also in the order in which they are usually corrected. Astigmatism, for example, is only corrected after both coma and SA have been eliminated.

5.5.2 Shapes of optical surfaces

A lens- or mirror-maker will find a spherical surface to be the easiest to produce and test. Other shapes, however, are frequently required, and the most

Table 5.4. *Third-order monochromatic aberrations*

Aberration	Functional dependence
Spherical aberration (SA)	ρ^4
Coma	$\rho^3 b \cos \phi$
Astigmatism	$\rho^2 b^2 \cos^2 \phi$
Curvature of field	$\rho^2 b^2$
Distortion	$\rho b^3 \cos \phi$

Table 5.5. *Conic section eccentricities*

Shape	Eccentricity	Conic constant
Sphere	0	0
Oblate ellipsoid	$0 < e < 1$	$-1 < K < 0$
Prolate ellipsoid	$e^2 < 0$	$K > 0$
Paraboloid	$e = 1$	$K = -1$
Hyperboloid	$e > 1$	$K < -1$

commonly used belong to the class generated by rotating a conic section around its axis of symmetry. Usually, this axis coincides with the symmetry axis of the telescope, as in Figure 5.27a, where we show the cross-section of such a surface (e.g. a mirror) in the y–z plane of our usual coordinate system. The cross-section satisfies the equation

$$y^2 = 2Rz - \left(1 - e^2\right) z^2$$

Here e is the eccentricity of the conic, and R is the radius of curvature at the vertex. In three dimensions, if this curve is rotated around the z-axis, the resulting surface satisfies the equation

$$\rho^2 = 2Rz - (1 + K) z^2$$

where $\rho^2 = x^2 + y^2$, and the conic constant, K, is just the value of $-e^2$ for the two-dimensional curve. Table 5.5 gives the values of e and K for specific conics.

Fig. 5.27 (a) Coordinate system for describing a mirror shaped like a rotated conic section. (b)–(f) Light rays originating at one focus of a conic of revolution will reconverge at the other focus. The degenerate cases are the sphere (f), where the two foci coincide, and the parabola (d), where one focus is at infinity. In the oblate ellipsoid, conjugate foci align diametrically on a ring around the axis of rotation. In the hyperboloid, either the object or the image is virtual.

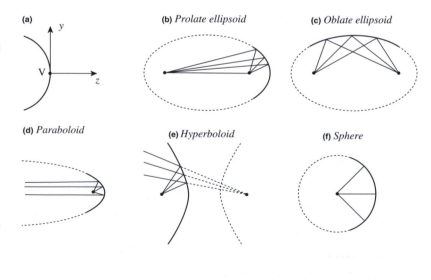

(a)

(b) *Prolate ellipsoid*

(c) *Oblate ellipsoid*

(d) *Paraboloid*

(e) *Hyperboloid*

(f) *Sphere*

As demonstrated by Descartes in 1630, each conic will perfectly focus a single object point onto a single conjugate point. These points, as you might guess, are the two geometric foci of each conic; see Figure 5.27.

5.5.3 Spherical aberration

Except for spherical aberration, all the wavefront errors in Table 5.4 vanish for sources on axis ($b = 0$). For visual astronomy, where one typically examines only on-axis images, SA is the only monochromatic aberration that is troublesome.

Figure 5.28 shows a spherical mirror and two rays from an on-axis source; one, near the axis, reflects to the Gaussian focus at F. The second, in violation of the paraxial approximation, strikes the mirror at an angle of 30 degrees, and is brought to focus at G, a distance of $0.845(R/2)$ from the vertex – considerably closer to the mirror than the Gaussian focus (see Problem 10).

For a mirror that is a conic of revolution, the focal length (not the wavefront error) of a ray parallel to the axis is

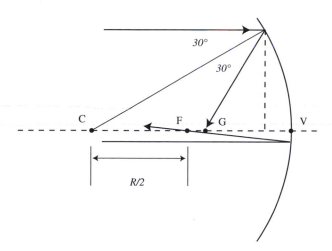

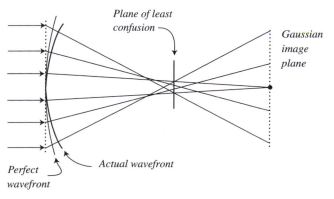

Fig. 5.28 A reflection from a spherical mirror that is not paraxial. A ray parallel to the axis strikes the mirror at a 30° angle of incidence. It does not pass through the paraxial focus at F.

Fig. 5.29 Spherical aberration. Note the best compromise focus is not at the Gaussian focus.

$$f(\rho) = \frac{R}{2} - (1 + K)\left\{\frac{\rho^2}{4R} + \frac{(3 + K)\rho^4}{16R^3} + \dots\right\} \qquad (5.16)$$

The first term in Equation (5.16) gives the Gaussian focus. The ρ^2 term is the third-order aberration, and the ρ^4 term is the fifth-order aberration. If SA is present in a mirror or lens, then the image will be blurred, and the best image is usually not at the Gaussian focus; see Figure 5.29. An equation similar to (5.16) exists for a lens, but is more complex because it accounts for two surfaces and the index of refraction.

It is possible to minimize, but not eliminate, SA in a spherical-surface single lens by minimizing the angles of incidence on every surface – a plano-convex lens with the curved side facing the sky is a common example. Likewise, any lens with a large enough focal ratio will approach the paraxial case closely enough that the blur due to SA can be reduced to the size of the seeing disk. Since a large focal ratio also minimizes chromatic aberration, very early (1608–1670) refracting telescope designs tended to have modest apertures and large focal lengths. The problem with these designs was reduced image brightness and, for large apertures, unwieldy telescope length.[4]

Another solution recognizes that a negative power lens can remove the SA of a positive lens of a different index, and an achromatic doublet can be designed to minimize both spherical and chromatic aberration. When large flint glass disks became available the 1780s, the stage was then set for the appearance of excellent refracting telescopes free of both spherical and chromatic aberrations. These grew in aperture, beginning with 3-cm aperture "spy glasses" used during the Napoleonic wars and advancing in the period 1812–25 with Fraunhofer's[5] superb instruments to the 10–24 cm range. The era of the refractor culminated with a 0.9-meter (36-inch) objective for Lick Observatory, California, in 1888 and a 1.02-meter (40-inch) objective for Yerkes Observatory, Wisconsin, in 1897. The legendary optical shop of Alvin Clark and Sons in Cambridge, Massachusetts, produced both objectives. A 1.25-meter refractor was on display

---

[4] Typical of these is the still-extant telescope that Giuseppe Campani produced in his optical shop in Rome, which has an aperture of 13 cm (5 inches) and a focal length of 10 meters (34 feet). King Louis XIV purchased this instrument in 1672 for Jean Dominique Cassini and the just-commissioned Paris Observatory, where Cassini used it to discover Rhea, the satellite of Saturn. Focal lengths of seventeenth-century telescopes approached the unusable – Cassini successfully employed a 41-meter (136-feet) telescope at Paris: the objective was supported by an old water tower and manipulated by assistants. "Aerial" telescopes – with no tube connecting objective and eyepiece – were not uncommon. Huygens used one 123 feet long, and Helvius operated one 150 feet in length.

[5] Friedrich Georg Wilhelm Struve used the largest of these, the 24-cm telescope at Dorpat (Tartu) Observatory, to measure the angular separations of 3000 double stars, as well as the parallax of Vega. Bessel used a 16-cm Fraunhofer telescope for the definitive measures of the parallax of 61 Cygni (1838). Argelander used a 9-cm Fraunhofer telescope for all his observations for the *Bonner Durchmusterung* (1862).

at the Paris exhibition of 1900, but never produced useful results, and a modern 1.0 m refractor was commissioned in 2002 for the Swedish Solar Telescope on La Palma. A diameter near 1 m is the upper limit for terrestrial refracting telescopes. Gravity would deform a lens larger than this to an unacceptable degree.

For mirrors, the removal of SA is simple. We have seen that a paraboloid reflector will display no on-axis aberrations whatever – rays from infinity parallel to the axis will all come to the same focus. This is consistent with Equation (5.16): SA is absent if the mirror shape is a paraboloid ($K = -1$). Isaac Newton constructed the first workable reflecting telescope in 1668. Popularized by William Herschel's spectacular discoveries, speculum-metal paraboloids were briefly fashionable from the 1780s until the superiority of the refractor became clear in the 1830s. Metal-film-on-glass reflectors gradually replaced the refractors in the first half of the twentieth century.

A strategy for correcting the spherical aberration of a *spherical* mirror recognizes that the amount of SA is independent of the off-axis distance of the source, i.e. the term b in Equation (5.15). This means a transparent corrector plate, when placed in front of a spherical mirror, can lengthen the optical path for rays of different ρ by the appropriate amount, whatever their direction, and thus cancel the mirror's aberration. Schmidt telescopes (Chapter 6) utilize such a corrector, and have found widespread use in astronomy.

5.5.4 Coma

Prior to the end of the nineteenth century, visual observers were concerned only with a telescope's on-axis performance. The advent of photography changed those concerns forever, and telescope design has since needed to satisfy more stringent optical criteria. Of the four off-axis aberrations, only the first two, coma and astigmatism, actually degrade the image resolution, while the other two only alter the image position.

Coma is the wavefront aberration that depends on $\rho^3 b \cos \phi$, which means, like spherical aberration, it is particularly a problem of large apertures. Unlike SA, coma increases with object distance from the axis. The $\cos \phi$ term means that even rays from the same radial zone of the objective will fail to come to a common focus. Figure 5.30 shows the ray paths and images typical of a system with coma.

If you imagine the rays from a source passing through a particular zone of the objective (i.e. a circle for which ρ = constant), then, in the presence of coma, those rays form a ring-shaped image offset from the Gaussian focus, as in the right of Figure 5.30. Comatic images of stars have a heavy concentration of light near the Gaussian focus, and become broader and fainter for rays in the outer circles – this comet-like appearance is the origin of the name for the aberration.

Fig. 5.30 The left-hand diagram is in the sagittal plane for a system with coma. The right-hand figure shows the image plane where each circle represents a bundle of rays from the same zone of the aperture. Dots on these circles mark the focus of each of the illustrated sagittal rays. Tangential rays from the same zone strike the tops of the corresponding circles.

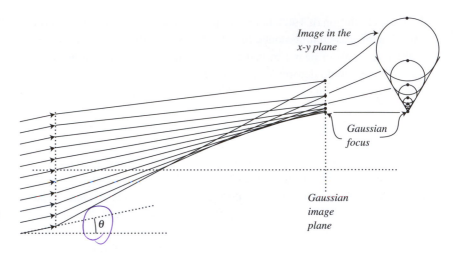

For a single surface, the length of the entire image – the angular size of the blur due to coma – is

$$L = A\frac{bD^2}{f^3} = A\theta\Re^{-2} \tag{5.17}$$

where D is the diameter of the aperture, f is the Gaussian focal length, $\Re$ is the focal ratio and A is a constant that depends on the shape of the surface (and index, if refracting).

An optical system with neither SA nor coma is called **aplanatic**. No single-element aplanatic telescope is possible, either in a refractor or reflector. As with SA, large focal ratios will reduce coma, but impose penalties in image brightness and telescope length. Otherwise, minimizing coma in refracting systems requires a system of lenses and, fortunately, achromatic doublet or triplet designs that minimize SA also reduce coma. Aplanatic reflecting telescope designs require two mirrors. Alternatively, a correcting lens system (usually with zero power) can also minimize coma in a single-mirror telescope. Such correcting optics may also aim to correct additional aberrations, but must take care to avoid introducing chromatic aberration.

5.5.5 Astigmatism

Astigmatism is an aberration whose wavefront distortion depends on the term $b^2\rho^2\cos^2\phi$. Therefore, it increases more rapidly than coma for off-axis images. The angular dependence suggests that this wavefront distortion is zero for rays in the sagittal plane (i.e. $\phi = 90°$, $\phi = 270°$), but an extremum for rays in the

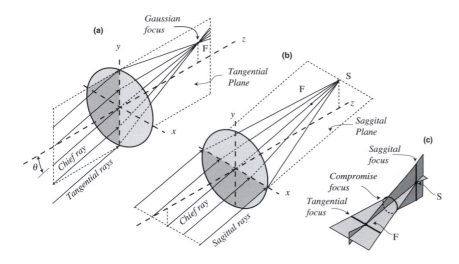

Fig. 5.31 Astigmatism for an off-axis source at infinity in direction θ. (a) Rays in the meridional plane suffer no aberration and come to a focus at the paraxial focus, F. (b) Rays in the sagittal plane come to a secondary focus at S. (c) All rays pass through two line segments that are perpendicular to the chief ray. The first, the tangential focus, is in the sagittal plane centered at F. The second, the sagittal focus, is in the tangential plane centered at S.

meridional plane. In the absence of other aberrations, astigmatic sagittal rays focus at the Gaussian image point, but astigmatic meridional rays focus at different a point; see Figure 5.31. All rays pass through two line segments. One line, called the sagittal focus, is in the meridional plane, and extends on either side of the Gaussian image point. The other line segment, called the meridional focus, is in the sagittal plane, centered on the focus point of the tangential rays.

For a single surface forming astigmatic images of objects at infinity, the length of each focal line is proportional to $\theta^2 \Re^{-1}$, and the best compromise focus position is midway between the two focal lines. Here, a star image has a circular shape and has a diameter equal to half the focal line length, so the blurred image length is

$$L_{\text{Astig}} = B\theta^2 \Re^{-1} \tag{5.18}$$

where B is a constant that depends upon the shape of the reflecting or refracting surface.

All uncorrected refractors and all practical two-mirror reflectors suffer from astigmatism. In many cases, the astigmatism is small enough to be ignored. In other cases, a corrector lens or plate removes the astigmatism. If a telescope design is free from astigmatism, coma, and SA, it is called an ***anastigmatic aplanat***. For large anastigmatic telescopes, the correction plate is usually located near the focal plane. For small telescopes, a popular design that can be made anastigmatic is the Schmidt–Cassegrain, a two-mirror telescope with a corrector plate located at the aperture (see below).

Fig. 5.32 Curvature of field. The mirror at right has a radius of curvature, *R*, with center of curvature at C. If all other aberrations are absent, off-axis images will focus on a spherical surface of radius *R*/2, also centered at C.

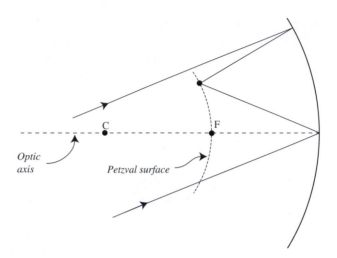

5.5.6 Field curvature

Two aberrations remain in an anastigmatic aplanat, but neither of them will cause images to blur. In the absence of other aberrations, third-order theory predicts that off-axis images will fall on a spherical surface known as the Petzval surface, not on the plane predicted in paraxial theory. The radius of the Petzval sphere depends on the curvatures and refractive indices of the mirrors and lenses in the optical system. For the simple case in Figure 5.32, a spherical mirror of radius of curvature *R* at its vertex, the Petzval surface has radius *R*/2 and has its center at the center of curvature of the mirror.

Detectors tend to be flat. A flat detector placed tangent to the Petzval surface, will necessarily record most of its images out of focus. For a small detector, this defocus will not exceed the seeing disk or diffraction limit and therefore not present a problem. A large enough detector, however, will produce blurred images because of field curvature. One solution is to bend the detector to match the Petzval surface. This has been done for many years with glass photographic plates, which can be forced, usually without breaking, into a considerable curve by a mechanical plate-holder. Large solid-state detectors like charge-coupled devices (CCDs) are mechanically quite fragile, so bending them is not an option. In many telescopes, a corrector plate or lens, which may also help remove other residual aberrations, serves to flatten the field.

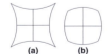

(a)   (b)

Fig. 5.33 (a) Pincushion distortion of an object that has a square outline centered on the optical axis. (b) Barrel distortion of the same object.

5.5.7 Distortion

Distortion relocates images in the focal plane so that the colinearity requirement is violated – straight lines on the sky become curved lines in the focal plane. There is no image blur. Figure 5.33 illustrates two kinds of distortion, "barrel" and "pincushion," either of which will increase more rapidly with distance from the axis than do the other Seidel aberrations.

Since distortion does not change image quality, an observer can remove it from an image if he has the calibrations needed. This is laborious for photographs, but relatively easy with digital images.

5.5.8 Spot diagrams

Figure 5.34 gives a qualitative summary of the distortions in the image of a point source that are introduced by the first four Seidel aberrations. Since more than one aberration may be present in an actual optical system, its image-forming behavior will generally exhibit some combination of the effects illustrated. For a mirror, the magnitude of a particular aberration will depend on the value of b, aperture, focal ratio, and conic constant. The figure shows **spot diagrams** for each aberration. Each spot is the focal-plane location of a single ray traced through the system. Rays are chosen to sample the entrance aperture in a uniform fashion, so the density of spots gives an indication of the brightness distribution of the final image.

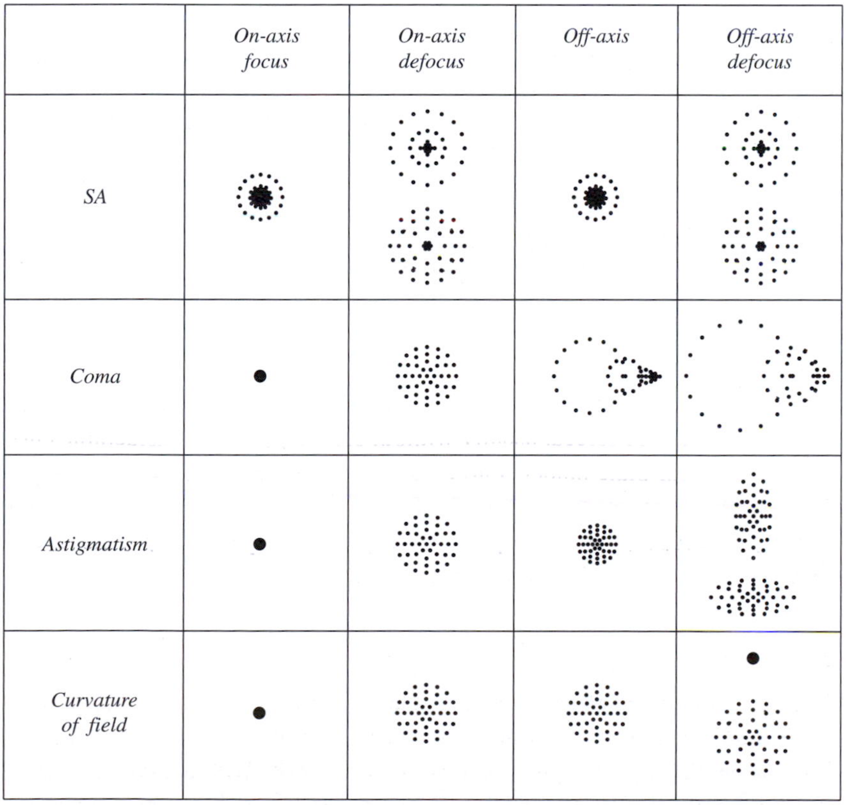

	On-axis focus	On-axis defocus	Off-axis	Off-axis defocus
SA				
Coma				
Astigmatism				
Curvature of field				

Fig. 5.34 Qualitative appearance of images of a point source in optical systems with a single aberration present. Discrete rays strike the entrance aperture in a pattern spaced around multiple. concentric circles. Actual sizes of the aberrations will depend upon details of the system. In the diagram, "focus" means the best compromise on-axis focus, which may differ from the Gaussian focus.

5.6 Summary

- Geometrical optics models light as a bundle of rays or, alternatively, as a sequence of geometrical wavefronts. Concepts:

index of refraction: $v(\lambda) = c/n(\lambda)$	*chromatic dispersion*
optical path length	***Fermat's principle***
Snell's law: $n_1\sin\theta_1 = n_2\sin\theta_2$	*total internal reflection*
paraxial approximation	*grazing incidence*
vertex focal length	***power***
optical axis aperture	***light-gathering power***

- In the paraxial approximation:

  power of a spherical surface ($n_1 = -n_2 = 1$ for a mirror):

$$P_{12} = \frac{n_2 - n_1}{n_2 R_{12}} = \frac{n_2}{s_2} - \frac{n_1}{s_1} = \frac{n_2}{f_2} = -\frac{n_1}{f_1}$$

  thick lens or two mirrors:

$$P = \frac{1}{f} = P_{12} + P_{23} - \frac{d}{n}P_{12}P_{23}$$

  thin lens or single mirror:

$$P = P_{12} + P_{23} = \frac{1}{f} = \frac{1}{s_2} - \frac{1}{s_1}$$

- There are simple rules for graphical ray tracing in the paraxial approximation. Ray-tracing computer applications employ much more exact rules and are important tools for optical design.
- A variety of special transmitting and mirror materials are available for astronomical applications. Special coatings can enhance reflection or transmission.
- Prisms are used for both the redirection of light and for angular dispersion.
- Astronomical telescopes gather light and enhance image detail. Concepts:

image scale: $s = 206\,265/f$	*focal ratio:* $\Re = f/D$
diffraction limit	*Airy disk*
Rayleigh criterion: $\alpha_A = 1.22\lambda/D$	*seeing*
speckle pattern	*scintillation*
differential atmospheric refraction	*chromatic aberration*
achromat	*apochromat*

- A system that fails to perform a collinear transformation between the object and its image is said to exhibit aberrations. Concepts:

Gaussian image	*3rd order ray theory*	*chief ray*
meridional plane	***sagittal plane***	***Seidel aberrations***

- Most optical surfaces used in astronomy are conics of revolution, characterized by a particular conic constant.
- Removing the Seidel aberrations produced by a spherical surface is a crucial concern in telescope design:
  - *Spherical aberration* is minimized in an achromatic doublet or absent in a paraboloid mirror.
  - *Coma* is also minimized in an achromatic doublet or triplet, and in certain aplanatic two-mirror systems.
  - *Astigmatism* requires additional optical elements for elimination
  - *Curvature of field* requires additional optical elements for elimination
  - *Distortion* is usually not corrected in telescopes

Exercises

1. Prove that the law of reflection follows from Fermat's principle by using the geometry illustrated in part (a) of the figure below. Let x be the unknown location of the point where the ray from P_1 to P_2 reflects from a horizontal surface. (Hint: write an expression for s, the total optical path length, set $ds/dx = 0$, and replace terms in the result with trigonometric functions of the angles of incidence and reflection.)

2. Prove that Snell's law of refraction follows from Fermat's principle by using the geometry illustrated in part (b) of the figure below. Let x be the unknown location of the point where the ray from P_1 to P_2 refracts at a horizontal surface. See the hint for Problem 1.

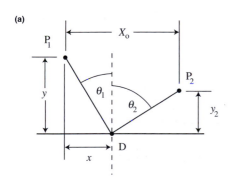
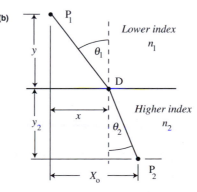

3. Two positive thin lenses with their optical axes aligned are separated by distance, d. If the lenses have focal lengths f_1 and f_2, use Equation (5.9) to find the value of d that will give the combination zero power. Illustrate your result with a graphical ray trace that shows rays that are incident parallel to the axis will emerge parallel to the axis. Will such a two-lens system have any effect at all on a beam of parallel rays?

4. A telescope has a focal ratio of $f/7.5$. You wish to use it with a spectrometer that requires an $f/10$ beam at its input. Compute the focal length of a 50mm diameter lens that, when inserted in the beam 150 mm in front of the unmodified focal plane, produces the required beam.

5. Compute the angular dispersion in red and ultraviolet light for a prism with apex angle 30 degrees and index given by $n(\lambda) \approx 1.52 + (0.00436 \, \mu m^2)\lambda^{-2}$. To do this, assume the prism is operated at minimum angular deviation for 600 nm light and compute the angular dispersion at that wavelength. Then repeat the calculation at minimum angular deviation for 350 nm light. Show that the ratio of the two dispersions is close to the value predicted by Equation (5.11).

6. Compute the frame width (in arcmin) and pixel scale (in arcsec per pixel) for CCD observations with the following telescope and instrument:

   > aperture 1.0 meter
   > focal ratio $f/6.0$.

   CCD is 1024×1024 pixels; each pixel is 18 μm square.

7. Show that in order for a ray to be transmitted by an optical fiber, its angle of incidence on the end of the fiber must satisfy the condition:

$$\sin \theta_1 \leq N.A. \equiv \sqrt{n_2^2 - n_3^2}$$

   where n_2 and n_3 are the indices of the core and cladding, respectively.
   (Hint: refer to Figure 5.13 and require that θ_W be greater than or equal to the critical angle.) The quantity $N.A.$ is known as the **numerical aperture** of the fiber.

8. On a particular night, the planet Mars has an angular diameter of 15 arcsec and an apparent brightness of $1.0 \times 10^{-7} \, W \, m^{-2}$. Two astronomers observe the planet, using identical CCD cameras whose pixels are 25 μm apart. Albert uses a telescope of 0.3 m aperture whose focal ratio is $f/8$. Bertha uses a telescope of 30 m aperture whose focal ratio is $f/4$. How much energy accumulates in a single pixel of Albert's CCD image of Mars in a 100 s exposure? How much in a single pixel of Bertha's image of Mars?

9. Compute the diffraction limit of the Hubble Space Telescope (2.4 m diameter) in the ultraviolet (300 nm) and near infrared (2.0 μm). Compare with the diffraction limit at 2.0 μm of a space telescope that has an 8 m diameter.

10. Show that for a spherical mirror, the focal length defined by a ray parallel to the optical axis is *exactly*

$$f = R - \frac{R}{2}\left[1 - \frac{\rho^2}{R^2}\right]^{-\frac{1}{2}}$$

   where R is the radius of curvature and ρ is the distance between the ray and the optical axis. Use this to verify the result quoted in the text for Figure 5.25.

Chapter 6
Astronomical telescopes

> The adventure is not to see new things, but to see things with new eyes.
>
> — Marcel Proust (1871–1922), *Remembrance of Things Past*, 1927

While I disagree with Proust about the thrill of seeing utterly new things (I'm sorry, that *is* an adventure), astronomers immediately come to mind if I wonder who might be obsessively concerned with the acquisition of new "eyes." No instrument has so revolutionized a science, nor so long and thoroughly dominated its practice, as has the telescope astronomy. With the possible exception of the printing press, no instrument so simple (amateurs still make their own) has produced such a sustained transformation in humanity's understanding of the Universe.

In this chapter, we examine the basic one- and two-mirror optical layouts of the preferred modern designs, as well as the layouts of a few telescopes that use both transmitting and reflecting elements. Schroeder (1987) provides a more advanced treatment.

Space-based telescopes have some pronounced advantages, disadvantages, and special requirements compared with their ground-based cousins, and we will consider these in some detail, along with recent advances in the construction of very large telescopes. Because it is such an important technology, we will take some trouble to understand the principles of adaptive optics, and its potential for removing at least some of the natural but nasty (for astronomy) consequences of living on a planet with an atmosphere. Finally, we will anticipate the future of the telescope, and consider astronomers' plans (and hopes) for advanced adaptive optics systems and — not just large — but *extremely* large telescopes.

We begin, however, not with the telescope, but with the apparatus that supports and points it.

6.1 Telescope mounts and drives

> My brother began his series of sweeps when the instrument was yet in a very
> unfinished state, and my feelings were not very comfortable when every moment
> I was alarmed by a crack or fall, knowing him to be elevated fifteen feet or more

on a temporary cross-beam instead of a safe gallery. The ladders had not even
their braces at the bottom; and one night, in a very high wind, he had hardly
touched the ground when the whole apparatus came down.

— Caroline Herschel, Letter, 1784

6.1.1 Altazimuth and equatorial mounts

For ground-based telescopes, the mount that supports the telescope has two
important functions. First, it moves the telescope to *point* at a specific position
on the celestial sphere. Second, the mount *tracks* the object pointed at — that is, it
moves the telescope to follow accurately the apparent position of the object.
Object position relative to the horizon changes rapidly due to diurnal motion (i.e.
the spin of the Earth) and less rapidly due to changing atmospheric refraction and
even proper motion (e.g. Solar System objects). In some circumstances, tracking
may need to compensate for effects like telescope flexure or image rotation.
Some specialized telescopes do not track: transit telescopes (see Chapter 3) point
only along the meridian. Some telescopes do not even vary their pointing: a
zenith tube points straight up at all times. In some specialized telescopes, the
tracking function is accomplished by moving the detector in the image plane
while the massive telescope remains stationary.

Most telescopes, however, are mounted so as to be mobile on two axes, and
track by moving the entire telescope. Figure 6.1 shows the two most common
forms of mount. The configuration in Figure 6.1a is called an *altazimuth* mount.
This mount points to a particular position on the sky by rotating the vertical axis
to the desired azimuth, and the horizontal axis to the desired elevation. To track,
the altazimuth mount must move both axes at the proper rates, rates that change
as the tracked object moves across the sky.

Fig. 6.1 (a) Altazimuth
and (b) equatorial
telescope mounts.

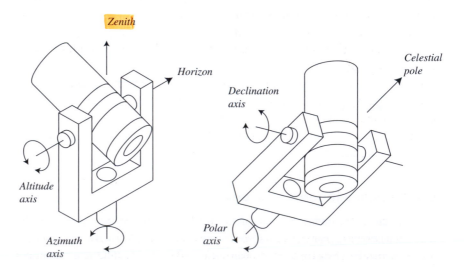

The ***equatorial mount***, illustrated in Figure 6.1b, has one axis (the polar axis) pointed directly at the celestial pole, and the second axis (the declination axis) at right angles to the first. This mount points to a particular position on the sky by rotating the polar axis to point the telescope at the desired hour angle, and rotating the declination axis to the desired declination. Tracking is simple with the equatorial – simply rotate the polar axis at the constant sidereal rate to match the rotation of the Earth.

The advantage of the equatorial mount is thus the simplicity of both pointing and tracking. Many manually operated equatorials, for example, are equipped with ***setting circles*** – graduated circles attached to each axis to indicate the declination and hour angle at which the telescope points. Since the hour angle is just the sidereal time minus RA of the object, astronomical coordinates translate directly into telescope position. Tracking is a simple matter of rotating the polar axis at the steady rate of one turn each sidereal day and for this reason the tracking mechanism of an equatorial is sometimes called a ***clock drive***. For very precise tracking, effects like atmospheric refraction, mechanical errors, and flexure require corrections to the sidereal rate as well as small movements in declination. Many of these can be computer-controlled, but an astronomer (or an instrument) often generates such small corrections by monitoring the position of a ***guide star*** near the field of interest and moving the telescope as needed.

The altazimuth mount is more compact (and therefore potentially less expensive) than the equatorial, and gravitational loading does not vary with pointing as it does in an equatorial. Virtually all the ground-based telescopes with apertures above 5 meters use altazimuth mounts.

The altazimuth has disadvantages. Neither of its axes corresponds to the axis of diurnal motion, so that pointing requires a complicated transformation of coordinates, and tracking requires time-varying rotation rates on both axes. In addition, as an altazimuth tracks, the image of the sky in the telescope focal plane rotates with respect to the altitude and azimuth axes, so any image-recording detectors must be mounted on a rotating stage in order to avoid trailing. Again, the rotation rate for this stage is not constant. Computer control of the drive axes and rotating detector stage more or less eliminates these disadvantages, but with additional cost and complexity. A final disadvantage arises because the coordinate transformation from hour angle–declination to altitude–azimuth becomes singular at the zenith. An altazimuth mount is therefore unable to precisely track objects within about five degrees of the zenith.

6.1.2 Telescope mounts in space

Telescopes in space must also point and track, but since gravity does not glue them to a spinning planet, at least some aspects of these tasks are less problematic. In general, two methods have been used to adjust and stabilize the orientation of a telescope in space: small rockets and spinning reaction wheels. Rockets require a supply of propellant. For example, bursts of a compressed gas escape from a nozzle to produce a force on the telescope. Reaction wheels require an on-board motor – when the rotation speed of the wheel is changed, the telescope begins to rotate in the opposite direction.

Since the resolution of a space telescope is generally much higher than a ground-based instrument, there are more stringent requirements for precision tracking. Space telescopes often rely on guide stars for this precision. To point the Hubble Space Telescope, for example, at a "fixed star" requires continuous telescope movement because of the aberration of starlight induced by the telescope's orbital velocity and because of torques induced by atmospheric drag and thermal effects. Some telescopes in space are not designed to produce steady images at all, but are mounted on spinning platforms, and data accumulates as objects drift repeatedly in and out of the fields of the detectors.

6.2 Reflecting telescope optics

All large-aperture and most small-aperture modern optical telescopes are reflectors. The great number of possible designs usually narrows to the few practical choices discussed in this section. Wilson (1996, 1999) gives a thorough treatment of reflecting telescope optics, as well the development of different designs in the historical context.

6.2.1 Prime focus and Newtonian focus

One very simple telescope design mounts a detector at the focus of a paraboloid ($K = -1$) mirror. This *prime focus* configuration has the advantage of both simplicity and minimum light loss (there is only one reflection), but has some limitations. First, any observer or apparatus at the prime focus will obstruct the mirror. For this reason, the prime focus configuration is generally only found in telescopes where the diameter of the aperture exceeds the diameter of the astronomer (or her observing apparatus) by a significant amount. In these cases (i.e. telescopes larger than around 3.5 meters in diameter), a prime focus *cage* is fixed on the optical axis of the telescope to carry an astronomer or remote-controlled instruments. Besides reducing the light-gathering power of the telescope and introducing a greater opportunity for scattering light, this central obstruction has minimal effect on properly focused images. Out-of-focus images will have a characteristic "doughnut" shape. The support structure that extends between the central obstruction and the side of the tube produces the artifact of radial diffraction spikes on very bright stars.

A second problem with the prime focus configuration is image quality. As described in Section 5.5, coma and astigmatism increase with distance from the optical axis. Applying aberration theory to a paraboloid, we can evaluate the

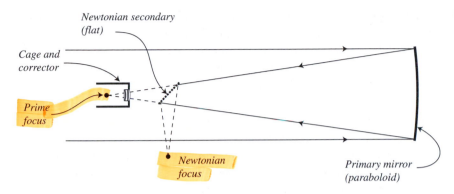

Fig. 6.2 Prime focus and Newtonian configurations.

Newtonian secondary (flat)

Cage and corrector

Prime focus

Newtonian focus

Primary mirror (paraboloid)

constants in Equations (5.17) and (5.18) to show that the angular diameters of blurred images due to coma and astigmatism are

$$L_{\text{coma}} = \theta / 16 \Re^2$$
$$L_{\text{astg}} = \theta^2 / 2 \Re$$

(6.1)

Here we recall that the terms on the right-hand side are the focal ratio, $\Re = f/D$, and θ, the angular displacement of the image from the optical axis in radians. Because coma can be severe even close to the optical axis for fast mirrors, prime focus cameras are usually equipped with compound corrector lenses that reduce aberrations over an acceptably wide field, but at the cost of some light loss.

For optical telescopes where the prime focus configuration is impractical, the **Newtonian** design, which uses a flat diagonal mirror to redirect the converging beam to the side of the telescope tube (see Figure 6.2) provides a more convenient access to the focus. The diagonal mirror introduces both a central obstruction and an additional reflection, but the Newtonian design is so simple that many homemade and inexpensive telescopes use this layout. Since many amateur observers only use images near the optical axis, the coma and astigmatism of the paraboloid need not be regarded as serious flaws. Although the wide-field performance of a Newtonian can be improved with corrector lenses near the focus, professional astronomers generally use a Cassegrain[1] design.

---

[1] Historically, James Gregory proposed the first reflecting telescope design in 1663, but was unable to acquire the concave mirrors to build a working model. (Robert Hooke implemented Gregory's design in 1674.) Meanwhile, Isaac Newton built a different two-mirror telescope in 1668, so the first reflector actually constructed was a Newtonian. A rather obscure Frenchman named Cassegrain proposed a third design in 1672. Newton promptly and unfairly criticized Cassegrain's design. Although early reflectors tended to be Newtonians, most modern telescopes use Cassegrain's design.

6.2.2 Cassegrain and Gregorian reflectors

An additional disadvantage of the Newtonian becomes serious at large focal length. Large weights, in the form of instruments or astronomers, need to be supported at the focus while tracking, usually at a great distance from both the primary and the observatory floor. The Newtonian is not only daunting for an astronomer with a touch of acrophobia, but any weight at the focus exerts a variable torque on the telescope tube.

Figure 6.3 shows two alternative two-mirror designs. Like the Newtonian, both the **Cassegrain** and the **Gregorian** utilize a paraboloid as the primary mirror. In Cassegrain's configuration, the secondary is a convex hyperboloid located on the optical axis, with one (virtual) focus coincident with the focus of the primary (point F in the figure). This means that rays converging to F will be redirected to the second focus of the hyperboloid at point F', which is usually located behind the primary. A hole in the center of the primary allows the rays to reach this secondary focus. In Cassegrain radio telescopes, the focus is more commonly just in front of the primary. The Gregorian design is similar to the Cassegrain, except the secondary is a concave ellipsoid.

In any two-mirror design, the multiple-lens formula results in the final power of the mirror combination as given by Equation (5.9):

$$P = \frac{1}{f} = P_1 + P_2 - dP_1P_2 \tag{6.2}$$

Here, P_1 and P_2 are the powers of the primary and secondary, respectively. It will be convenient to define three dimensionless parameters that describe,

Fig. 6.3 (a) Cassegrain and (b) Gregorian mirror configurations.

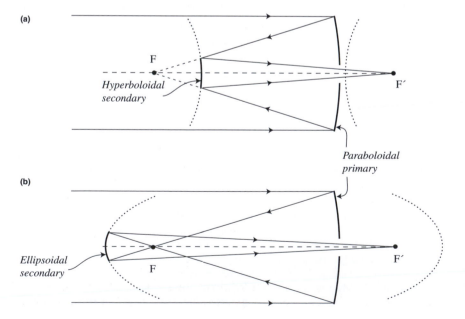

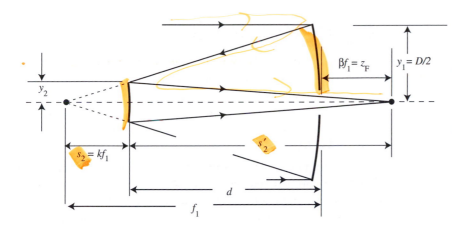

Fig. 6.4 Parameters for a two-mirror telescope. The sign convention measures s and z_F positive to the right from the mirror vertex. The value of d is always positive.

respectively, the final focal length, the distance between the primary and secondary, and the back focal distance (the distance from the vertex of the primary to the final focus) — see Figure 6.4:

$$m = \frac{P_1}{P} = \frac{f}{f_1} = -\frac{s_2'}{s_2}$$

$$k = \frac{y_2}{y_1} = 1 - \frac{d}{f_1} \tag{6.3}$$

$$\beta = \frac{z_F}{f_1}$$

The sign convention is that β is positive if the focus is behind the primary, and that m and k are both positive for a Cassegrain and negative for a Gregorian. Because the foci of the conics must coincide, a telescope designer may choose only two of these parameters, with the third given by

$$k = \frac{(1 + \beta)}{(1 + m)} \tag{6.4}$$

The overall focal length, f, is positive for a Cassegrain and negative for a Gregorian. Substitution into Equation (6.2) requires that

$$\frac{f_2}{f_1} = -\frac{km}{m - 1} \tag{6.5}$$

In both designs, the primary is a paraboloid by definition (so $K_1 = -1$). The requirement that the spherical aberration (SA) remains zero determines the conic constant, K_2, of the secondary. Specifically:

$$K_{1C} = -1, \qquad K_{2C} = -\left(\frac{m + 1}{m - 1}\right)^2 \tag{6.6}$$

Both Cassegrain and Gregorian designs locate the focus conveniently — behind the primary. Clearly, the tube of the Cassegrain will be much shorter than a Newtonian of the same focal length, and the overall cost of the observatory, which includes the mount, dome, and building, will thus be much lower. A Gregorian will be longer than a Cassegrain, but will still be much shorter than a Newtonian of the same effective focal length. For telescopes of even moderate size, these advantages easily outweigh the trouble caused by the increased optical complexity of two curved surfaces.

6.2.3 Aplanatic two-mirror telescopes

The classical versions of the Cassegrain and Gregorian just described begin with the assumption that the primary must be a paraboloid. The resulting telescopes suffer from coma and astigmatism. One can, in fact, choose any conic constant for the primary, and find a secondary shape that preserves freedom from SA. The proper choice K_1 and K_2, in fact, will remove both SA *and* coma, producing an *aplanatic reflecting telescope*. The aplanatic Cassegrain is called a *Ritchey–Chrétien* or *R–C*, and consists of a hyperbolic primary and hyperbolic secondary. The aplanatic Gregorian has no special name and utilizes an ellipsoidal primary and secondary. The required conic constants are

$$
\begin{aligned}
K_1 &= K_{1C} - \frac{2(1 + \beta)}{m^2(m - \beta)} \\
K_2 &= K_{2C} - \frac{2m(m + 1)}{(m - \beta)(m - 1)^3}
\end{aligned}
\tag{6.7}
$$

The remaining aberrations are smaller for the aplanatic Gregorian than for the R–C. However, for a given f, both the central obstruction due to the secondary and the overall length of the tube are greater for the Gregorian, and the resulting increase in expense and decrease in optical efficiency have usually been decisive for the R–C. Most modern telescopes also tend to favor the Ritchey–Chrétien over the primary focus. In addition to tube length, convenience, and weight considerations, corrector lenses for the prime focus must have many elements to remove aberrations, and thus tend to lose more light than does the single reflection from the R–C secondary. For wide-field applications, R–C telescopes frequently carry refracting optics to correct aberrations like curvature of field.

6.2.4 Nasmyth and coudé foci

Astronomers can mount heavy equipment at the Cassegrain focus, on the strong part of the telescope tube that also supports the primary mirror cell. There are limits, though. The *Nasmyth focus*, illustrated in Figure 6.5, can carry even heavier instruments on an altazimuth mount. (James Nasmyth (1808–90) was a British machine tool inventor and amateur astronomer.) In this arrangement, the secondary produces a longer final focal length than in the Cassegrain (the absolute value of the parameter m in Equation (6.3) becomes larger), and a flat

(a)

(b)

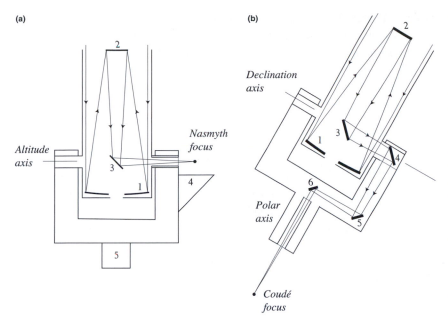

Declination axis

Altitude axis

Nasmyth focus

Polar axis

Coudé focus

Fig. 6.5 (a) The Nasmyth configuration. Light from the secondary mirror (2) is redirected by a tertiary flat (3) along the hollow altitude axis and reaches a focus above the Nasmyth platform (4). The platform rotates around the azimuth axis (5) as the telescope points and tracks. (b) A coudé configuration. Light from the secondary is redirected by flat 3 to a series of flats (4–5–6) that bring the beam to the polar axis. Similar arrangements can direct the beam along the azimuth axis of an altazimuth.

mirror intercepts the beam from the secondary and directs it horizontally along the altitude axis. As the telescope tracks, this focus remains fixed relative to the mount. Equipment at the Nasmyth focus thus exerts a gravitational stress on the telescope and mount that will not change over time.

If an instrument is very massive or very delicate, the ***coudé focus*** (French for "bent like an elbow") is superior to the Nasmyth. Figure 6.5b gives an example of this arrangement, implemented in an equatorial mount. A flat mirror redirects light from the secondary along the declination axis, and then a series of flats conducts the beam to emerge along the polar axis, where it reaches focus at a point that does not move with respect to the Earth. Here astronomers can locate elaborate instruments, like very large spectrographs, sometimes housed in stabilized and ultra-clean enclosures. A similar arrangement of flats will establish a coudé focus by directing the beam down the vertical (azimuth) axis of an altazimuth mount.

Both the Nasmyth and coudé have some disadvantages, so are usually implemented as temporary modifications of a general-purpose telescope. Compared with the Cassegrain, both Nasmyth and coudé require additional reflections, so there is some light loss. As the telescope tracks, the image field rotates at both these foci. A third problem concerns aberrations. Suppose you design a telescope to be aplanatic in the R–C configuration with the focus behind the primary, and with conic constants K_1 and K_2 given by Equations (6.7). To switch from the R–C to the Nasmyth or coudé, swap in a new secondary to get a longer focal length, and therefore get different values for m and β. However, the existing K_1 no longer be satisfies Equations (6.7). The resulting combination, called a ***hybrid two-mirror telescope***, cannot be aplanatic. In the hybrid, you can choose a value for K_2 so that SA is zero, but coma will still be present. Some modern telescopes can adjust the conic constant of the primary through active optics, and thereby reduce or eliminate coma when the secondary is changed. In other cases, a correcting element can be added to reduce the aberrations in the hybrid.

6.2.5 Schmidt telescopes

> I shall now show how completely sharp images can be obtained with a spherical mirror . . . . If the correcting plate is now brought to the center of curvature of the mirror, . . . the spherical aberration is abolished, even over the whole field.
>
> – Bernhard Schmidt, Ein lichtstares komafreies Spiegelsystem,
> *Mitteilungen der Hamburger Sternwarte*, vol. 7, no. 15, 1932

All telescope designs discussed so far give good images over relatively small fields of view – diameters of slightly less than one degree of arc for an RC, and perhaps up to three degrees with a corrector lens, depending on the focal ratio. The Schmidt[2] telescope produces good images over a much larger field – six to eight degrees. Schmidt telescopes became the standard instrument for many important photographic surveys during the mid and late twentieth century.

The Schmidt telescope exploits the symmetry of a spherical mirror to avoid off-axis aberrations. It consists of three elements: a spherical primary mirror, an aperture stop located at the center of curvature of the primary, and a refracting corrector plate designed to remove spherical aberration.

Figure 6.6 shows the layout. The aperture stop insures there can be no distinction between on-axis and off-axis sources: wave fronts from different directions will illuminate slightly different parts of the mirror, but since it is a spherical mirror, all will experience an identical change in wavefront shape upon reflection. Because of the aperture stop, the chief ray from every source always passes through point C, the center of curvature. This means that points B and V in Figure 5.25 always coincide, and there can be therefore no third-order coma or astigmatism. The absence of these off-axis aberrations means that the Schmidt offers the possibility of a fast focal ratio and large field of view. The stop does not affect the curvature of field, so the Petzval surface will be the one expected for a spherical mirror.

---

[2] Bernard Schmidt (1879–1939), an Estonian, lost his right arm in a boyhood experiment with gunpowder. At a time when almost all optical work was done by hand, he nevertheless became internationally recognized as a master lens- and mirror-maker. He constructed the first Schmidt camera (36-cm aperture, *f*/1.7, with a 16° field) at Hamburg Observatory in 1930 and described the design in the 1932 paper quoted. Schmidt never divulged his method for the very difficult task of grinding the surface of the corrector.

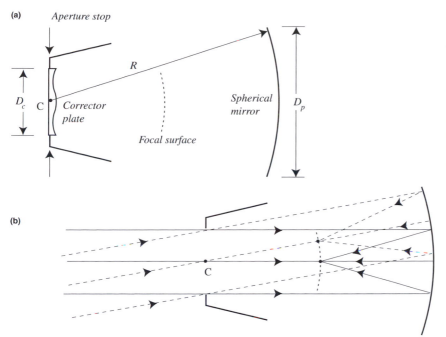

Fig. 6.6 The Schmidt telescope: (a) shows the arrangement of aperture stop, corrector plate, primary mirror, and focal surface; (b) shows how the aperture stop located at the center of curvature results in identical optics for beams from different directions.

The corrector plate, located in the plane of the aperture stop, is designed to remove SA. If you review Figure 5.28 and Equation (5.16) you can see that SA in a spherical mirror means the marginal rays (the ones near the edge of the aperture, ρ large) converge more strongly than the axial rays (the ones near the center). A Schmidt corrector, then, should be a refracting element whose power is larger (more positive) for the axial rays and smaller for the marginal. An entire family of shapes can do the job. Two possible shapes are sketched in Figure 6.7. The shape labeled (b), which is thickest at center and thinnest at 86.6 % radius, is the one usually chosen, since it minimizes the chromatic aberration introduced by the corrector plate. It is possible to further minimize chromatic aberration by using a two-element achromatic corrector. Unlike the spherical mirror, the corrector plate *does* have an optical axis, and introduces some off-axis aberrations, which are of concern in systems with very fast focal ratios ($<f/2$). The refracting corrector plate limits apertures to modest values, and the focal surface is inaccessible to a human observer, so the instrument is often called a Schmidt camera.

The location and curvature of the focal surface is the main inconvenience of the design. Until recently, the usual observing method was photographic. A special frame mechanically flexed a large[3] glass photographic plate to match

---

[3] The UK Schmidt, for example, uses square glass plates that measure 356 mm (14 inches) on a side and are 1.0 mm thick. Each of these plates will photograph an area $6.4° \times 6.4°$ on the sky. A large detector, of course, also vignettes (blocks) the beam.

the "diameter" of the image of a star will be something like the diameter of the central part of the Airy disk (Equation 5.13):

$$\theta = 2\alpha_A = \frac{2.44\lambda}{D}$$

This equation only applies in the absence of aberrations, and we know that the design and quality of the telescope optics determine image quality and useful field size. The precision and alignment of optical surfaces thus becomes especially critical in space, where seeing will not mask small errors. With excellent optics, the higher resolution of a space telescope produces smaller stellar images and more detail in the images of extended objects like planets and galaxies.

Detection limits

For stellar objects, freedom from wavefront distortions due to the atmosphere also means that a space telescope can detect fainter objects than an identical ground-based telescope, because the same light can be concentrated in a smaller image. Consider the simple problem of detecting the presence of a star. Assume we have a telescope with aperture, D, that the star produces flux f_λ, and we observe for time t. This means that in a narrow wavelength band, the telescope (either in space or on the ground) will collect and count a number of photons from the star given by

$$S = \text{signal} = \frac{\pi D^2}{4} \frac{\lambda}{hc} f_\lambda t Q$$

Here Q is a factor that depends on the bandwidth and counting efficiency of the detector, and the factor λ/hc converts from energy units to number of photons. We will say that the star is *detectable* if this signal is about the same size as its uncertainty. What is the uncertainty in measuring the signal? For faint stars like this, the measurement is of the combined signal and background (sky), B:

$$\text{measurement} = S + B$$

We compute the signal by subtracting the sky level from the measurement. The background is the count due to sky within the boundaries of the stellar image. If the image of the star has angular diameter θ, and the sky has average surface brightness b_λ, then

$$B = \text{background} = \frac{\pi D^2}{4} \frac{\lambda}{hc} \frac{\pi \theta^2}{4} b_\lambda t Q$$

For purposes of estimating the uncertainty in the measurement, we note that since the star is faint, we can ignore its contribution to the measurement, and approximate the uncertainty of the combination as the uncertainty in the background. This uncertainty is often called the noise, and if one is counting photons, Poisson statistics gives:

$$N = \text{noise} = \sigma(S) = \left[(S + B) + B\right]^{\frac{1}{2}} \approx \sqrt{2B} = \frac{\pi D\theta}{4}\left(\frac{2\lambda b_\lambda tQ}{hc}\right)^{\frac{1}{2}}$$

The signal-to-noise ratio then is

$$\frac{S}{N} \approx \left(\frac{\lambda Qt}{hcb_\lambda}\right)^{\frac{1}{2}}\frac{D}{\theta}f_\lambda$$

For a star that is just detectable, we set the signal-to-noise ratio equal to one, and find that the flux of such a star is

$$f_d = \left(\frac{hc}{\lambda Q}\right)^{\frac{1}{2}}\left(\frac{b_\lambda}{t}\right)^{\frac{1}{2}}\frac{\theta}{D} \tag{6.8}$$

Therefore, on the ground, where θ is set by seeing independent of telescope size, the detection threshold decreases only as the first power of D, even though light-gathering power increases as D^2. But in space, a telescope is diffraction limited, so substituting the Airy disk diameter for θ in Equation (6.8)

$$f_{d,\text{space}} = \left(\frac{hc\lambda}{Q}\right)^{\frac{1}{2}}\left(\frac{b_\lambda}{t}\right)^{\frac{1}{2}}\frac{2.44}{D^2} \tag{6.9}$$

With perfect optics in space, then, the payoff for large apertures is superior, since the detection threshold depends on D^{-2}.

Low Background

Equation (6.8) suggests another advantage to the space environment. The Earth's atmosphere is itself a source of background light, so from space, the value of b_λ is lower than from the ground. On the ground at visible and near-infrared (NIR) wavelengths the atmosphere contributes light from several sources: *airglow* (atomic and molecular line emission from the upper atmosphere), scattered sunlight, starlight, and moonlight, and scattered artificial light. Further in the infrared, the atmosphere and telescope both glow like blackbodies and dominate the background. From space, the main contribution to the background in the visible and NIR comes from sunlight scattered from interplanetary dust (visible from dark sites on the surface as the *zodiacal light*). In the V band, the darkest background for the HST (near the ecliptic poles) is about 23.3 magnitudes per square arcsec, while at the darkest ground-based site, the sky brightness is about 22.0 magnitudes per square arcsec. In the thermal infrared, the sky from space can be much darker than the sky from the ground because it is possible to keep the telescope quite cold in space. Plans for the JWST suggest the sky at 5 μm should be on the order of 12 magnitudes darker in space.

Atmospheric transmission

A fourth advantage of a space telescope is freedom from the absorbing properties of the Earth's atmosphere. This means, of course, that those parts of the electromagnetic spectrum that never reach the surface of the Earth are observable from space, and it is only here that gamma-ray, X-ray, and far ultraviolet astronomy, for example, are possible. But there is a further benefit in space. Even in the visible and infrared, the atmosphere is not completely transparent. From the ground, a major observational problem arises from variations in atmospheric transmission. Not only does the amount of absorbing material vary with zenith distance, but the atmosphere itself is dynamic – clouds form; the concentration of aerosols fluctuates; weather happens. All this variation seriously limits the accuracy one can expect from ground-based astronomical photometry, where astronomers are often pleased to achieve 1% precision. From space, weather never happens, and photometry precise to one part in 10^5 is possible.

Access to sky

A fifth advantage of a space telescope is its improved access to all parts of the celestial sphere. From the ground, half the sky is blocked by the Earth at all times, and for more half the time – daytime and twilight – atmospheric scattering of sunlight makes the sky too bright for most observations. For most locations, part of the celestial sphere is never above the horizon. Even nighttime has restrictions, since, for a substantial fraction of each month, moonlight limits the kinds of observation that can be made. In space, a telescope far enough away from the Earth and Moon has access to most of the sky for most of the time. The HST, in a low Earth orbit with a period of about 97 minutes, has somewhat greater restrictions. Many objects are occulted by the Earth once each orbit. Because of scattering by residual atmosphere and zodiacal dust, the telescope cannot point within 50° of the Sun, within about 25° of the illuminated Earth, or within 10° of the illuminated Moon. The JWST (see below) will be in an orbit that avoids most of the HST constraints.

Perturbing forces and environment

A telescope on the ground will experience changing gravitational stresses as it points in different directions, and will respond by changing shape – it is impossible for large telescopes to maintain the figures and alignments of optical surfaces without careful and expensive engineering. Stresses induced by wind or by temperature changes generate similar problems. A whole other set of difficulties arise from the toxic environment for telescopes on Earth – optical coatings get covered with dirt and degraded by atmospheric chemicals; abundant oxygen and high humidity promote corrosion of structures. Most of the expense of a large modern telescope is not in the optics, but in the systems needed to move and shelter the optics, while maintaining figure and alignment.

A telescope in space is in an ultra-clean environment, in free-fall, and experiences relatively tiny gravitational and wind forces. The forces needed to point, track, and maintain optical integrity can be relatively small, and the large mechanical bearings, massive mounts, and protective buildings of ground-based observatories are unnecessary.

6.3.2 Disadvantages of space telescopes

From the previous section, it might seem that telescopes belong in space, and that it would be foolish to build any serious ground-based astronomical facilities. This is not the case. At the present time, the total optical/infrared aperture on the ground exceeds that in space by a factor of at least 200, and that factor is likely to increase in the near future. The disadvantages of a space observatory are epitomized by its enormous cost compared to a ground-based observatory of similar aperture. For example, the two 8-meter Gemini telescopes had a construction budget of about $100 million per telescope. The 2.4-meter HST cost $2000 million to construct and launch.

Part of the great expense is space transportation – boosting a large telescope (or anything else) into orbit requires an enormous technical infrastructure. The transportation requirements also place severe constraints on telescope design – the instrument needs to be lightweight, but also sturdy enough to survive the trauma of a rocket launch. Once in space, the telescope must function automatically or by remote control, and needs to communicate its observational results to the ground. This technical sophistication entails substantial development costs and investment in ground stations and staff.

Although the space environment has many benefits, it also harbors hazards. Any spacecraft will be vulnerable to intense thermal stresses since its sunward side receives a high heat flux, while its shadowed side sees only cold space. Low Earth orbits pose additional thermal problems. The HST, for example, experiences intense thermal cycling as it passes in and out of the Earth's shadow once each orbit. Radiation is also physically destructive – X-rays and ultraviolet light from the Sun can damage electronic and structural components. Although a spacecraft can be shielded from sunlight, it is impossible to avoid energetic particles, either from cosmic rays or from the solar wind, especially for orbits that encounter particles trapped in the Earth's magnetosphere. Detectors on the HST, for example, periodically experience high background counts when the spacecraft passes through the South Atlantic Anomaly – an inward extension of the van Allen radiation belts which contain the trapped particles.

6.3.3 The James Webb Space Telescope

This replacement for the HST, built by NASA, the European Space Agency (ESA) and the Canadian Space Agency (CSA), should launch sometime before

Fig. 6.9 Schematic design for the JWST, showing major elements.

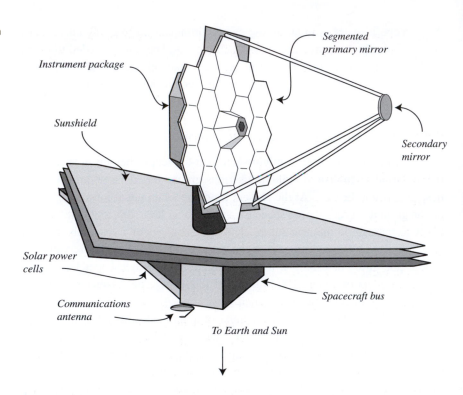

the year 2015. The design, illustrated in Figure 6.9, should produce substantially advanced observational capabilities compared to HST. The main features are:

- **A heliocentric orbit at the L_2 libration point**. This orbit, about 1.5 million kilometers from the Earth, keeps the Sun, Earth and Moon all in roughly the same direction. It avoids some of the troublesome features of a low Earth orbit: repeated occultation of targets by the Earth, periodic thermal cycling, the terrestrial radiation belts, and dynamic drag and scattering of light by residual atmosphere.
- **A stable thermal environment**. A highly reflective and heavily insulated sunscreen will keep the telescope and instruments in perpetual shadow. Under these conditions, the telescope should reach a relatively constant temperature of around 40 K. This should substantially reduce the background level in the mid infrared (MIR), so that it is dominated by emission from the zodiacal dust, rather than from the telescope.
- **Large aperture and high resolution**. The JWST has a lightweight beryllium primary mirror with a diameter of 6.5 meters, designed to produce images that are diffraction-limited at a wavelength of 2 μm ($\theta = 150$ mas) over a field of 20 minutes of arc. Even in the largest available spacecraft cargo shroud, the JWST must launch in a "folded" configuration, and then deploy the segments of the primary mirror and secondary support once in space. This means that the final shape of the mirror must be adjustable, and requires an active optics system (see below) if it is to achieve the diffraction limit.
- **Advanced instrumentation**. The telescope will have four instruments: a fine-guidance camera, a NIR (0.6–5 μm) imager, a NIR spectrograph, and a MIR (5–28 μm) imaging

spectrograph. Because of the large aperture and low background, the JWST should be between 100 and 1 million times more sensitive to faint sources in the NIR and MIR region than any existing telescope, including the HST.

The JWST promises revolutionary advances in observational capabilities. Meanwhile, however, a new species of telescope gives every indication that equally spectacular advances can be expected on the ground.

6.4 Ground-based telescopes

By the end of the nineteenth century, the most successful species of astronomical telescope, in the form of the achromatic aplanatic refractor, had evolved to the 1.0-meter aperture limit set by the strength of the Earth's gravity and the fluidity of glass. Telescopes long before this time encountered the limit on resolving power set by the Earth's atmosphere. There was barely a pause, however, before a new species of telescope, the reflector with a massive silvered-glass primary, shattered the aperture limit. The resolving-power limit was more stubborn, but did yield a bit as astronomers realized that atmospheric turbulence was minimized at certain locations.[4] The evolution of this new species culminated in the 5-meter Hale[5] reflector at Palomar in 1948. Image quality from the Palomar site was mediocre, and telescopes like the 4-meter Blanco Telescope (1976) on Cerro Tololo in Chile and the 3.6-meter Canada-France-Hawaii Telescope on Mauna Kea in Hawaii (1979) produced considerably better images (median seeing of 0.7–0.8 arcsec) using the classical Hale telescope design.

At mid-twentieth century, however, telescopes had again reached an aperture and resolving-power limit, and for a generation, mountaintop 4- and 5-meter instruments were the best astronomers had to work with. Beginning in the 1980s, however, a spectacular series of technological advances produced a third species of reflecting telescope with 6- to 10-meter apertures, capable of HST-quality resolution over narrow fields of view. Appendix A6.2 gives the current list of the largest telescopes on Earth. This new ground-based species has by no means evolved to its fundamental limits, and we expect to see greater resolving power on even larger telescopes and in the near future. A 25–45-meter class

---

[4] Until late in the 1800s most of the great observatories of the world used telescopes conveniently located in university towns or near national capitals: Paris, Greenwich, Potsdam, Chicago. This gradually changed with the realization that better conditions existed at remote mountaintop locations like Mt. Hamilton (Lick Observatory, 1888) and Mt. Wilson (1904), both in California, Lowell Observatory (1894) in Arizona and Pic du Midi Observatory (first large telescope, 1909) in the French Pyrenees.

[5] George Ellery Hale (1868–1938) an astronomer and extraordinary administrator, founded the Kenwood, Yerkes, Mt. Wilson, and Palomar Observatories, and four times raised funds and supervised the construction of the largest telescopes in the world: the Yerkes 1-meter (40-inch) refractor, the Mt. Wilson 1.5-meter (60-inch) reflector, the 2.5-meter (100-inch) Hooker reflector, and the 5-meter (200-inch) Palomar reflector.

seems to be a realistic near-term goal, with some more speculative plans for 100-meter apertures.

6.4.1 Large mirrors

The mirror for the Hale 5-meter telescope has a finished weight of 14 tons. Made of Pyrex glass, it was designed to be rigid enough to maintain its paraboloidal figure in all orientations, resulting in a support structure (tube and mount) that has a moving weight of 530 tons.

The Hale mirror set the standard for an entire generation of telescopes. In this classical fabrication method, the mirror-maker pours a pool of molten glass into a cylindrical mold with a diameter-to-thickness ratio of about 6:1. The mold sometimes impresses a ribbed pattern on the back of the glass to reduce mirror weight while retaining stiffness, and to provide a method for attaching the mirror-support structure. After casting, the mirror must be annealed – gradually cooled at a rate slow enough to avoid stress in the glass due to thermal gradients. Improperly annealed glass can chip or shatter during the later stages of figuring, and the first 5-meter blank had to be rejected in 1934 because of poor annealing. The second 5-meter blank required 10 months to anneal, during which time the lab survived both a flood and an earthquake.

The front surface of the blank is next ground into a spheroid, and then polished into a conic with the desired focal ratio. For the Hale mirror (an $f/3.3$ paraboloid), this process required about 3 months, removed 5 tons of glass, and consumed 10 tons of abrasive.

A mirror larger than the Hale mirror cannot be made rigid enough, even with a massive support structure, to retain its shape in a moving telescope. Moreover, the classical fabrication method becomes very costly and time-consuming with increasing size (cooling and grinding time should scale as the second or third power of the mirror diameter). The current generation of large telescopes employs a new strategy of telescope design, and new methods of mirror fabrication. The strategy recognizes that any large mirror will be "floppy", and uses techniques known as ***active optics*** to adjust and maintain the mirror shape. Since mirrors are expected to be flexible, they can be low mass, which cuts cost, fabrication time, mount bulk, and thermal response time.

Large modern telescopes have used three different types of mirror:

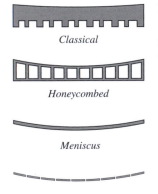

Classical

Honeycombed

Meniscus

Segmented

Fig. 6.10 Cross-sections of types of large astronomical mirrors.

> **Honeycombed monolithic mirrors** are an extension of the classical design, but with reduced mass and stiffness. Fabrication of these mirrors is greatly facilitated by a technique known as ***spin casting***. The idea here is to rotate the glass mold at a constant rate, so that the centrifugal effect forms the surface of the molten glass into a paraboloid. Spin casting greatly reduces or even eliminates the grinding phase, saving months or years of work, and makes it possible to fabricate very fast ($f/1.2$) surfaces simply by selecting the correct spin rate for the mold. The two 8.4-meter mirrors of the Large Binocular Telescope are spun-cast honeycombed mirrors.

Segmented mirrors are mosaics of several easily fabricated smaller mirrors arranged to produce a single large aperture. The primaries of the two 10-meter Keck telescopes on Mauna Kea, for example, each consist of 91 hexagonal segments. Each 1.8-meter segment is held in place by an active control system that constantly adjusts mirror positions to compensate for misalignments. The JWST will have a segmented primary

Thin meniscus mirrors have a diameter-to-thickness ratio of something like 40:1. They are usually spun-cast in a bowl-shaped mold. Unlike the honeycombed monolith or the individual segments of a mosaic, these mirrors have no ability to retain their shapes unless supported by an active cell. The Gemini 8.4-meter telescopes and the four 8.2-meter elements of the Very Large Telescope (VLT) use meniscus primaries.

6.4.2 Computers

I wake and feel the fell of dark, not day.
What hours, O what black Hours we have spent
This night! What sights you, heart, saw; ways you went!
And more must, in yet longer light's delay....
– Gerard Manley Hopkins (1844–1889),
I wake and feel the fell of dark, not day

The advent of inexpensive and powerful digital computers completely transformed the practice of observational astronomy. Without computers to monitor and adjust mirror shape, the large "floppy" mirrors of modern telescopes are useless. Without computers in control of the fabrication process, it is doubtful these mirrors could be made in the first place. Without computers to manage pointing, tracking, and instrument rotation, an altazimuth mount becomes a very tricky proposition. Without computers to command instruments and gather the data from a modern camera or spectrograph, the flow of information from even the largest telescope would choke off to a trickle. Without computers, elimination of the effects of atmospheric seeing on the ground would be impossible. Without computers, the HST and JWST would be utterly unthinkable.

Astronomers were quick, in the 1950s and 1960s, to utilize early "mainframe" electronic computers for the reduction and analysis of data, and for constructing theoretical models of astrophysical phenomena. Then, in 1974, the 3.9-meter Anglo-Australian Telescope became the first large telescope to use computer-controlled pointing and tracking. As the price-to-power ratio of mini- and micro-computers fell, digital electronics moved into observatories. The advent of CCD detectors in the 1980s meant that computers not only reduced data and moved telescopes and but also controlled instruments and acquired data.

In 1870 (or 1670) an observational astronomer woke to spend the night in the cold and dark, eye to ocular through the black hours, making occasional notes or calling measurements to an assistant. By 1970, little had changed, except things were sometimes a bit more gymnastic: still in the cold and dark, the astronomer used the ocular only to guide the telescope, perhaps while exposing a photographic plate. A frantic rush to the (blissfully warm) darkroom provided an occasional interlude when the photograph needed to be developed.

Today, a night at the telescope differs little from a day at the office: the warm room is brightly lit, and the astronomer types an occasional command at a computer console: move the telescope; change the filter; expose a CCD frame; start a pre-programmed sequence of moves, changes, and exposures. Digital data accumulate in computer storage, and if things are moving slowly, the astronomer can start reducing data as they arrive. Rewards are immediate, right there on the monitor. The telescope is in another room: cold, dark, and open to the sky. That room could be next door, but could also be on another continent. Or there might be no room at all, as the telescope orbits above. The old methods are exhausted, and discovery and adventure come with the new, but the price is reduced acquaintance with the fell of dark, and the exchange of exotic photons from the depths of space for mundane emissions from a monitor screen.

6.4.3 Observatory engineering

Several important principles govern modern telescope design. Except for the first listed below, none of these were part of the thinking that produced "classical" telescopes like the 5-meter Hale (1948) or the 4-meter Mayall (1973).

1. **The location of an observatory is crucial to its success**. Seeing, for example, is substantially better at high altitude on isolated islands like Mauna Kea and La Palma, and at the various sites in northern Chile. Better, at least, than it is at most other inhabitable places on Earth. Remote sites at high altitude have dark skies. Dry climates are important because clouds are so detrimental to optical work and because atmospheric transmission in the infrared is closely linked to total atmospheric water vapor content.

2. **Lightweight mirrors in a compact structure are cost effective**. Modern mirrors for large telescopes are lightweight, which means a smaller moving mass, which translates into lower cost and easier control. The moving weight of one of the Keck 10-meter telescopes is 300 tons, while the moving weight of the Hale 5-meter is 530 tons. Modern primary mirrors have fast focal ratios ($f/1.75$ for Keck, $f/3.3$ for Hale), which means a shorter telescope length and a smaller building. Modern mounts are altazimuth, and occupy less space than the classical equatorial, again producing a smaller (cheaper) enclosure. Although one Keck telescope has four times the light-gathering power of the Hale telescope, the observatory domes are about the same size. Finally, the reduction in mass makes it easier to maintain the telescope at the same temperature as the outside air.

3. **Active optics can improve image quality**. Active optics systems use motorized push-pull attachments to adjust mirror shape and position to optimize image quality. Such systems are a required part of any large telescope with a non-rigid mirror. Even a rigid mirror that does not flex in response to changing gravitational loads can benefit from an active system. Such a mirror will be subject to shape changes due to differential thermal expansion, and may well have optical imperfections polished into the glass itself.

4. **Local climate control can improve natural seeing**. Appreciable turbulence can exist inside the telescope shelter. A "dome" (in fact, many telescope enclosures are not dome-shaped, but astronomers use the term generically) and its contents will warm up during daylight hours. At night, when the air temperature drops, the dome itself or objects inside – like the telescope mirror, the tube, the mount, the floor – generate turbulence as they cool convectively. Even worse, a poorly designed observatory might have artificial heat sources near the telescope – a leak from another room, or non-essential power equipment in the dome itself. Structures that permit substantial airflow (e.g. with fans, louvers, retractable panels) while still protecting the telescope from wind buffeting can often improve seeing appreciably. For example, the WIYN 3.5-meter telescope, installed on Kitt Peak in 1994, has a low-mass mirror and active optics. The WIYN enclosure was designed to maximize airflow, and minimize seeing induced by thermal effects by the mirror, mount, and other structures. The WIYN delivered a median 0.7 arcsec seeing during 1995–1997. During this same period, the Kitt Peak Mayall 4-meter Telescope (dome and telescope both of the classical Palomar design), delivered median seeing of 1.1 arcsec.

5. **Novel focal arrangements can reduce costs of specialized telescopes**. Both 10-m telescopes of the Keck Observatory can combine their beams at a common focus. The four 8.2-m unit telescopes of the Very Large Telescope (VLT) at the European Southern Observatory can do the same, for a light-gathering power equivalent to a 16-m aperture. The Hobby-Eberly Telescope, in another example, has an 11-meter diameter spherical primary made of identical 1.0-meter hexagons, and is intended for spectroscopic use only. The optical axis is permanently fixed at an elevation angle of 55 degrees. To observe, the entire telescope structure rotates to the desired azimuth and stops. The system tracks during the exposure (limited to a maximum of 2.5 hr, depending on declination) by moving an SA corrector and the detector along the focal surface. Most, but not all, of the sky is accessible. This design produces substantial savings due to the tiny moving weight during exposures and the invariance of the gravitational load on the primary.

6. **Adaptive optics can eliminate some of the effects of atmospheric seeing**. With a perfect telescope at a very good site, if local turbulence in minimized, image quality is largely determined by uncontrollable turbulence in the upper atmosphere. It is possible to reduce the effects of this kind of seeing by quickly adapting the shape of an optical element to undo the distortions caused by the atmosphere. Ground-based telescopes with adaptive optics have in fact attained image quality to match the HST over narrow fields of view.

Fig. 6.11 Circulation and turbulence in the Earth's atmosphere. (a) The basic circulation pattern set up by convection above a warm surface. (b) The circulation modified by turbulence. On Earth, turbulence is usually strongest within 1 km of the ground, and has a weaker peak near the tropopause – the upper limit of the convective region. Large local values of wind shear in the boundary layer can also produce appreciable turbulence.

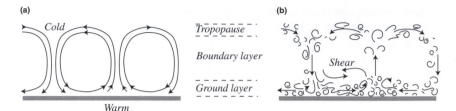

6.5 Adaptive optics

Long telescopes may cause Objects to appear brighter and larger than short ones can do, but they cannot be so formed as to take away that confusion of Rays which arises from the Tremors of the Atmosphere. The only Remedy is a most serene and quiet Air, such as may perhaps be found on the tops of the highest Mountains above the grosser Clouds.

— Isaac Newton, *Opticks*, 1704

6.5.1 Atmospheric wavefront distortion

For nearly four centuries, Newton's judgment about astronomical seeing held true, but there are now technological remedies for unquiet air. We discussed wavefront distortions and atmospheric seeing in Chapter 5, but now wish to examine these in a more quantitative way. The index of refraction of air depends on its density (and the wavelength of light – see Table 5.3). In a perfectly serene and quiet atmosphere, the density and index will depend only on altitude, and every point at the same height will have the same index.

In the real and imperfect atmosphere, however, solar heating drives convective cells in the lowest layer of the atmosphere, a region about 10–12 km thick called the *troposphere*. Here, one mass of air in contact with the surface can become slightly hotter and thus more buoyant than its neighbors. That mass rises. Another moves horizontally to fill its place; cold air from above drops down to make room for the rising mass and completes the circulation around a cell. Many cells are established, and the air, especially at the boundaries of the flow, tends to break up into ever smaller eddies and lumps of different density (i.e. temperature) – this break-up of the flow is turbulence. See Figure 6.11.

Now consider a wavefront from a distant star that passes through the Earth's turbulent atmosphere. The front arrives as a plane, but different parts will encounter slightly different patterns in the index of refraction. Each ray from the front will traverse a slightly different optical path, and the wavefront after passage will no longer be a plane, but will be distorted with dents and bumps corresponding to larger or smaller total optical path lengths through the atmosphere. Since the turbulent lumps and eddies at each altitude move at the local wind speed, the distortion in the wavefront changes very quickly.

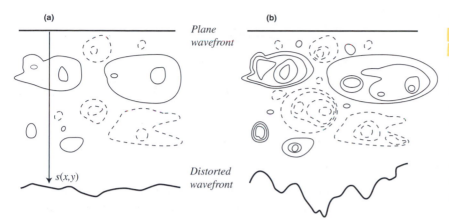

Fig. 6.12 Distortion of wavefronts by a turbulent atmosphere. The figure shows plane wavefronts incident on turbulent layers, and the distortion that results. Solid contours indicate lower temperatures; dotted contours indicate higher temperatures. The distortion in (b) is much greater than in (a).

Figure 6.12 gives an illustration, where we follow two downward-traveling fronts in cross-section. The fronts pass through turbulent layers, represented by contours mapping lower-than-average or higher-than-average temperatures and indices of refraction. The left-hand wavefront, the weaker-turbulence case, experiences a mild distortion, while the right-hand front becomes rather badly crumpled. Each part of a wavefront traverses some optical path length, $s(x, y)$. If s does not change rapidly with x and y, there is little distortion; if it does change rapidly, there is great distortion.

In Figure 6.13, we suggest a method for quantifying the wavefront distortion. First consider a method that works in one dimension (the x-direction). Start at one end of the wavefront ($x = 0$) and fit a straight line to a segment of the front. How long can this segment be before the fit becomes "poor?" We need a criterion for judging goodness of fit, and we choose the root-mean-square difference (in the z-direction) between the front and the fit. If this quantity becomes greater than $\lambda/2\pi n$ (about 1/6 of a wavelength), then the fit is poor. This is equivalent to saying that the RMS deviation of the phase, ϕ, of the wave is less than one radian. The maximum length that can be fitted well is r_1, which we will call the **coherence length** of the first segment. Now move along the front, fitting successive segments with "good fit" straight lines of length r_i. The statistical mean of all the r_i values s is r_{avg}, the **coherence length of the wavefront**. Each segment has a different slope, so each will propagate in a slightly different direction, and each will focus at a different spot in the image plane of a telescope (review Figure 5.21). The shorter the coherence length, the more speckles in the image.

Now we extend the idea of coherence length to two dimensions. Suppose we select a random point on a two-dimensional (2-D) wavefront, and ask how large a 2-D patch of the front we can expect to be coherent. The answer is given by **Fried's parameter**, $r_{0\lambda}$.

$r_{0\lambda}$ = the expected diameter over which the root-mean-square optical phase distortion is 1 radian

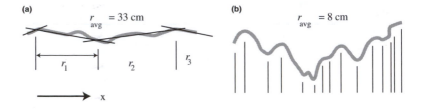

Fig. 6.13 The coherence length of a 1-meter section of a wavefront. Within each segment of length r_i, the variation in optical phase is less than one radian. Fronts (a) and (b) are the ones sketched in Figure 6.12. The scale is greatly exaggerated in the x-direction.

Note that Fried's parameter is a statistical description of how the optical path length varies across the wavefront. Nevertheless, you will find it most useful to regard $r_{0\lambda}$ as a measure of how large a segment of the wavefront you can expect to treat as a plane wave. Fried's parameter is a good indicator of image quality. In particular, the full-width at half-maximum (FWHM) of the seeing disk is given by

$$\theta(\text{seeing, in arcsec}) \approx 0.2 \frac{\lambda[\mu m]}{r_{0\lambda}[m]} \qquad (6.10)$$

If the diameter of a telescope, D, is larger than $r_{0\lambda}$, then Equation (6.10) gives the image size. If $D < r_{0\lambda}$, then the telescope is **diffraction limited** and the image size is given by the Airy disk, Equation (5.13). (Seeing can still move the diffraction-limited image location around in the focal plane, however.)

Since the index of refraction of air depends only weakly on wavelength, the same must be true of the optical path length, $s(x, y)$. The variation in phase produced by a variation in s is inversely proportional to wavelength. The Fried parameter is therefore a function of wavelength:

$$r_{0\lambda} = r_0 \left(\frac{\lambda}{0.5\mu m} \right)^{\frac{6}{5}} (\cos \zeta)^{\frac{3}{5}} \qquad (6.11)$$

Here ζ is the angle between the direction observed and the zenith. The Fried parameter is usually quantified simply by quoting its value at a wavelength of 500 nanometers, r_0. Values for r_0 generally range from a few centimeters (poor seeing) to 15 or 20 cm (superb seeing). Whatever the value of r_0 at a site, $r_{0\lambda}$ will be larger (and the size of the seeing disk will be somewhat smaller) for observations made nearer the zenith as well as for observations made at longer wavelengths. The Fried parameter can fluctuate by large factors over time, on scales of seconds or months.

6.5.2 The idea of adaptive optics

Adaptive optics (AO) aims at removing wavefront distortions by inserting one or more adjustable optical elements into the path between source and detector. In practice, the adjustable elements are usually reflective, and usually located near the telescope focal plane. The shape of the reflecting surface is

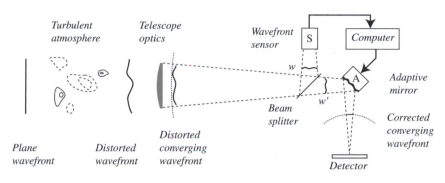

adjusted to exactly cancel distortions generated by atmospheric turbulence. Figure 6.14 is a schematic representation of the AO concept. In the figure, a partially reflecting mirror splits the distorted wavefront in the telescope into two fronts, w and w'. These fronts have identical distortions. Front w proceeds to a sensor, S, which detects the magnitude of its distortion at some number of locations on the front. The other front, w', is reflected from an adjustable mirror, A, onto the detector. Meanwhile, the computer has read the distortions sensed by S, and commands A to adjust the shape of its surface so as to exactly cancel them. If all goes perfectly well, the compensated image formed at the detector will be diffraction-limited, with all effects due to the atmosphere removed.

Things seldom go perfectly well. One measure of how well an AO system succeeds is the **Strehl ratio**, R_S. For a point source detected by a particular telescope, the run of intensity with angular distance from the image center is called the **point-spread function**. If I_{PSF0} is the peak intensity of the point-spread function, and I_A is the peak intensity of the Airy function for the same source, then

$$R_S = \frac{I_{PSF0}}{I_A}$$

The idea here is that an AO system should transfer intensity from the outer part of the seeing disk to the core, and increase R_S over the uncompensated value. A perfect AO system will produce $R_S = 1$, and should improve the Strehl ratio by a factor of $(D/r_0)^2$.

6.5.3 The Greenwood time delay

The idea of adaptive optics is simple. Its execution is not. The books by Roddier (1999) and Hardy (1998) provide a full discussion. We can only examine a few concepts here. One has to do with the time element.

Clearly, any time delay between sensing the wavefront and adjusting the deformable mirror is a problem. The maximum delay that can be tolerated will depend on the size and velocity of each distortion in the atmosphere. Statistically,

Fig. 6.15 Anisoplanatism. Plane wavefronts a and b arrive from sources separated by angle θ. They experience different distortions, a' and b'. An AO compensation for a' will be incorrect for b'. Turbulence in the upper atmosphere limits isoplanism more severely than does turbulence in the lower atmosphere.

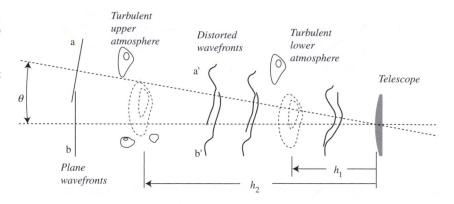

if $\bar{v}$ is the weighted average velocity of turbulent features, then the amount of delay, in seconds, before a wavefront error of 1 radian accumulates is (λ in μm)

$$\tau = 0.559 \frac{r_0}{\bar{v}} \lambda^{\frac{6}{5}} (\cos \zeta)^{\frac{3}{5}}$$

For typical values of the parameters, the Greenwood time delay is several milliseconds. Practical AO systems must not only respond within the Greenwood time, but must also update the shape of correcting elements at at least the Greenwood frequency, $1/\tau$. The wavelength dependence of τ is one (of many) reason(s) why AO is a lot easier at longer wavelengths.

6.5.4 Anisoplanatism

Figure 6.15 illustrates a serious limitation of simple AO systems. A turbulent layer is some height h above a telescope. Rays from two sources, separated in the sky by angle θ, traverse the turbulence along different paths. The layer introduces different wavefront distortions for the two sources. For very small values of θ, the distortions will not differ greatly, but if $\theta \geq r_{0\lambda}/h$ the phase distortions will be uncorrelated. The isoplanatic angle, θ_i, in radians, is the largest angle for which the expected distortions differ by less than one radian:

$$\theta_i = 0.314 \frac{r_{0\lambda} \cos \zeta}{\bar{h}}$$

Since turbulence generally occurs at different heights, $\bar{h}$ represents a weighted mean height. For a typical site, the isoplanatic angle ranges from around two seconds of arc in the blue to around 30 seconds of arc at 10 μm. This means that the compensated field of view – the isoplanatic patch – is very limited. Note also that ground-level turbulence (height h_1 in the figure) does not restrict the size of the compensated field as seriously as motions in the upper atmosphere. Again, the wavelength dependence means that AO systems will be more useful in the infrared than in the optical.

6.5.5 Guide stars

Another serious limitation for AO arises from the obvious condition that AO only works if it can sense the distortions in a wavefront. This wavefront is usually (but not always) from a point source – the *guide star*. Since exposure times for the sensor must be less than the Greenwood delay time, the guide star must be bright. If the science source is faint, and if there is no *natural guide star* of sufficient brightness closer to it than the isoplanatic angle, then AO cannot compensate the image of the source. Guide stars can be much fainter in the infrared than in the optical (τ is longer in the infrared), and the isoplanatic angle is much larger. Once again we see that AO is easier in the infrared than in the optical – at longer wavelengths it is much more likely that an acceptable natural guide star will fall in the same isoplanatic patch as the science source.

Even in K band, though, it is difficult to find suitable natural guide stars. One possible solution is to use an artificial Earth satellite beacon that can be positioned in the isoplanatic patch, but this has severe practical difficulties. A more promising technique is the *laser guide star*. The basic idea is to use a laser to illuminate a small spot in the upper atmosphere well above the turbulence layer.

There are two current methods for implementing a laser guide star. The first uses a pulsed laser to illuminate a narrow column of air and observe the back-scattered light (Rayleigh scattering). The wavefront sensor observation is also pulsed, so that the altitude of the illuminated spot can be selected by adjusting the time between pulsing the laser and making the observation. Maximum altitude for Rayleigh laser beacons is about 20 km because of the exponential drop in air density with height.

The second method depends on a curiosity of the Earth's atmosphere – the presence of a 10-km thick layer in the mesosphere (90 km up) with an unusually high concentration of neutral sodium and potassium atoms, probably of meteoritic origin. A laser on the ground near the telescope is tuned to one of the sodium D lines (589.00 or 589.59 nm) and fired to pass through the mesospheric layer at the desired position. The laser light excites sodium atoms, which in turn emit line radiation by spontaneous emission (after about 10^{-8} seconds), with most of the emission concentrated in the sodium layer. For astronomy, sodium beacons are superior to Rayleigh beacons in that their higher altitude permits a more accurate replication of the ray path from the science object.

Systems to produce and utilize a laser beacon are expensive and difficult to implement, but they do allow AO to operate where natural guide stars are unavailable.

6.5.6 Wavefront correctors

Wavefront correctors in current astronomical AO systems are almost always small mirrors. These deformable mirrors have been of several types, but all must

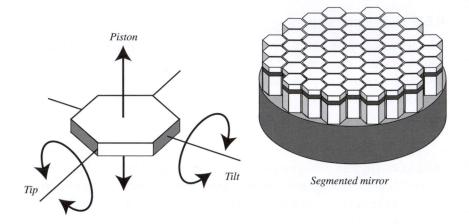

Piston

Tip

Tilt

Segmented mirror

have surfaces whose shape can quickly (within milliseconds) adjust to compensate for sensed distortions in the incoming wavefront. For example, Figure 6.16 shows a segmented mirror made up of 54 independent flat reflectors. In operation, the AO system senses the wavefront distortion, then very quickly positions each of the segments to exactly cancel the pathlength errors in each portion of the wavefront. Each hexagonal segment is a low-mass mirror that can execute three types of motion: piston, tip, and tilt. The number of segments needed in a deformable mirror depends on the coherence length of the wavefront, and should scale roughly as $(D/r_0)^2$.

One or more ***actuators*** adjust each segment. An actuator converts an electrical signal into a change in position, and most current actuators rely on the piezoelectric effect. Certain polarized ceramic materials respond to an imposed electric field by changing dimension, and have a relatively fast response time. In the pure piezoelectric effect, the change in dimension is directly proportional to the voltage drop across the material, as illustrated in Figure 6.17a. An array of actuators (three per segment) can define the piston, tip, and tilt for each component of a segmented mirror, as suggested in Figure 6.17b.

Segmented mirrors have the advantage of being made up of many identical and easily replaceable components. They have the disadvantage of diffraction and scattering effects produced by the gaps between segments, as well as requiring a relatively large number of actuators. Continuous-surface deformable mirrors eliminate these problems, and have seen the most use in astronomical AO. One design bonds a flexible thin reflecting glass, quartz, or silicon face-sheet to an array of actuators (Figure 6.17c). A second design is the bimorph mirror, which consists of two piezoelectric disks with local electrodes bonded to the back of a flexible reflecting surface (Figure 6.17d). Voltage applied at a particular location causes local material in one disk to expand while the material in the other disk contracts, producing surface curvature.

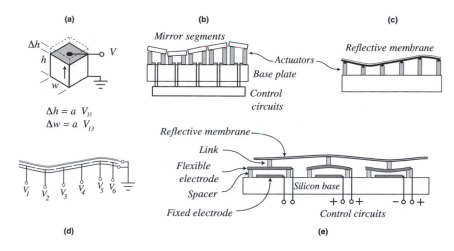

Fig. 6.17 Mirror actuators. (a) A piezoelectric stack, which will change its vertical dimension in response to an applied voltage. (b) Piezoelectric actuators control the tip-tilt of each element of a segmented mirror. (c) A smaller number of actuators are bonded to a thin monolithic reflector. (d) A bimorph actuator whose curvature is controlled by adjusting the local tension/compression of the bimorph pair. (e) A cross-section of a MEMS deformable mirror.

Deformable mirrors that are micro-electronic machined systems (MEMS) are potentially very inexpensive. Employing the lithography methods of the electronics industry, MEMS technology fabricates electrodes, spacers, and electrostatic actuators on a silicon chip, and bonds a flexible reflecting membrane to the surface of the device. The primary disadvantage is the limited size of the resulting mirror.

A deformable mirror can be the secondary mirror in a reflecting telescope – an advantage in simplicity. Mirrors this large, however, would be very expensive. More often, the deformable element is located behind the focus, often at an image of the primary.

Most practical AO systems correct the wavefront in two stages. One flat mirror corrects the overall **tip–tilt** of the entire wavefront, and a deformable mirror corrects the remaining distortions. Although this arrangement is more complex, it has important advantages, including a reduction in the overall piston motion required by the elements of the deformable mirror. (Both piezoelectric actuators and MEMS electrostatic actuators have limited piston motion.)

6.5.7 Wavefront sensors

A wavefront corrector is only useful if the AO system is able to measure the distortions that need compensation. There are at least a half-dozen techniques for performing this function. Here we will describe only one, the **Shack–Hartmann sensor**, because it has seen wide application in astronomy and is relatively easy to understand.

The principle of the Shack–Hartmann sensor, illustrated in Figure 6.18, is that rays propagate perpendicular to the surface of the wavefront, so that sensing the direction of a ray is equivalent to sensing the slope of the wavefront. In Figure 6.18a, we sketch a Shack–Hartmann sensor operating on a wavefront that

Fig. 6.18 The Shack–
Hartmann sensor. (a) The
beam from a perfect
point source – all images
on the sensors are in the
null position. (b) A
distorted wavefront and
the resulting image
displacements from tilted
segments.

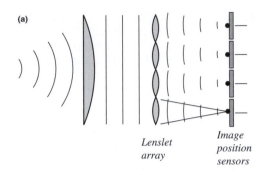
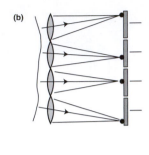

*Lenslet
array*

*Image
position
sensors*

was a perfect plane when it was incident on the aperture. The device is located behind the focus, where a ***field lens*** converts the diverging spherical (we assume the telescope optics are perfect) wavefront back into a plane. An array of lenses then separates the rays from each segment of the front into isolated bundles, called ***sub-apertures***, each of which is brought to a different focus on a detector array.

Now, in Figure 6.18b, consider what happens if the original wavefront is distorted. The distortion causes local changes in the wavefront slope and corresponding changes in directions of the rays. Each bundle of rays now comes to a focus whose location depends on the slope of the wavefront. The detector array senses the position of the focused spot behind each small lens, and a computer can determine the corresponding value for the wavefront slope in the x and y directions in each sub-aperture. Note that the guide object need not be a point source: the system only requires some image whose relative displacement can be determined in each sub-aperture.

But one more step remains before we can command a deformable mirror. Figure 6.19 shows the process of ***wavefront reconstruction***. Our computer must convert the array of slopes for each sub-aperture into the actual shape of the distorted front. Obviously some integration-like algorithm is necessary, and the computations generally require a mathematical matrix inversion. For a large number of sub-apertures, this requires a very fast computer, since adaptive corrections must be made within the Greenwood time.

6.5.8 Practical AO systems

Designing and building a practical AO system for astronomy is a very complex undertaking, one so expensive that astronomers can usually justify it only for the largest of telescopes, where the cost of the AO system shrinks to a reasonably small fraction of the overall budget. Figure 6.20 is a very schematic view of the optical layout of a practical AO system. The system is located behind the telescope focus, where a parabolic mirror converts the diverging spherical wavefront to a plane before the un-compensated front encounters the

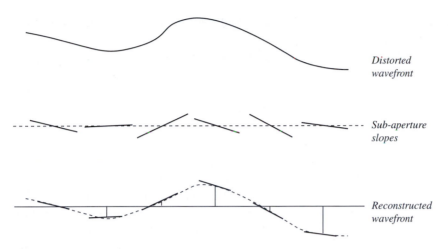

Fig. 6.19 Wavefront reconstruction

Distorted wavefront

Sub-aperture slopes

Reconstructed wavefront

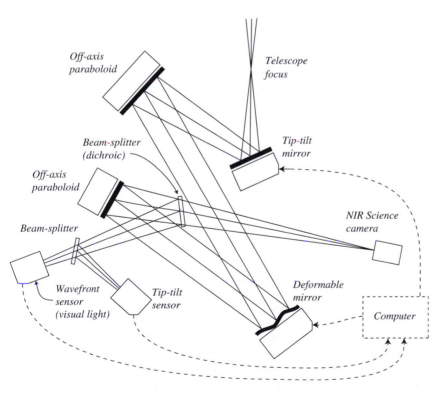

Fig. 6.20 Schematic of a practical AO system.

Off-axis paraboloid

Telescope focus

Beam-splitter (dichroic)

Tip-tilt mirror

Off-axis paraboloid

NIR Science camera

Beam-splitter

Deformable mirror

Wavefront sensor (visual light)

Tip-tilt sensor

Computer

deformable mirror. A second paraboloid then focuses the compensated front on the detector.

The illustration shows a separate system to correct for errors in the overall position of the image – errors that can arise not only from seeing, but also from telescope drive imperfections. A sensor commands the tip and tilt orientation of a conventional mirror to maintain the centroid of the image at the same spot on the detector. This separate tip–tilt correction minimizes the amount of correction (and therefore actuator motion) that the deformable mirror needs to make.

Another important feature of the practical system is the location of the wave-front sensor. The sensor examines the front *after* the tip–tilt and the deformable mirrors have corrected it, an arrangement termed ***closed-loop*** operation. This means that the imperfections in the image should be relatively small if the adaptive cycle is shorter than the Greenwood time. The task of the AO computer is to null out only wavefront errors introduced (or residual) since the last command cycle. This requires that the null point be well calibrated, but the range of motions directed by the feedback loop will be relatively small, and any errors in compensation will tend to be corrected in the next cycle.

Also note the use of a dichroic beam-splitter – this is a special mirror that passes infrared light but reflects the optical. Since optical path-length deviations introduced by the atmosphere are independent of wavelength, this means that the NIR image can be corrected by sensing the distortions in the optical image. This allows all the infrared light to be directed to the image, while all the optical light is used for wavefront sensing.

6.6 The next stage: ELTs and advanced AO

> Perhaps more excellent things will be discovered in time, either by me or by others with the help of a similar instrument, the form and construction of which . . . I shall first mention briefly. . . .
>
> – Galileo Galilei, *Sidereus Nuncius*, 1610

If adaptive optics can produce Strehl ratios close to unity, then the payoff in detection and resolution limits for large apertures is governed by the equations for space telescopes (Section 6.3), so a 40-meter telescope should detect stellar sources 100 times fainter than an AO-equipped 4-meter, and resolve six times finer detail than the JWST. With AO, an extremely large telescope (ELT) on the ground thus becomes a very attractive proposition.

As of this writing (2010) three multinational groups worldwide are construct-ing single-aperture telescopes much larger than the current crop of 8–11 meter instruments. All three projects are expected to attain first light in around the year 2018, and Table 6.1 gives some details for each. A half-dozen other equally

Table 6.1. *ELT projects currently under way*

Project	Effective aperture (m), number of segments, and aperture (m)	Name, major partners, URL
E-ELT	42, 1000 × 1.4	European Extremely Large Telescope, European Southern Observatory (13 member nations), http://www.eso.org/sci/facilities/eelt/
GMT	24.5, 7 × 8.4	Giant Magellan Telescope, nine partner institutions from the USA, Australia, and Korea, http://www.gmto.org/
TMT	30, 492 × 1.40	Thirty Meter Telescope, California Institute of Technology, University of California, Association of Canadian Universities, National Astronomical Observatory of Japan, http://www.tmt.org/

ambitious and more ambitious projects are in the early planning stages. Note that no technology exists for casting monolithic mirrors larger than 8.4 m, so any ELT must use segmented mirrors.

Each of the three ELTs in progress will cost in the range of one billion dollars – a price that essentially requires multi-institutional, multinational cooperation. Doing ELT astronomy is similar to doing astronomy with space missions like the JWST and quite unlike the historic single astronomer/single telescope practice. In addition to funding and cooperation, an ELT's success will depend on solving two key technological problems:

- A telescope structure with sufficient stiffness to withstand wind loading, equipped with an active optics system that will maintain segment positioning and alignment in the face of inevitable deformations.
- An advanced adaptive optics system that is capable of producing a large Strehl ratio over a large isoplanatic patch.

Some of the advances needed in AO systems are well under way. First, it is obvious that simply scaling up current AO techniques would require improvements in deformable mirrors (the number of elements needed in a deformable mirror depends on the telescope aperture squared) and processing speed. The MEMS deformable mirrors seem to be the most promising technology here, since multiplying the number of actuators with photolithographic methods is

Fig. 6.21 The MCAO (multiple conjugate adaptive optics) concept. For clarity, the corrected beam for the science image is omitted. Two separate guide stars – either natural or laser – provide beams that probe atmospheric turbulence in different directions. Corresponding wave-front sensors (WFS) measure the wave-front errors in each of these directions. The locations of the two deformable mirrors (DMs) determine the layer (conjugate) corresponding to the distorted wavefront (WF) they should compensate. Practical systems may use more than two guide stars and DMs.

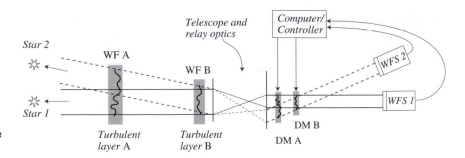

relatively inexpensive. Refer to the websites of the major MEMS manufacturers for the most current capabilities (e.g. www.bostonmicromachines.com, www.alpao.fr, and www.irisao.com)

Planning for ELTs has spurred progress on methods for increasing the size of the isoplanatic patch. The use of multiple guide stars in a technique called multiple conjugate adaptive optics (MCAO) would allow for the computation of the wavefront distortion present at different layers in the atmosphere – a kind of tomography of the turbulence structure. With this information separate deformable mirrors can apply different corrections for beams incident from different directions. Figure 6.21 illustrates a simple example, and the ELT websites given in Table 6.1 provide introductions to some of the more elaborate advanced AO systems in planning and development stages.

Summary

- Ground-based telescopes usually utilize either an equatorial or an altazimuth mount. Each has advantages. Concepts:
  setting circle   *clock drive*   *guide star*
  instrument rotator
- Telescopes in space use rockets and/or reaction wheels to point and guide.
- Prime focus and Newtonian optical layouts with a parabolic primary mirror suffer from coma.
- Cassegrain and Gregorian layouts use two curved mirrors, are generally preferred because of compactness and convenience, and can be made aplanatic. Concepts:
  back focal distance   *Ritchey–Chrétien*
- Simple algebraic relationships (Equations (6.2)–(6.7)) govern the design of Cassegrain and Gregorian telescopes.
- The Nasmyth focus location is fixed relative to the mount, and the coudé focal location is fixed relative to the ground.

- Catadioptric telescopes utilize both reflecting and refracting optics. The Schmidt telescope, a very useful wide-angle camera, combines a spherical mirror, aperture stop, and a transmitting corrector plate to remove all third-order aberrations except curvature of field. The Schmidt–Cassegrain is a modification popular in the amateur market.

- A space telescope is generally superior to ground-based telescopes of the same aperture in its ability to resolve detail and to detect faint objects (Equations (6.8) and (6.9)). Concepts:

Airy disk	*seeing*	*airglow*
zodiacal light	*HST*	*JWST*

- Since 1980, ground-based telescopes have achieved very large apertures (6 to 10 meters) and an ability to compensate for atmospheric seeing. Concepts:

5-m Hale telescope	*floppy mirror*	*active optics*
honeycombed mirror	*segmented mirror*	*spin casting*
meniscus mirror	*local climate control*	

- Adaptive optics (AO) technology is based on three components: a wavefront sensor, an adjustable mirror, and a computer that quickly interprets the sensor output and adjusts the mirror shape to cancel atmospheric distortions in the wavefront. Concepts:

turbulence	*tropopause*	*boundary layer*
coherence length	*Fried's parameter*	*point-spread function*
Strehl ratio	*Greenwood time*	*isoplanatic angle*
natural guide star	*laser guide star*	*tip–tilt correction*
closed-loop operation		

- Practical deformable mirrors are either segmented or continuous-surface. Actuators may be based on piezoelectric elements or MEMS technology.

- One form of wavefront sensor is the Shack–Hartmann device, which depends on intercepting the guide object's wavefront with multiple apertures.

- Astronomers expect telescopes with apertures in the 24- to 42-meter range before the year 2020. Efficient use of telescopes of this size requires significant advances in adaptive optics.

Exercises

1. Describe the kind of motion an altazimuth mount must execute to track a star through the zenith. Describe the motion of an instrument rotator at the Cassegrain focus during such a maneuver. Assume an observatory latitude of 45 degrees, and only consider tracking close to the zenith.

2. Investigate and provide an image of, or sketch, the different configurations of the equatorial mount known as (a) German (universal or Fraunhofer), (b) English,

(c) horseshoe, and (d) open fork. The book by King (1979) is a good source for older illustrations.

3. Investigate and describe the operation of the telescope mountings known as (a) a siderostat and (b) a heliostat.

4. A prime focus telescope with a parabolic mirror will be used at a site where the best seeing is 0.9 arcsec. What is the limiting radius of the field of view for which the comatic blur is smaller than the best seeing disk? Your answer will depend on the focal ratio of the primary, so compute the radius for $f/2.5$, $f/8$, and $f/10$. Show that for the $f/8$ telescope, the blur due to astigmatism at the edge of this coma-limited field is much smaller than the seeing disk.

5. Compare the tube lengths of the following telescopes, all of which have 1.0-meter apertures.
   (a) an $f/10$ refractor,
   (b) an $f/10$ Schmidt,
   (c) an $f/2.5$ Schmidt,
   (d) an $f/10$ classical Cassegrain with an $f/3$ primary mirror and final focus that is 20 cm behind the vertex of the primary (length is the distance from the secondary to the final focus),
   (e) an $f/10$ Cassegrain with an $f/2$ primary mirror and final focus that is 20 cm behind the vertex of the primary.

6. A Gregorian telescope and a Cassegrain telescope have identical primary mirrors, back focal distances, and final focal lengths (i.e. $\beta_{Cass} = \beta_{Greg}$, $m_{Cass} = -m_{Greg}$) Using the definitions in Equations (6.3) show that the difference between the lengths of two tubes is proportional to

$$\frac{|m|}{m^2 - 1}$$

7. An $f/7.5$ 4-meter RC telescope has an $f/2.5$ primary and a final focus 25 cm behind the vertex of the primary. Compute conic constants of the primary and secondary mirrors, and the diameter of the secondary.

8. A classical Schmidt telescope is being designed to have a 1-meter diameter aperture stop and an un-vignetted field of view of 10 degrees diameter.
   (a) Compute the diameter of the primary mirror, D_p, if the focal ratio of the system is $f/3.5$.
   (b) Compute the diameter of the primary if the focal ratio of the system is $f/1.7$.
   (c) Ignoring the effect of any central obstruction, show that the image of a star will be dimmer at the edge of the field that at center by a factor of cos 5°.

9. The detection threshold of the HST (aperture 2.4 meters) for a certain application is $m = 26.0$. What is the magnitude threshold for the same application for the JWST (aperture 6.5 meters)? Assume both telescopes are diffraction limited and the background for the JWST is 1.0 magnitudes per square arcsec fainter than for the HST. (Caution: recall the relation between magnitude difference and flux ratio.)

10. Again, assume the detection threshold of the HST (aperture 2.4 meters) for a certain application is $m = 26.0$. What is the magnitude threshold for the same application of

an ELT with a 24-meter aperture? Assume the background for the ELT is 2 magnitudes per square arcsec brighter than for the HST.

11. A particular site has a median Fried parameter of 10 cm during the month of August.

   (a) Estimate the expected FWHM of the uncompensated seeing disk and the Strehl ratio, R_S, in U band, I band and K band. (Hint: how should the Strehl ratio scale with seeing disk diameter?)

   (b) Above what wavelength will the images in an optically perfect 1-meter telescope be unaffected by turbulence?

   (c) For an AO system in K band, for this situation, compute the Greenwood time delay imposed by ground-layer turbulence alone (average velocity 10 km hr^{-1}) and by tropospheric turbulence alone (average velocity 120 km hr^{-1}).

12. On the Internet, investigate the technique called "ground-level conjugate adaptive optics," and explain why removing the effects of only ground-layer turbulence should improve the seeing disk over a relatively wide field of view.

Chapter 7
Matter and light

Because atomic behavior is so unlike ordinary experience, it is very difficult to get used to, and it appears peculiar and mysterious to everyone – both to the novice and to the experienced physicist. Even experts do not understand it the way they would like to, and it is perfectly reasonable that they should not, because all of direct human experience and of human intuition applies to large objects.

– Richard Feynman, *The Feynman Lectures on Physics*, 1965

Chapter 1 introduced the situations that produce line and continuous spectra as summarized by Kirchhoff's laws of spectrum analysis. This chapter descends to the microscopic level to examine the interaction between photons and atoms. We show how the quantum mechanical view accounts for Kirchhoff's laws, and how atomic and molecular structure determines the line spectra of gasses.

To understand modern astronomical detectors, we also turn to a quantum mechanical account – this time of the interaction between light and matter in the solid state. The discussion assumes you have had an introduction to quantum mechanics in a beginning college physics course. We will pay particular attention to some simple configurations of solids: the metal oxide semiconductor (MOS) capacitor, the p–n junction, the photo-emissive surface, and the superconducting Josephson junction. Each of these is the physical basis for a distinct class of astronomical detector.

7.1 Isolated atoms

7.1.1 Atomic energy levels

A low-density gas produces a line spectrum, either in absorption or emission, depending upon how the gas is illuminated (review Figure 1.7). The formation of lines is easiest to understand in a gas composed of single-atom molecules, like helium or atomic hydrogen. Consider the interaction between a single atom and a single photon. In either the absorption or the emission of a photon, the atom usually changes the state of one of its outermost electrons, which are therefore termed the *optical electrons*. The same electrons are also called the *valence electrons*, since they largely influence an atom's chemical properties by participating in covalent and ionic bonds with other atoms.

Observations of atomic spectra and the theory of quantum mechanics both demonstrate that the energy states available to any bound electron are *quantized*. That is, an electron can only exist in certain permitted energy and angular-momentum states. In theory, these permitted states arise because an electron (or any other particle) is completely described by a *wave function*. In the situation in which the electron is bound in the potential well created by the positive charge of an atomic nucleus, the electron's wave function undergoes constructive interference at particular energies, and destructive interference at all others. Since the square of the wave function gives the "probability density" of the electron existing at a certain location and time, the electron cannot have energies that cause the wave function to interfere destructively with itself and go to zero. Physicists call these *forbidden* states. In the isolated atom, most energies are forbidden, and the energies of the rare *permitted* states are sharply defined.

Figure 7.1a illustrates the permitted energy levels for a fictitious atom, which appear as horizontal lines. In the figure, energy, in units of electron-volts, increases vertically. (One electron-volt (eV) is the energy gained by an electron accelerated by a potential difference of one volt. 1 eV = 1.6022×10^{-19} J.) There are seven *bound states*, labeled a – g, . . . , in this particular atom. (Real atoms have an infinite number of discrete states – see below.) These different energy levels correspond to different configurations or interactions of the outer electrons. The idea here is that the atom must exist in one of these permitted energy states. The lowest energy state, the one assigned the most negative energy, is called the *ground state* (level a in the figure). This is the configuration in which the electrons are most tightly bound to the nucleus,

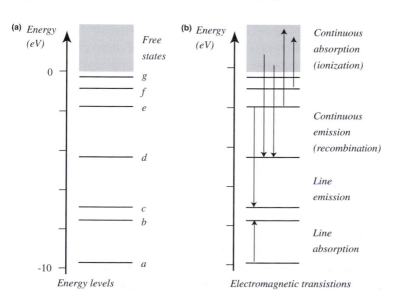

Fig. 7.1 (a) Permitted energy levels for an electron in a hypothetical atom that has seven bound states (a–g). The most tightly bound states (lowest energy) correspond to an electron location closer to the nucleus. (b) Absorption or emission of photons. The probabilities of different transitions can be vastly different from one another.

and would be the state of an undisturbed atom at zero temperature. Above the ground state are all other permitted *excited* states, up to the *ionization level*. The ionization level, conventionally assigned zero energy, corresponds to an atom that has so much internal energy that an electron is just able to escape. In that situation, the free electron is no longer part of the atom, and the remaining positive ion will have internal energy states described by a completely different diagram. Because the wave function of the free electron in a vacuum is not restricted by variations in potential energy, the energy of the free electron is not quantized.

You can think of bound states with higher energies as situations in which the optical electrons are *on average* physically further away from the nucleus. Be aware, though, that the vision of electrons orbiting the nucleus like planets in the Solar System (i.e. the early Bohr theory) is limited in its usefulness. The best answer to the question "where is this electron?" is a function that says certain locations are more likely than others, but, unlike the energy situation, a rather broad and sometimes complicated range of positions is possible for each bound state.

7.1.2 Absorption of light by atoms

Even though we can't see the positions of an atom's electrons, we *can* measure their energies when light interacts with atoms. Remember that a photon carries a specific amount of energy that is directly proportional to its frequency, v:

$$E = hv = \frac{hc}{\lambda} \tag{7.1}$$

The atom can make a transition from one bound state to another by either absorbing (the process is called *photo-excitation*) or emitting a photon of the correct frequency or wavelength, as illustrated in Figure 7.1b. In the process of photo-excitation, the photon is truly absorbed, and ceases to exist. The figure shows a photo-excitation transition from the ground state (level a, which has energy E_a) to the first excited state (level b, which has energy E_b). The photon responsible for this transition must have wavelength

$$\lambda_{ab} = \frac{hc}{\Delta E_{ab}}$$

where

$$\Delta E_{ab} = E_b - E_a$$

This explains why a beam of light with a continuous spectrum that passes through an atomic gas will emerge exhibiting an absorption line spectrum. Since only photons with energies corresponding to the energy difference between bound electron states, ΔE_{ij}, can be absorbed, only lines with the corresponding wavelengths of λ_{ij} will be present as absorption features in the spectrum that emerges.

As Figure 7.1 illustrates, photons capable of *ionizing* the atom can have any wavelength, so long as they are energetic enough to move an electron from a bound to a free state. This is observed in the spectrum as a feature called an ***absorption edge*** – a drop in the intensity of the transmitted continuum at wavelengths shorter than the ionization wavelength.

7.1.3 Emission of light by atoms

An isolated hot gas produces an emission line spectrum. Again, you can understand why by considering the quantized energy levels. In Figure 7.1, for example, an atom changing from state e to state c must lose energy. It can do so by creating a photon with energy ΔE_{ec}. This process of ***de-excitation*** by photoemission can occur spontaneously, or it can be stimulated to occur by an incoming photon of exactly the transition energy. This latter process is the equivalent of negative absorption: one photon collides with the atom and two identical photons emerge. Stimulated emission is the basis for the operation of lasers and masers.

If there is a significant number of free electrons in a hot gas, then the gas will emit continuous radiation along with the usual emission lines. As illustrated in Figure 7.1, a photon is emitted if a free electron loses energy and recombines with a positive ion, forming the bound state of the neutral atom. The resulting radiation will be continuous, since the energy of the free electron is not quantized. Transitions from one free state to another are also possible, and will also contribute to a continuous spectrum.

7.1.4 Collisions and thermal excitation

Atoms prefer to exist in the lowest possible energy state, the ground state. An isolated atom in any excited state will spontaneously decay to a lower state. The length of time an atom can expect to remain in a particular excited state depends on the rules of quantum mechanics, but if there is a quantum-mechanically "permitted" transition to a lower state, the half-life of the excited state usually is on the order of 10^{-8} seconds. How do atoms get into an excited state in the first place? One way, of course, is by absorbing electromagnetic radiation of the proper wavelength. A second path is via collisions with other atoms or particles. Atom-on-atom collisions can convert kinetic energy into internal energy in the form of optical electrons in excited states. In the very eventful environment of a hot gas, atoms that want to stay in the ground state have little chance of doing so for long, because they are kicked up into higher states by collisions. A hot gas glows because the resulting excited atoms will decay back to lower energy levels, emitting photons in the process.

Collisions can transfer energy out of an atom as well as into it. With many collisions, at constant temperature, the population and de-population rates for

Matter and light

one level due to all processes are equal, and the expected number of atoms in a particular bound state is well defined. The **Boltzmann distribution** describes the number of atoms in each energy state in such a situation of **thermodynamic equilibrium**. Consider any two bound states, i and j, having energies E_i and E_j. The Boltzmann equation gives the ratio of the number of atoms in these two states as

$$\frac{n_i}{n_j} = \frac{g_i}{g_j} \exp\left\{\frac{E_j - E_i}{kT}\right\} \tag{7.2}$$

Here g_i and g_j are the **statistical weights** of each level (g is the number of distinct quantum mechanical states at the specified energy – see the next section). Boltzmann's constant, k, has the value 1.381×10^{-23} J K^{-1} = 8.62×10^{-5} eV K^{-1}.

7.1.5. Specification of energy levels

In the terminology of quantum mechanics, the state of every bound electron is specified by four **quantum numbers**:

> n, the principal quantum number, can take on all positive integer values 1, 2, 3,.... This number is associated with the radial distribution of the probability density of the electron as well as with its energy, and in the terminology used by chemists, specifies the **shell**.
>
> l, the azimuthal quantum number, can take on values $0, 1, \ldots, (n-1)$. It can be associated with the angular distribution of the probability density, and can have a secondary effect on the energy of the state.
>
> m, the magnetic quantum number, can take on values $0, \pm 1, \ldots, \pm l$. It describes the possible interaction between the electron and an imposed magnetic field. It can have an effect on the energy of the electron only if a field is present.
>
> s, the electron spin quantum number can have only two values, $+1/2$ or $-1/2$. It can affect the electron energy by interacting with the angular momenta of other parts of the atom.

In particle physics, a **fermion** is a particle like the electron, proton or neutron, whose spin quantum number has a half-integer value like $\pm 1/2$, $\pm 3/2$, etc. Any particle's intrinsic angular momentum has the value $(h/2\pi)\sqrt{s(s+1)}$, where h is Planck's constant and s is the spin quantum number. Particles with integer spin $(0, \pm 1$, etc.) are called **bosons**.

The **Pauli exclusion principle** states that no two identical fermions may occupy the same quantum state. This demands that **no two electrons bound in an atom may have the same four quantum numbers (n, l, m, s)**. Table 7.1 lists all possible values of the four quantum numbers for electrons in the first

Table 7.1. *Quantum numbers of the first 30 bound atomic states (up to the ground state of zinc). In the periodic table, the 4s states are usually filled before the 3d states, 5s before 4d, etc. See Figure 7.2*

Quantum numbers				Name of configuration	Number of states
n	l	m	s		
1	0	0	±1/2	1s	2
2	0	0	±1/2	2s	2
2	1	−1	±1/2		
		0	±1/2	2p	6
		+1	±1/2		
3	0	0	±1/2	3s	2
3	1	−1	±1/2		
		0	±1/2	3p	6
		+1	±1/2		
3	2	−2	±1/2		
		−1	±1/2		
		0	±1/2	3d	10
		+1	±1/2		
		+2	±1/2		
4	0	0	±1/2	4s	2

few levels. Each of the states listed must be either empty or occupied by a single electron. The ground state of an atom with atomic number Z will have the lowest-energy configurations occupied, up to the Zth available state, and all other states empty. The actual energy of a particular state depends not only on the atomic number and the values of the four quantum numbers for the occupied states, but also on other details like the atomic weight, and magnetic interactions between the electron, nucleus, and electrons in other states.

The energy of an electron will depend most strongly upon both n and l quantum numbers. The ***configuration*** of electrons in an atom is therefore usually described by giving these two numbers plus the number of electrons in that n, l level. The ***spectroscopic notation*** for a configuration has the form:

$$ny^x$$

where

n is the principle quantum number,

x is the number of electrons in the level — many electrons can have the same n, l so long as they have different m and/or s values, and

y codes the value of the l quantum number according to the following scheme:

l	0	1	2	3	4	5	6	7, 8, …
Designation	s	p	d	f	g	h	i	k, l, etc.

Lithium, atomic number 3, for example, has the ground-state configuration $1s^2\,2s^1$; that is, two electrons in the $n = 1$ state, one with quantum numbers $(1, 0, 0, -1/2)$, the other with $(1, 0, 0, 1/2)$. The third lithium electron (this is the valence electron) is in the $n = 2$ level with quantum numbers either $(2, 0, 0, -1/2)$ or $(2, 0, 0, 1/2)$. Table 7.2 gives some further examples of electron configurations.

Figure 7.2 is a schematic energy-level diagram that shows the relative energies of the electron configurations in atoms. As one moves from element to element in order of increasing atomic number, electrons are added from the bottom up in the order suggested by Figure 7.2. (There are minor exceptions.)

The **periodic table**, one of the triumphs of human learning, summarizes our knowledge of the chemical properties of the elements, and recognizes that chemical behavior is periodic in atomic number. The table is organized according to similarities in optical electron configurations. Each row or period contains elements with identical values of n for outer electrons. In chemical terminology, the valence electrons of atoms in the same period are all in the same **shell**. The

Table 7.2. *Examples of a few electron configurations*

Element	Atomic number	Electron configuration of the ground state
Hydrogen	1	$1s^1$
Helium	2	$1s^2$
Boron	5	$1s^2\,2s^2\,2p^1$
Neon	10	$1s^2\,2s^2\,2p^6$
Silicon	14	$1s^2\,2s^2\,2p^6\,3s^2\,3p^2$
Argon	18	$1s^2\,2s^2\,2p^6\,3s^2\,3p^6 = [\text{Ar}]$
Potassium	19	$1s^2\,2s^2\,2p^6\,3s^2\,3p^6\,4s^1 = [\text{Ar}]\,4s^1$
Scandium	21	$[\text{Ar}]\,3d^1\,4s^2$
Germanium	32	$[\text{Ar}]3d^{10}\,4s^2\,4p^2$
Krypton	36	$[\text{Ar}]3d^{10}\,4s^2\,4p^6 = [\text{Kr}]$
Rubidium	37	$[\text{Kr}]\,5s^1$

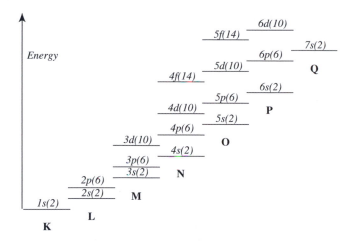

Fig. 7.2 Schematic energy levels of electronic configurations. Each level is labeled with the spectroscopic designation, including the number of electrons needed to fill the configuration in parentheses. Chemical shell designations (K, L, etc.) are at the bottom of each column. The diagram indicates, for example, that the two 5s states will fill before the ten 4d states. Energy levels are illustrative only of the general order in which configurations or sub-shells are filled and are not to scale. There are several exceptions to this overall scheme.

atomic properties of elements, like electro-negativity, ionic radius, and ionization energy all trend generally in one direction along the row. Chemical behavior likewise trends from one extreme to the other along a row. Period three, for example, ranges from the reactive metal sodium, through the less reactive metals magnesium and aluminum, the semi-metal silicon, the increasingly reactive non-metals phosphorus, sulfur, and chlorine, and the inert gas, argon. Elements in the same *column* of the table, in contrast, have the same electron configuration in their outer shells, and therefore all have very similar chemical properties. The noble gases, for example, (helium, neon, argon, krypton, xenon, and radon – column 18 or group VIIIA) all exhibit chemically inert behavior, and all have a filled outer shell with eight electrons in the s^2p^6 configuration. Similarly, the halogens in column 17, all highly reactive non-metals like fluorine and chlorine, have outer shells with the s^2p^5 configuration. There is also a secondary trend in properties moving down a column: the chemical reactivity of the halogens, for example, decreases steadily from fluorine, the lightest, to astatine, the heaviest.

Because of the order in which configurations are filled (see Figure 7.2) many elements have identical valence electron configurations and differ only in their inner electron shells. For example, the rare-earth elements, or lanthanoids – cerium ($Z = 58$) through ytterbium ($Z = 70$) – are all in period 6 of the table, but have chemical properties that are almost indistinguishable from one another. This is because they have identical outer shells ($6s^2$), and differ only in the configurations of their inner (primarily 4f) sub-shells.

For atoms with multiple valence electrons, the energy level of an excited configuration may depend not only on the quantum numbers of the electrons, but upon the interactions between the electron spins and angular momenta. For example, the excited state of helium that has configuration $1s^1\,2p^1$ has *four* possible energies spread over about 0.2 eV. States differ because of different relative orientations of the two electron spins and the $l = 1$ angular momentum of the p electron (directions are quantized and thus limited to four possibilities). The details of how multiple electrons interact are beyond the scope of this book, but for now, it is sufficient to recognize that such interactions can cause the energy level of a configuration to split into multiple values.

7.2 Isolated molecules

The outermost electrons of a molecule see a more complex binding potential due to the presence of two or more positively charged nuclei. Generally, this results in a greater number of electronic energy states. Each electronic state is still characterized by four quantum numbers, but in the molecule, the value of the m quantum number has an important effect on the energy level. More importantly, the molecule itself has internal degrees of freedom due to its ability to rotate around its center of mass, as well as its ability to vibrate by oscillating chemical bond lengths and angles. These internal rotational and vibrational modes are quantized as well, and they vastly increase the number of energy states permitted to the molecule.

Quantum mechanical theory approximates the total internal energy of a molecule as the sum of three independent terms:

$$E = E_{\text{electron}} + E_{\text{vibration}} + E_{\text{rotation}}$$

In addition to the quantum numbers specifying the electronic state, a diatomic molecule like CO or TiO will have one quantum number, J, to specify the rotational state, and one, v, for the vibrational state. Specification of the vibrational mode for molecules with more than two atoms becomes quite complex.

Figure 7.3 is a schematic energy-level diagram for a fictitious diatomic molecule. The energy levels in the figure are not scaled precisely. Transitions between the ground state and the first excited electronic state are usually in the

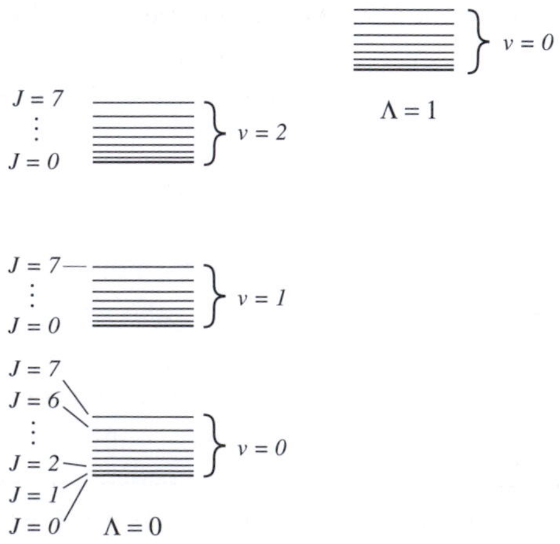

Fig. 7.3 Energy levels in a simple molecule. Right and left columns are different electronic states, as indicated by the quantum number Λ. Quantum numbers J and v specify the rotational and vibrational states, respectively. We show only three rotation states and seven vibration states in the lower electronic level.

range 0.5 to 100 eV. Transitions between adjacent vibrational states are about 100 times smaller than this, and between adjacent rotational states, about 10^6 times smaller yet.

The spacings between the rotational levels at different electronic and vibrational states are similar. As a result, the spectra of even simple diatomic molecules show a complicated pattern of lines consisting of extensive *bands*, with each band composed of many closely packed lines.

7.3 Solid-state crystals

7.3.1 Bonds and bands in silicon

A crystal is a mega-molecule in which the pattern of atoms and bonds repeats periodically with location. Many of the detectors we discuss in the next chapter are made of crystalline solids, so we now describe in detail the electronic structure of silicon, the most important of these materials. The silicon atom, located in column IVa of the periodic table, has 14 electrons, four of which are in the outer shell, with configuration $3s^2\,3p^2$. The outer shell will be filled when it contains eight electrons, not four. According to the theory of chemical valence, the component atoms of a molecule try to attain the electron structure of an inert gas (eight outer-shell electrons) by an appropriate sharing or transfer of electrons. Shared or transferred electrons produce, respectively, covalent or ionic bonds between atoms.

Consider the formation of a silicon crystal. Figure 7.4a shows what happens to the energy levels of an isolated silicon atom when a second silicon atom is brought closer and closer to it. As the electron wave functions begin to overlap, the levels split into two, outermost first. The nearer the neighbor, the greater is its influence, and the greater the splitting of levels. The outer electrons of both atoms can enter those levels since their wave functions overlap.

If we construct a crystal atom by atom, new energy states appear with each addition. For five atoms in a row, we expect something like Figure 7.4b. As crystal construction continues, more and more electron states become available as more and more atoms are added to the structure. Since even a tiny crystal

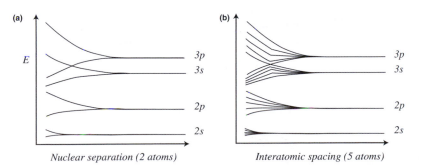

Fig. 7.4 (a) Changes in the electron energy levels in a silicon atom as a second atom is brought into close proximity. (b) The same diagram for the case of five atoms in a linear matrix.

Fig. 7.5 Schematic diagram of the bands in silicon crystals. The diagram at right shows the bands formed at the preferred inter-atomic spacing. Dark-gray bands are occupied, light-gray are empty but permitted. Energies in the band gaps (white) are forbidden to electrons.

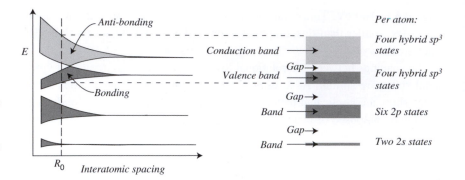

contains on the order of 10^{20} atoms, each causing a split in the energy levels, the spacing between levels must be on the order of 10^{-20} eV. These levels are so closely spaced that for practical purposes we treated them as a continuous **band** of available energies. If bands do not overlap, they will be separated by energy **band gaps**. An electron anywhere in the crystal lattice is permitted an energy anywhere in a band, and is forbidden an energy anywhere in a gap.

Figure 7.5 shows the energy situation in crystalline silicon. The preferred inter-atomic spacing between nearest neighbors is R_0 (0.235 nm at room temperature). Note that at this spacing, the 3p and the 3s energy levels overlap. The result is called a crossover degeneracy, and energies in the crossover region are forbidden by quantum mechanics. The permitted states in the quantum mechanical view are certain linear combinations of s and p states, not the separate s and p states of the isolated atom. The periodic potential pattern of the regularly spaced nuclei in the crystal causes electron wave functions to interfere constructively at particular locations, and produces a set of states called **sp^3-hybrid orbitals**. Each silicon atom contributes eight such states to the bands. Four of the sp^3 hybrid orbitals, the ones with lowest energy, correspond to an electron having its most probable location midway between the atom and one of its nearest neighbors. The nearest neighbors are at the four vertices of a tetrahedron centered on the nucleus. These four sp^3 hybrid orbitals all have energies that lie in the **valence band** and constitute the **bonding states**. Four other sp^3 hybrid states have energies in the **conduction band** and locations away from the bonding locations. These are the **anti-bonding states**.

From now on, we will use band diagrams, like the right side of Figure 7.5, to account for all the electrons in the entire crystal. At zero temperature all the anti-bonding states are empty and make up the **conduction band**. The difference between the energy of the top of the valence band, E_v, and the bottom of the conduction band, E_c, is called the **band gap energy**:

$$E_G = E_c - E_v$$

In silicon, the band gap is 1.12 eV at room temperature.

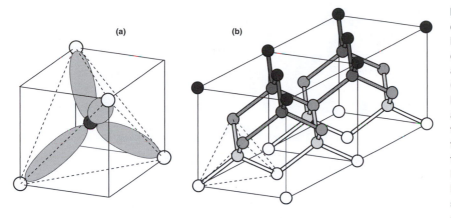

Fig. 7.6 (a) Tetrahedral covalent bonds for sp³ hybridized orbitals for one atom (black) pictured at the center of a cube (solid lines). Its nearest bond-forming neighbors are at the four corners of the cube. These define the vertices of a tetrahedron (dashed lines). The electron bonding states are shown as light-gray ellipsoids – the regions where there is the highest probability of finding a valence electron. (b) A stick-and-ball model of the diamond lattice. Darker grays represent atoms higher up in the vertical direction. Each cube outlines a unit cell of the crystal, and a complete crystal is built by assembling many identical adjoining unit cells in three dimensions.

The most probable physical location of the valence electrons is on the line joining neighboring nuclei. In a silicon crystal, one would find two electrons in each of the light-gray region of Figure 7.6a – one from each atom – and this pair of shared electrons constitutes a covalent bond. Each atom forms four such bonds, symmetrically placed, and each atom therefore "sees" eight outer electrons – a complete shell. The bonds arrange themselves in a tetrahedral structure about each nucleus. The symmetry here reflects the transformation of the two s state and two p state electrons of the isolated silicon atom into the four hybrid sp³ state electrons of the silicon crystal. X-Ray diffraction studies confirm that this tetrahedral structure repeats throughout the crystal in a three-dimensional pattern called a ***diamond lattice***, as sketched in Figure 7.6b.

7.3.2 Conductors, semiconductors, and insulators

It is instructive, although overly simple, to explain the differences between electrical conductors, semiconductors, and insulators as arising from differences in the size of the band gap and in electron populations within the bands. The important principle is that a material will be a good conductor of electricity or heat if its electrons can accelerate (i.e. change quantum state) easily in response to an applied electric field.

An analogy may help you understand the effect of band structure on conductivity. Imagine that you are standing in an almost empty room. You are free, in this environment, to respond to a whim to run across the room at top speed. On the other hand, if the same room is packed shoulder-to-shoulder with people, running is out of the question, no matter how strong your desire. Indeed, in a sufficiently dense crowd, moving at all is impossible.

Similarly, an electron in relative isolation can help conduct electricity or heat because it can (and must) accelerate when a global electric field is imposed, or when a strong local field arises during a collision with another particle. In a

Matter and light

Fig. 7.7 Band structure of insulators, conductors, and semiconductors: (a) an intrinsic semiconductor at zero temperature; (b) the same material at a higher temperature; and (c) an extrinsic semiconductor.

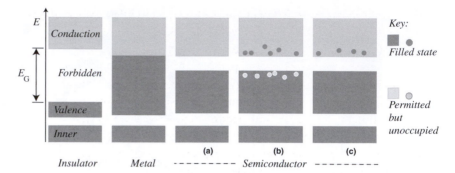

crystalline solid, however, options are more restricted. The Pauli exclusion principle explains that an electron can only accelerate (i.e. change quantum states) if it can move into a new state that is (a) permitted and (b) not occupied by another electron. In a perfect silicon crystal at zero temperature, these two conditions are difficult to satisfy: every electron is in the valence band and every electron part of a covalent bond. Every permitted state in the band is occupied. The electrons, in effect, are packed shoulder-to-shoulder. Although we have ignored movement of the nuclei (which can oscillate around their mean positions), as well as surface effects, the basic conclusion is: electron crowding makes cold silicon a poor conductor of electricity, heat, and sound.

There *are* available states at much higher energies – the anti-bonding states in the conduction band. If an electron can acquire at least enough energy to jump the band gap, then it finds itself in the relatively empty conduction band where it is able to move around. In the crowded-room analogy, you might have enough energy to climb up a rope through a trap door to the empty room on the next storey, and there you are free to run.

Silicon conductivity will improve at higher temperatures, because an electron in a hot crystal might gain enough energy from a thermal collision to reach a state in the conduction band.

Figure 7.7 shows simplified diagrams of the band structures typical of metals, insulators, and semiconductors. In a typical insulator, the valence band is completely filled. The band gap is large compared to both the thermal energy, kT (at room temperature (300 K), $kT = 0.026$ eV), and any other energy sources. Because of the large gap, valence electrons cannot reach any permitted states in the conduction band. The exclusion principle forbids any electron to move into an already occupied state, so electrons cannot move at all – the material is a non-conductor.

A metallic conductor, in the second panel of the figure, has unoccupied permitted states immediately adjacent to the occupied valence states. If an electron near the top of the valence band absorbs even a tiny amount of energy, it may move into the conduction band, and from there to virtually anywhere in the material. The horizontal coordinate in these diagrams represents position in the material.

The figure shows three different views of materials called ***semiconductors***. The first, (a) an ***intrinsic semiconductor***, looks like an insulator, except it has a small band gap. This is similar to the band diagram of silicon at zero temperature. A valence electron can jump the gap into the conduction band by absorbing a modest amount of energy, either from thermal excitation or from some other energy source. Illustration (b), for example, shows the material in (a) at a high temperature. A few electrons have absorbed sufficient thermal energy to rise to the conduction band. This material will conduct, but the size of the current is limited because only these few conduction band electrons can easily change states. More electrons, of course, will rise to the conduction band to improve the conductivity if the temperature is increased further, and materials of this kind, in fact, can be used to make temperature gauges (thermistors).

The other thing to notice in illustration (b) is that whenever an electron is boosted into the conduction band, it must leave behind an empty state in the valence band. Another valence electron can shift into this vacated state and create a new empty state in the location it vacates. Since yet another electron can now move from a third location to fill this second empty state, it is clear that valence electrons can move through the crystal by occupying and creating empty states. It is easier to concentrate on the motion of the empty states, and to think of these ***holes*** as the entities that are moving. Holes thus behave like mobile positive charges in the valence band, and will contribute to the overall electrical conductivity. In intrinsic semiconductors, holes are usually less mobile than conduction-band electrons.

The third semiconductor (c) also has a few electrons in the conduction band, but without any corresponding holes in the valence band. Materials of this kind, called ***extrinsic semiconductors***, are extremely important in the construction of most electronic devices. A second class of extrinsic semiconductors has valence-band holes without corresponding conduction-band electrons.

7.3.3 Intrinsic semiconductors

Semiconductor crystals

A pure silicon crystal forms by linking all atoms with the tetrahedral covalent bond structure pictured in Figure 7.6. Part (b) of that figure shows the arrangement of a few dozen silicon atoms and bonds, and although not all bonds have been drawn, you can assure yourself that, as the pattern repeats, each atom will end up with four bonds. This geometry, called the ***diamond lattice***, insures that each atom shares eight electrons, completely filling the outer shell and producing a chemically stable structure. Indeed, the regularity of the diamond-lattice structure is tightly enforced, even if impurities are present in the silicon.

Matter and light

Table 7.3. *Periodic table of the elements near column IVA*

IIB s^2	IIIA s^{2}p^1	IVA s^{2}p^2	VA s^{2}p^3	VIA s^{2}p^4
	B	^{6}C	N	O
	Al	^{14}Si	P	S
Zn	Ga	^{32}Ge	As	Se
Cd	In	^{50}Sn	Sb	Te
Hg	Tl	^{82}Pb	Bi	Po

Elements with similar outer-electron configurations form similar diamond lattice crystals. These are in column IVA (also called column 14) of the periodic table, and include carbon, germanium, and tin.[1] Similar bonds also form in binary compounds of elements symmetrically placed in the table on either side of column IVA. For binary compounds, the crystal structure is called the "zinc blend" structure, which resembles Figure 7.6 except for alternation of the chemical identity of the nuclei on either end of each bond. Most useful semiconductors exhibit the diamond or zinc blend crystal structure. (Exceptions include lead sulfide and zinc oxide.) Table 7.3 shows part of the periodic table containing elements that combine to form important semiconductors.

Examples of binary-compound semiconductors are gallium arsenide (GaAs, a III–V compound) and cadmium telluride (CdTe, a II–VI compound). Some ternary compounds, notably $(Hg_xCd_{1-x})Te$, and quarternary compounds like $In_xGa_{1-x}As_yP_{1-y}$, also form useful semiconductors. Commercially, silicon is by far the most commonly used semiconductor. Germanium and gallium arsenide also find important commercial applications.

Semiconductor materials generally have a resistivity in the range 10^{-2} to 10^9 ohm cm, midway between that of a good conductor (10^{-6} ohm cm) and a good insulator ($>10^{14}$ ohm cm). As we have already seen, resistivity depends critically on both temperature and the size of the band gap. Table 7.4 lists the band-gap energies for several semiconductors. Note that since the lattice spacing in a crystal is likely to change with temperature, so too will the band gap. Carbon in the diamond allotrope is an insulator because its band gap is so large that very few electrons can be thermally excited to the conduction band at room temperature; other carbon allotropes (graphite, carbon nanostructures) are conductors.

---

[1] The most common allotrope of tin, white tin, the familiar metal, has a tetragonal crystal structure. Gray tin, a less common allotrope, crystallizes in the diamond lattice. Lead, the final member of column IVA, crystallizes in a face-centered-cubic lattice.

Table 7.4. *Some common semiconductors. Forbidden band-gap energies and cutoff wavelengths at room temperature. A more complete table is in Appendix H1. Data from Sect. 20 of Anderson (1989)*

Material		Band gap (eV)	λ_c (μm)
IV			
Diamond	C	5.48	0.23
Silicon	Si	1.12	1.11
Germanium	Ge	0.67	1.85
Gray tin	Sn	0.0	
Silicon carbide	SiC	2.86	0.43
III–V			
Gallium arsenide	GaAs	1.35	0.92
Indium antimonide	InSb	0.18	6.89
II–VI			
Cadmium sulfide	CdS	2.4	0.52
Cadmium selenide	CdSe	1.8	0.69
Mercury cadmium teluride	$Hg_xCd_{1-x}Te$	0.1–0.5 ($x = 0.8$–0.5)	12.4–2.5
IV–VI			
Lead sulphide	PbS	0.42	2.95

Conductivity and temperature

At zero temperature, all the materials in Table 7.4 are non-conductors. As temperature increases, thermal agitation causes ionizations: electrons are promoted to the conduction band, free of any one atom; similarly mobile holes are created in the valence band. The material thus becomes a better conductor with increasing temperature. At equilibrium, we expect the rate of electron-hole recombinations to exactly equal the rate of thermal ionizations.

How, exactly, does an electron in a bonding state receive enough energy to jump the band gap? Optical electrons can collide with one another, of course, but it is important to note also that the lattice itself is an oversized molecule that can vibrate by oscillating bond length or angle. Just as with molecules, lattice vibration states are quantized with respect to energy. Solid-state theory often associates each discrete lattice vibration energy with a particle, called a **phonon**, an entity analogous to the photon. Changes in electron state may thus involve the absorption or emission of a phonon. An electron can jump the band gap because it absorbs a phonon of the correct energy, and can lose energy and momentum by creation of, or collision with, a phonon.

At a particular temperature, the density of electrons at any energy within the bands will depend upon the product of two functions, (a) the probability,

Fig. 7.8 Electron and hole density in an intrinsic semiconductor. (a) The locations of the band edges and the Fermi level midway between them. (b) The probability of finding an electron in a permitted state as a function of energy, $P(E)$. The solid line shows P at zero temperature, and the broken lines at two higher temperatures. (c) The density of permitted states as a function of energy, $S(E)$. (d) The density of electrons (the product of (b) and the highest temperature curve in (c)) and holes as functions of energy. The horizontal scale of plot (c) has been expanded to show detail.

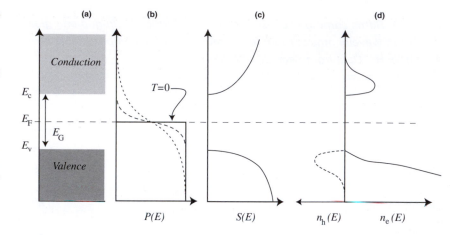

$P(T, E)$, of an electron having that energy, and (b) the number density of available states at each energy, S:

$$n_e(T, E) = P(T, E)S(E)$$

With respect to the probability of a fermion having a particular energy, recall that the exclusion principle causes important restrictions on occupancy. This is certainly the case for the electrons in the bands of a semiconductor, where most of the valence states are fully occupied. In such a situation of **electron degeneracy** the probability per unit energy that an electron has energy, E, is given by the **Fermi–Dirac** distribution:

$$P(T, E) = \frac{1}{[1 + \exp\{(E - E_F)/kT\}]} \tag{7.3}$$

This expression reduces to the Boltzmann distribution, Equation (7.2), at high temperatures. At the limit of zero temperature, the Fermi–Dirac distribution requires that all of the lowest energy states be occupied, and all of the higher states (those with energies above E_F) be empty. That is, at $T = 0$,

$$P(E) = \begin{cases} 1, E < E_F \\ 0, E > E_F \end{cases} \tag{7.4}$$

The parameter E_F is called the **Fermi energy**, and might be defined as that energy at which the probability for finding an electron in a permitted state is exactly one half. According to this definition, the Fermi energy will itself be a function of temperature for some systems at high temperature. However, for all cases we are concerned with, the Fermi energy can be treated as a constant equal to the energy of the highest permitted state at $T = 0$.

Figure 7.8a shows the energy bands for silicon at absolute zero, where electrons will fill all available states in the permitted bands up to the Fermi level, which falls midway between the valence and conduction bands. Figure 7.8b plots Equation (7.3) at three different values of temperature.

Figure 7.8c shows a schematic representation of the **density of permitted states**, $S(E)$, for the valence and conduction bands of silicon. Note that $S(E)$, which gives the number of quantum states that are available at a particular energy (per unit energy and volume), is approximately quadratic near the permitted band edges and vanishes in the band gap.

The product $P(E)S(E)$ gives $n_e(E)$, **the number density of electrons** at energy E. The **number density of holes** at energy E in the valence band is just

$$n_h = [1 - P(E)]S(E)$$

Figure 7.8d shows these two functions, n_e and n_h, for a non-zero temperature. The total number densities of charge carriers of each kind (negative or positive) are given by the integrals of these functions in the appropriate band – conduction electrons and valence holes:

$$
\begin{aligned}
n_N &= \int_{E_F}^{\infty} n_e dE \\
n_P &= \int_{-\infty}^{E_F} n_h dE
\end{aligned}
\tag{7.5}
$$

In intrinsic semiconductors, the number density of these two kinds of charge carriers in equilibrium must be equal, so $n_P = n_N$. The temperature dependence in Equation (7.5) follows from the Fermi distribution, and has the form

$$n_P = n_N = AT^3 e^{-\frac{E_G}{kT}} \tag{7.6}$$

The number density of charge carriers in intrinsic semiconductors, and therefore properties like resistivity and conductivity, should vary approximately exponentially with temperature.

7.3.4 Intrinsic photoabsorbers

An electron can leave a covalent bond if it is given sufficient energy (at least the value of E_G) to jump the band gap into an anti-bonding state in the conduction band. The required energy could be supplied by a photon, if the photon has a wavelength less than the **cutoff wavelength** for the material:

$$\lambda_c = \frac{hc}{E_G} = \frac{1.24 \, \mu m}{E_G[eV]} \tag{7.7}$$

The band gap for silicon, for example, corresponds to a cutoff wavelength λ_c of 1.1 μm.

Photo-absorption changes a material's properties. For example, Figure 7.9a shows a simple device that utilizes such a change to measure light intensity. Photons absorbed by a block of semiconductor material produce ionization

Fig. 7.9 Simple
photoconductors. In (a),
light strikes the exposed
surface of a
semiconductor linked to
a simple circuit by two
metal contacts. Photo-
ionization produces
charge carriers that
reduce semiconductor
resistance. Current
through the device will
increase with increasing
illumination, and output
is the voltage across a
load resistor. In (b), a
three-pixel device
registers three different
voltages in response to
local illumination. Here
photons pass through
upper (transparent)
contacts. The lower
contact is reflective so
that photons passing
through the device are
redirected for a second
pass.

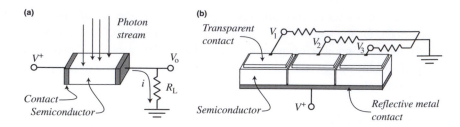

events – electrons in the valence band promoted to the conduction band, leaving
an equal number of holes. The greater the flux of incoming photons, the greater
the equilibrium concentration of charge carriers, and the greater the conductivity of the detector. If you maintain a constant voltage across the semiconductor,
as in the figure, then the electrical current through the circuit should increase
with the number of photons absorbed per second. A measurement of the voltage
at V_0 thus monitors light intensity. Figure 7.9b shows an alternative illumination
strategy that facilitates a close-packed array of detectors.

Notice that this **photoconductor** responds to the *number* of photons per
second absorbed, not, strictly, to the rate of *energy* absorbed. Of course, if
you know their spectral distribution, it is an easy matter to compute the energy
flux carried by a given number of photons.

There are at least three reasons why a photon incident on the top of the device
in Figure 7.9a will fail to generate an electron–hole pair. First, we know that
those with frequencies below the band-gap frequency, E_G/h, cannot move an
electron from the valence to conduction band, and thus cannot be detected.

A second failure is due to reflection of photons from the top surface of the
device. As we saw in Chapter 5, minimal reflection occurs at normal incidence,
and depends on the refractive index of the material:

$$R = \frac{(n_1 - n_2)^2}{(n_1 + n_2)^2}$$

The refractive index (and thus reflectivity) for silicon and most other semiconductors is very high in the ultraviolet, decreases through visible wavelengths, and is low (3.5 to 4) in the red and infrared. Reflectivity is also low
in the X-ray band. Anti-reflection coatings can considerably reduce reflectivity
for a particular wavelength.

A third reason for detection failure is that photons above the band-gap frequency might pass completely through the device without interaction. We now
examine this transmission phenomenon in greater detail.

Once entering a semiconductor, the distance a photon can travel before being
absorbed depends very strongly on its wavelength, as well as the quantum
mechanical details of available electron states in the material. If a beam of
photons enters material in the z-direction, its intensity at depth z will be

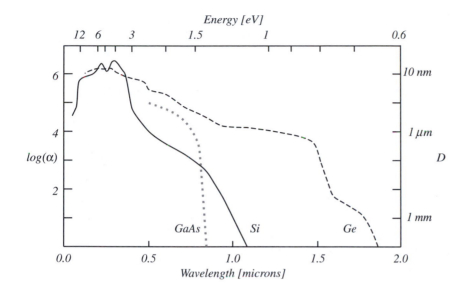

Fig. 7.10 The absorption coefficient, α, measured in m^{-1}, as a function of photon wavelength or energy. The absorption depth, D = 1/α, is on the right axis. Two indirect transition materials, silicon and germanium, show a much more gradual change with energy than does gallium arsenide, a direct transition material.

$$I(z) = I_0 \mathrm{e}^{-\alpha z}$$

where I_0 is the intensity at $z = 0$, and α is the ***absorption coefficient***. A large absorption coefficient means light will not travel far before being absorbed. Figure 7.10 shows the absorption coefficient as a function of wavelength for silicon, germanium, and gallium arsenide. These curves illustrate an important divergence in behaviors caused by the details in the available energy states in two classes of materials. Notice that GaAs absorbs strongly right up to the cutoff wavelength, whereas Si and Ge very gradually become more and more transparent approaching that wavelength. Materials with an abrupt cutoff, like GaAs and InSb, are called ***direct transition*** semiconductors. Materials of the second kind, like Si and Ge, are called ***indirect transition*** semiconductors.

For both direct and indirect materials, when photons of the proper energy are absorbed, they almost always produce electron–hole pairs. The exceptions are usually due to flaws in the material. In some cases, a photon can interact with the lattice (particularly defects in the lattice) and deposit its energy as a phonon, not as a photo-ionization. For this reason, light-detecting devices require a semiconductor material that has been crystallized with strict controls to assure chemical purity and lattice integrity.

7.3.5 Extrinsic semiconductors

In practical devices, crystals inevitably have some chemical impurities and mechanical imperfections. These alter the energies and momenta of the states available near the sites of the defects, usually in undesirable ways. Curiously, though, some of the most useful semiconductor devices are made by intentionally introducing impurity atoms into the lattice.

(a) *Pure crystal*

(b) *Acceptor impurities*

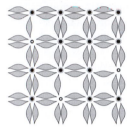

(c) *Donor impurities*

Fig. 7.11 A schematic of the bond structure in intrinsic and extrinsic semiconductors. In an actual crystal, the bond arrangement is three-dimensional – see Figure 7.5: (a) a pure intrinsic material; (b) three p-type impurity atoms in an extrinsic material; and (c) three n-type impurity atoms.

Figure 7.11a shows a flattened schematic of the positions of the atoms and outer electrons in an intrinsic semiconductor like silicon. Each atom shares eight valence electrons, forming four complete bonds. All atoms and bonds in the lattice are identical. Diatomic semiconductors like GaAs have a similar structure, except the chemical identity of the atoms alternates along rows and columns.

Now we intentionally introduce an impurity into the lattice, as in Figure 7.11b, where a few of the silicon atoms have been replaced by atoms that have only three valence electrons, like boron, gallium, or indium. Each impurity creates a vacancy in the electron structure of the lattice – a "missing" electron in the pattern.

The crystal, in fact, will try to fill in this "missing" electron. The impurity creates what is called an ***acceptor state***. It requires relatively little energy (on the order of the room temperature thermal energy, kT) to move a valence electron from a silicon–silicon bond elsewhere in the lattice into this gap at the impurity site. This creates a hole at the site that donates the electron. Such a hole behaves just like a mobile hole in an intrinsic semiconductor – a positive charge carrier that increases the conductivity of the material. Semiconductors in which impurities have been added to create positive charge carriers are termed ***p-type extrinsic*** semiconductors.

Figure 7.12a is an energy-band diagram for a p-type semiconductor. At zero temperature, a small number of (unoccupied) acceptor energy states exist within the band gap of the basic material. The energy difference, E_i, between the top of the valence band and the acceptor states is typically on the order of 0.05 eV in silicon (see Table 7.5). At a finite temperature, excitation of electrons from the valence band into these intermediate states creates valence-band holes (Figure 7.11b). Because the electrons in the intermediate states are localized at the impurity sites, they are immobile and cannot contribute to the conductivity. The mobile holes in the valence band, of course, can contribute, and are termed the ***majority charge carriers***. In contrast with intrinsic semiconductors, $n_P > n_N$ in p-type materials. Adding impurities to create an extrinsic semiconductor is called ***doping***, and the more heavily doped the material is, the higher is its conductivity. The transparent conductors used as contacts in Figure 7.9b, for example, are often made of highly doped silicon.

There is a second kind of doping. Figure 7.11c illustrates intrinsic material doped with atoms that have five valence electrons, like arsenic or antimony. The result is an ***n-type extrinsic*** semiconductor. Here, the "extra" electrons from the ***donor*** impurities are easily ionized into the conduction band. This ionization restores the bond structure to that of a diamond crystal (only eight shared outer-shell electrons, not nine) and consequently produces conduction-band electrons that constitute the majority carriers. Figure 7.12c shows the band structure of an n-type material, with occupied impurity states at energy E_i below the bottom of the conduction band at zero temperature. At higher temperatures, as shown in

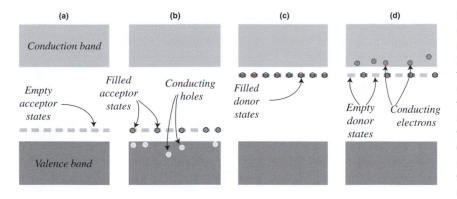

Fig. 7.12 Band structure of extrinsic semiconductors. (a) A p-type material at zero temperature. The energy difference between the top of the valence band and the acceptor states is typically on the order of 0.05 eV. (b) The same material at a higher temperature. Electrons excited into the acceptor states have created valence-band holes. (c) An n-type material at zero temperature; (d) shows the same material at a higher temperature, where electrons from the donor states have been ionized into the conduction band.

Figure 7.9d, some of these donor states are ionized to add electrons to the conduction band.

Extrinsic semiconductors respond to light in nearly the same way as intrinsic material. In fact, because the concentration of impurity atoms is always quite small (typically one part in 10^3 or 10^4), the presence of dopants does not appreciably modify intrinsic photo-absorption *above* the band-gap energy. The important difference occurs with photons whose energies lie *below* the intrinsic band-gap energy but above the dopant ionization energy, E_i.

Suppose, for example, a sample of boron-doped silicon (usually symbolized as Si:B), a p-type material, is kept so cold that the acceptor states, which lie 0.045 eV above the top of the valence band, are mostly empty. Intrinsic absorption in silicon cuts off at wavelengths longer than 1.12 μm. Short-ward of this cutoff wavelength, our sample absorbs as if it were intrinsic silicon. However, because photons with wavelengths shorter than $\lambda_i = hc/E_i = 26$ μm can ionize electrons from the valence band into the acceptor states, the number of majority carriers in the extrinsic material will increase in proportion to the number of photons short-ward of 26 μm. In effect, extrinsic absorption moves the cutoff to the longer wavelength. The implication for the construction of detectors for infrared light is obvious.

The absorption coefficient for extrinsic operation depends upon the dopant concentration. An important difference, then, between intrinsic and extrinsic photo-absorption is that the coefficient for extrinsic absorption can be adjusted in the manufacturing process. However, there are limits to the amount of impurity that can be added, so the absorption coefficient for extrinsic operation will always be low. One set of limits arises because, at high concentrations, the dopant atoms are so close together that their electron wave functions overlap, producing an ***impurity band***. If the states in this band are partially occupied, then the material will be conducting – with charge carriers "hopping" from one impurity state to another, effectively short-circuiting any photoconductive

Table 7.5. *Ionization energies, in eV, for different impurity states in silicon and germanium. Data from Kittel (2005) and Rieke (1994)*

Acceptors	Si	Ge
B	0.045	0.0104
Al	0.057	0.0102
Ga	0.065	0.0108
In	0.16	0.0112
Tl	0.26	0.01
Be	0.146	0.023
Cu	0.23	0.039
Donors		
P	0.045	0.0120
As	0.049	0.0127
Sb	0.039	0.0096
Bi	0.069	

effect. Extrinsic detectors therefore tend to be rather thick (1 mm) to provide adequate depth for photo-absorption.[2]

7.4 Photoconductors

Both intrinsic and extrinsic semiconductors, employed in a circuit like the one illustrated in Figure 7.9, in principle make excellent light detectors. If the voltage across the semiconductor is maintained as a constant (i.e. if $R_L = 0$), then the current will be directly proportional to the number of charge carriers in the material, which (at a sufficiently low temperature) will be directly proportional only to the rate at which it absorbs photons. More precisely, the electric current due to photons will be something like

$$I_{\text{photo}} = nqP_aP_e$$

where n is the number of photons incident per second and q is the electron charge. The quantity P_a is the probability that an incident photon will generate a pair of charge carriers in the detector; P_a depends upon the factors discussed above: surface reflectivity, the absorption coefficient, and the thickness of the sensitive layer. (The fraction of the photons entering a layer of

---

[2] There are some important exceptions, e.g. the *blocked impurity band (BIB) detector*, also called the *impurity band conduction (IBC) detector*, in which a thin layer of highly doped material is bonded to a layer of intrinsic material, so that the intrinsic material breaks the continuity (and conductivity) of the impurity band. See McLean (2008).

thickness z that are absorbed is just $1 - e^{-\alpha z}$.) The absorption coefficient, in turn, will depend upon wavelength and (for extrinsic materials) impurity concentration.

The quantity P_e is the probability that a charge carrier, once created, will actually move from the semiconductor to the appropriate electrode. It depends on a number of factors: electric field strength, thickness of the material, charge carrier mobility, and the presence of flaws in the crystal that might promote recombination.

Although is desirable to make I_{photo} as large as possible, there are limits. For example, increasing the voltage across the electrodes increases P_e, but at large potential differences electrons will gain considerable kinetic energy. At high energies, electrons will ionize atoms by collision to create new charge carriers, and these secondaries will in turn be accelerated to produce additional carriers. This avalanche of charge produces a spike in the output current. At high enough voltages, in a condition called **breakdown**, the avalanche becomes constant, destroying the resistance of the material and making it useless as a detector. Similarly, increasing the thickness, z, of the material increases the probability that a photon will be absorbed. At the same time, however, increasing z in Figure 7.9b reduces the probability that a charge carrier will be able to move to an electrode before it recombines.

7.5 The MOS capacitor

The metal-oxide-semiconductor (MOS) capacitor is the basic element of an important class of astronomical detectors. Figure 7.13 illustrates the essentials of the device, which is a three-layer sandwich. In the figure, the left-hand layer is a block of p-type semiconductor. The left-hand face of this block is connected to electrical ground. A thin layer of insulator forms the middle of the sandwich. The semiconductor is usually doped silicon, and the insulator layer is usually silicon dioxide. The right-hand layer is a thin coating of metal, which is held at a positive voltage. If the insulating layer is not made of SiO_2 (silicon nitride,

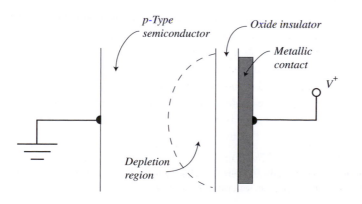

Fig. 7.13 Cross-sectional view of the physical structure of a MOS capacitor. Positive voltage (usually a few volts) applied to the metal layer creates a depletion region in the semiconductor.

Fig. 7.14 An energy-band; diagram for the MOS capacitor. Majority carriers are swept out of the depletion region. Minority carriers are swept towards the boundary with the insulator.

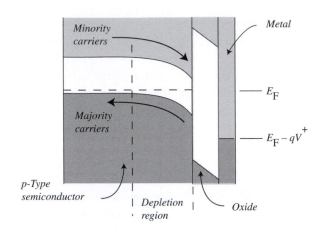

Si$_3$N$_4$, is the usual alternative), then the device is called an **MIS (metal-insulator-semiconductor) capacitor**.

Figure 7.14 shows the band structure of the device. The positive voltage of the metal layer distorts the energies of the bottom and top of the semiconductor forbidden gap. The tilt of the band reflects the strength of the electric field. In the diagram, the electric field forces electrons to move to the right and holes to the left. If a hole loses energy, it moves upwards in the diagram.

The large band gap in the insulator prevents minority electrons from crossing into the oxide layer. The flow of majority holes to ground in the valence band, in contrast, is not impeded. The result is that, in equilibrium, a **depletion region** devoid of the majority charge carriers develops in the semiconductor adjacent to the insulator. The minority carriers here are immobile – trapped in the potential well formed by the bottom of the semiconductor valence band and the band gap of the insulator.

The MOS capacitor is especially useful because it will *store* electrons that are generated by ionization. Referring to Figure 7.15a, it is clear that if an electron–hole pair is created in the depletion region, the pair will be swept apart before they can recombine: the electron goes into the well, and the hole leaves the material. Electrons in the well remain there indefinitely, since they sit in a region depleted of holes. Ionizations outside the depletion zone are less likely to produce stored electrons since charges there move by diffusion, and the longer it takes for the electron to reach the depletion zone, the greater are its chances of encountering a hole and recombining.

For ionizations in the depletion zone, however, charge storage can be nearly 100% efficient. Eventually, if enough electrons accumulate in the zone, they will neutralize the effect of the positive voltage and remove the potential well for newly generated electrons. Figure 7.15b illustrates this saturated situation. Saturation destroys the depletion zone and allows generated charge carriers to move only by diffusion, eventually recombining in equilibrium just as in an

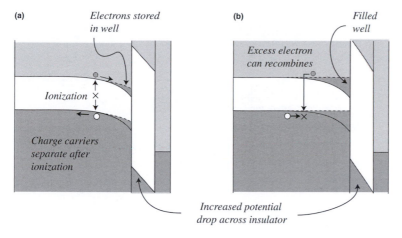

(a) Electrons stored in well

Ionization ×

Charge carriers separate after ionization

(b) Filled well

Excess electron can recombines

Increased potential drop across insulator

Fig. 7.15 (a) The movement of charge carriers created by ionization in the semiconductor layer of an MOS capacitor. Conduction-band electrons will move into the potential well, while valence-band holes move out of the material to ground. There is a net increase in the negative charge in the semiconductor layer. (b) In a saturated device, there is no longer a potential gradient in the semiconductor, so recombination and ionization will be in equilibrium, and there will be no further gain in net negative charge.

ordinary semiconductor. Newly created electrons are no longer stored. The capacitor has exceeded its ***full-well capacity***.

Short of saturation, the MOS capacitor is a conceptually simple detector of light. For every photon below the cutoff wavelength absorbed in the semiconductor layer, the device stores something like one electron. Making a photometric measurement then consists of simply counting these electrons. For an astronomer, this is a wonderful characteristic. It means that a very weak source can be detected by simply exposing the capacitor to light from the source for a time long enough to accumulate a significant number of electrons.

7.6 The p–n junction

A very significant situation arises if a p-type material and an n-type material are brought into contact. Junctions of this sort are the basis for many solid-state electronic devices and for some astronomical detectors. Figure 7.16 illustrates the behavior of charge carriers at a p–n junction. In the figure, we imagine that a block of n-type material has just been brought into contact with a block of p-type material.

Figure 7.16a shows the non-equilibrium situation immediately after contact. The majority charge carriers start to flow across the junction. Electrons in the n-side conduction band will move across the junction and drop down in energy to fill the available acceptor states on the p side (broken lines). Likewise, mobile holes in the valence band of the p-type material will move across the junction to neutralize any electrons in donor states in the n-type material. Opposite charges build up in the doping sites on either side of the junction – excess negative charge on the p side, excess positive charge on the n side. Electrostatic repulsion eventually halts further transfer of carriers across the junction.

Fig. 7.16 The p–n junction; (a) shows the flow of charge carriers immediately after contact between the two regions. Majority carriers (n-side electrons and p-side holes) recombine, fill acceptor sites, and ionize donor sites. The band structure in equilibrium is shown in (b). The accumulation of charges near the junction creates a built-in field, which alters the energy levels of available states so that the Fermi energy is the same everywhere in the crystal.

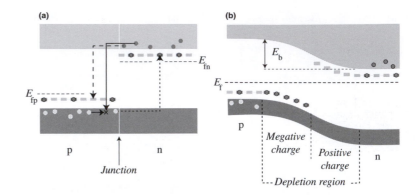

7.6.1 Generation and recombination

Figure 7.16b shows the situation once equilibrium is established. As in the MOS capacitor, a ***depletion region***, constantly swept clear of mobile charge carriers, has formed in the volume surrounding the junction. The lack of charge carriers means this region should have very high electrical resistance. In equilibrium, the energy of the bottom of the conduction band and the top of the valence band changes across the depletion region – it requires work to move an electron from the n region to the p region against the electrostatic force. The potential difference across the depletion zone, E_b, is just sufficient to bring the Fermi energy to the same level throughout the crystal. In equilibrium, charges do move through the depletion region, but the two electric currents here cancel:

$$I_r = -I_g$$

The first current, the ***recombination current***, I_r, is due to the majority carriers that are able to overcome the potential barrier, cross the depletion region, and undergo recombination. This I_r is a positive current that flows from p to n; it has two components: one caused by n-side electrons, the other by p-side holes. Figure 7.17 illustrates the flow of the recombination current, whose magnitude *will depend on the size of the barrier* and on the temperature.

The second current, I_g, the ***generation current***, is due to minority carriers and flows in the opposite direction (from n to p). The minority carriers are thermally ionized conduction-band electrons on the p side and valence-band holes on the n side, which diffuse away from their creation sites. If such a carrier reaches the depletion region, it will be swept across. Diffusion speed outside the depletion region depends on the temperature and the impurity concentration, but is (to first order) *independent* of E_b. Thus, I_g depends on temperature, but in contrast to I_r is virtually independent of the size of E_b.

7.6.2 p–n Junction diodes

The different behaviors of the two currents mean that the p–n junction can function as a ***diode***: it will carry (positive) current in the direction

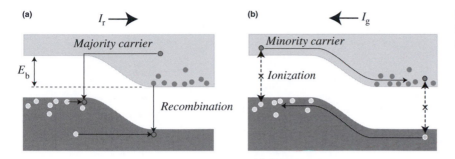

Fig. 7.17 Recombination (a) and generation (b) currents at a p–n junction.

p to n, but not in the reverse direction. Figure 7.18 illustrates the basic process.

In the condition known as **_forward bias_**, a positive voltage connected to the p side of the junction reduces the size of the potential barrier E_b. The recombination current, I_r, will flow more strongly. (That is, more electrons will have energies greater than the barrier, and can move from n to p.) The size of this current will depend in a non-linear fashion on the size of the applied voltage, V_{ext}. The applied voltage, however, does not affect the generation current in the opposite direction, I_g, due to minority carriers. The relatively poor conductivity of the depletion region guarantees that almost all of the potential drop will occur here, and the applied voltage will have little influence on the diffusion rate outside the depletion region. Thus, in the forward bias case, $I_r > -I_g$, and current flows from p to n.

A negative voltage connected to the p side of the junction – a condition known as **_reverse bias_** – increases the size of the potential barrier E_b. This chokes off the flow of majority carriers and lowers I_r from its equilibrium value. Again, the minority carrier current, I_g, remains little changed, so the result of the reverse bias circuit is a very small current in the direction n to p. Boltzmann's law and the above arguments suggests a **_diode equation_** that gives the voltage–current relationship for an "ideal" diode:

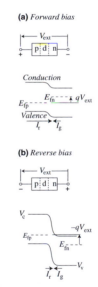

(a) *Forward bias*

Conduction

(b) *Reverse bias*

$$I_{\text{TOTAL}} = I_r + I_g = I_s\left(e^{\frac{qV_{ext}}{kT}} - 1\right) \tag{7.8}$$

Here, q is the electron charge, and current and voltage are assumed to be positive in the p to n direction. You can verify that this formula corresponds to the behavior seen in an actual diode illustrated in Figure 7.19. The formula does not describe the phenomenon of diode breakdown at large reverse biases.

Fig. 7.18 Biased diodes: (a) forward bias reduces the size of the barrier, so the recombination current increases; (b) reverse bias increases the barrier and decreases the recombination current. In both cases, the generation current remains unchanged.

7.6.3 Light detection in diodes

Figure 7.20 shows the result of photo-absorption in a p–n diode. Each absorption of a photon causes an ionization and the creation of a conduction electron and

Fig. 7.19 The current–voltage relation for an ideal p–n diode. The solid line is the relation given by Equation (6.8). The dotted line shows the phenomena of breakdown in real diodes, which become conducting at very negative external voltages.

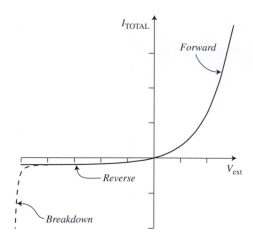

Fig. 7.20 Photo-absorption in a p–n junction diode.

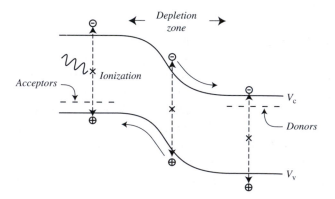

valence hole. This adds a new contribution to the generation current, this one dependant on ϕ, the number of photons that enter the detector per second. The inclusion of a photocurrent modifies Equation (7.8):

$$I_{\text{TOTAL}} = I_{\text{ph}} + I_{\text{r}} + I_{\text{g}} = -q\phi\eta + I_{\text{s}}\left(e^{\frac{qV_{\text{ext}}}{kT}} - 1\right) \tag{7.9}$$

Here η is a factor that depends on the fraction of incident photons absorbed as well as the probability that a generated charge carrier will cross the junction before recombining. Note that charge pairs created in the n or p material must move by diffusion to the junction, as discussed above, and have a finite probability of recombining before crossing the junction. Electron–hole pairs created in the depletion zone, on the other hand, are immediately swept apart by the strong electric field there, and have little chance of recombining. Majority

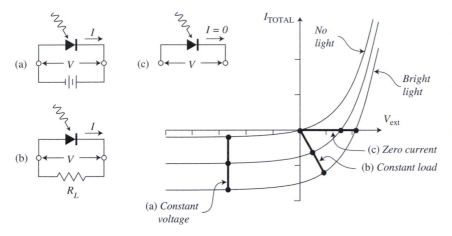

Fig. 7.21 Current–voltage relations for a photodiode at three different levels of incident photon flux. Heavy lines show electrical properties as a function of photon intensity for three different modes of operation.

carriers will thus tend to accumulate on either side of the depletion zone, and the junction will behave like a charge storage capacitor if an external circuit does not remove the carriers.

There are different strategies for employing the light sensitivity of a photo-diode. Figure 7.21 is a plot of Equation (7.9) for three different light levels, as well as three different modes of operation: (a) In the **photo-conductor** mode, a battery holds the external voltage to a constant value, and the current is a linear function of the incident photon flux. (b) In the **power-cell** mode, the diode is connected to a constant-load resistance, and the power output depends on the incident photon flux. This is the principle of operation for solar power cells. (c) In the **photovoltaic** mode, current from the diode is held at zero (making it a storage capacitor by connecting it to a very high impedance voltmeter, for example), and the voltage across it is a non-linear function of the photon flux.

7.6.4 Variations on the junction diode

Some modifications of the simple p–n junction can improve the device's response to light. Several are important in astronomy.

The **PIN diode** sandwiches a thin layer of intrinsic (undoped) silicon between the p material and the n material of the junction. This increases the physical size of the depletion zone, and the resulting p–intrinsic–n (PIN) diode has larger photosensitive volume, higher breakdown voltage, lower capacitance, and better time response than the simple p–n device.

The **avalanche photodiode** is both a physical modification and a mode of operation. Consider a photodiode (a modified PIN type) that is strongly back-biased at close to its breakdown voltage. Because of the large voltage drop across the intrinsic region, charge carriers created by photo-absorption will accelerate to high kinetic energies – high enough to produce impact ionization of additional charge carriers. These secondaries will in turn accelerate to produce further ionizations. The resulting avalanche of carriers constitutes a current pulse that is easy to detect. At low levels of illumination, counting the pulses is equivalent to counting photons. At higher illuminations, the pulses are too close together to count, but the resulting current, although noisy, is very large and therefore easy to detect.

The **Schottky photodiode** is especially useful as a detector in the near- and mid-infrared. Figure 7.22 shows a junction between a metal and p-type silicon as well as the corresponding electron energy bands. At the junction, electrons will spontaneously flow from the metal to neutralize majority holes in the semiconductor until the Fermi levels in the two materials match. The resulting charge transfer sets up a potential barrier at the junction, as well as a depleted zone on the semiconductor side. Light detection occurs because holes created by photo-absorption in the metal layer, if energetic enough, will move across the potential barrier into the depleted region, where they will be swept out into the semiconductor (remember, our band diagrams show energies for electrons, so holes will move upwards in the diagram). The barrier height for holes, E_{ms}, determines the long-wavelength cutoff of the diode; E_{ms} depends on the metal used. Common choices are all metallic silicides, which form easily when a very thin metallic layer is deposited on silicon by evaporation. The most useful is PtSi, with $E_{ms} = 0.22$ eV, and $\lambda_c = 5.6$.

Fig. 7.22 A Schottky photodiode. The upper diagram shows a cross-section of the material structure, and the lower shows the energy bands. Photo-ionized holes produced in the metal move upward in the diagram, and will be detected if they overcome the potential barrier between metal and semiconductor. The reflector sends photons back to the PtSi layer for a second pass.

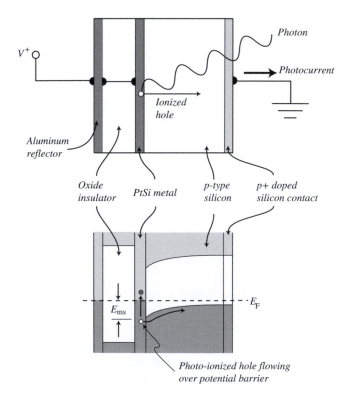

Schottky diodes have relatively low efficiencies, but they also have several virtues: they are very easy to manufacture and mate to read-out electronics. They also tend to have uniform responses and therefore are a good choice for the elements in an array. Their sensitivities extend into the infrared, where intrinsic silicon is useless.

7.7 The vacuum photoelectric effect

The vacuum photoelectric effect depends on the ejection of electrons from the surface of a solid, and has important applications in astronomical detectors. Figure 7.23 illustrates the effect, which is simplest in metals. A thin slab of the metal cesium occupies the left side of the figure, which shows the band structure. We use cesium as an example because it has a relatively loose hold on its surface electrons. The surface of the metal runs vertically down the center. If the potential energy of an electron at rest well away from the metal is zero, then the **work function, W,** of the material is the difference between this free electron energy and the Fermi energy of the solid. In the case of cesium, the work function is 2.13 eV.

We would like to use the energy of one photon to move one electron from the metal to the vacuum. This operation has two requirements: the electron must be given a positive energy, and it must be located at the surface. In general, the absorption of a photon with energy $hv \geq W$ will take place in the interior of the metal, and will promote an electron there into the conduction band. This electron *might* have positive energy. If, after diffusing to the surface, the electron still has both positive energy and an outward-directed momentum (*case* B in the figure) it can move into the vacuum.

A simple device called a **photocell** (or more properly, a **vacuum photodiode**), illustrated in Figure 7.24, uses this effect to measure the intensity of light. In the

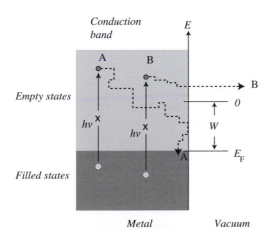

Fig. **7.23** The vacuum photoelectric effect in a metal. Photoelectron B reaches the vacuum with positive energy, while photoelectron A does not. Both photoelectrons make collisions with the lattice, and execute a random walk to the surface. Photoelectrons gradually become thermalized – if the metal is cold, they tend to lose energy on each lattice collision.

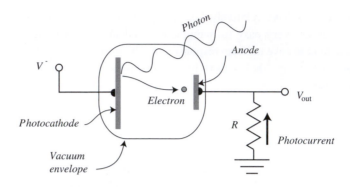

diagram, two conductors are sealed in an evacuated cell with a transparent
window. One conductor, the ***photocathode***, is made from some material (e.g.
cesium) that exhibits an efficient vacuum photoelectric effect. The photocathode
is held at a negative voltage. The other conductor, the anode, is connected
through a load resistor to the ground as illustrated. Illumination of the photocathode ejects electrons into the vacuum. These accelerate to the anode. The
result is an output current and voltage across the resistor that is proportional to
the photon arrival rate at the cathode.

Metals actually make rather poor photocathodes. For one thing, they exhibit
large work functions. (Cesium, the metal with one of the smallest values for W,
will only detect photons with wavelengths shorter than 580 nm.) A second, even
more serious, disadvantage is that metals are highly reflective. Semiconductors
usually make better photocathodes since they are much less reflective. The
photoelectric effect is slightly more complex in a semiconductor, as illustrated
by the band diagram in Figure 7.25a. The zero of energy and the work function
are defined as in a metal, and a new variable, the ***electron affinity***, χ, is defined
as the difference between the zero point and the energy at the bottom of the
conduction band. For a simple semiconductor, as in Figure 7.25a, the electron
affinity is a positive number. Since there are no electrons at the Fermi level in a
semiconductor, the energy required to eject an electron is

$$h\nu \geq E_{\mathrm{G}} + \chi$$

This can be relaxed by creating a p–n junction near the emitting surface.
In Figure 7.25b, the displacement of the energy of the bottom of the conduction
band in the n material near the surface means that a photon with energy
greater than just E_{G} can cause electron emission if it is absorbed in the p-type
material. In this case, the effective electron affinity of the p-type material is a
negative number. The n-type layer is so thin and transparent that its more
stringent energy requirements are not a serious detriment to the cathode's sensitivity to long wavelengths. Materials of this type, termed ***NEA photocathodes***
(negative electron affinity), are usually fabricated with a III–V semiconductor as

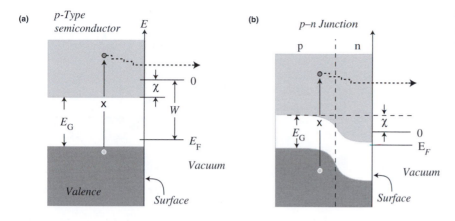

Fig. 7.25 The vacuum photoelectric effect in semiconductors

the p-type material and oxidized cesium as the n-type material. For example, an NEA photocathode made from p-doped gallium arsenide (E_G = 1.4 eV) with a surface layer of n-doped Cs_2O (E_G = 2.0 eV, χ = 0.6 eV) is sensitive out to 880 nm and has been important for some astronomical applications.

We have assumed that emitted photoelectrons will leave from the surface that is illuminated. This need not be the case, and many photocathodes are **semi-transparent**: photons enter on one side and electrons emerge from the opposite side.

7.8 Superconductivity

Superconducting material has an electrical conductivity that falls to zero at and below a critical transition temperature, T_c. The simplest superconductors, and the first investigated, are all metallic elements with very low critical temperatures (T_c < 10 K). These are called type I superconductors. More complex materials (alloys, ceramics, and various exotic compounds) may have higher transition temperatures. Currently (2010), the highest claimed T_c is 242 K.

Type I superconductors are the basis of some potentially important light detectors in astronomy, so we briefly describe their behavior here. The website superconductors.org or the modern physics text by Harris (1998) gives a more complete introduction, and chapter 10 of Kittel (2005) provides a more advanced treatment, as does Blundell (2009).

7.8.1 The superconductor band gap

Above the critical temperature in a superconducting metal like lead, the Fermi–Dirac formula describes the energy distribution of the valence electrons in the conduction band. At T_c (7.19 K for lead) a lattice-mediated force between electrons makes new energy states available below the Fermi level — two spatially separated electrons can form a **Cooper pair** of exactly canceling

Fig. 7.26 Energy bands in
a superconductor. The
band-gap energy is the
energy required to break
apart two electrons
bound in a Cooper pair,
placing them in an
excited quasiparticle
state (dotted arrow). The
density of states just
below and just above the
band gap is very high,
although there are no
states in the gap itself.

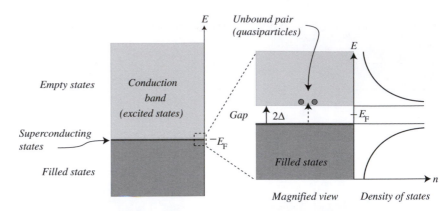

momenta and spins. Each pair has a binding energy well below the thermal
energy of the lattice and, with zero spin, behaves in many ways like a boson –
the Pauli exclusion principle does not apply to these states, and all pairs have the
same momentum (zero, when there is no current). It is the Cooper pair states that
are responsible for superconductivity and many resultant behaviors – including
perpetual electric currents and magnetic levitation.

Our concern, however, is the manner in which a superconductor interacts with
light. Figure 7.26 shows the special energy-band diagram for a superconductor.
At temperatures below T_c, an unlimited number of superconducting states exist at
an energy Δ below the Fermi level. Single electrons will therefore occupy only
states of energy $(E_F - \Delta)$ or lower. The value of Δ is a strong function of
temperature, rising from zero at T_c to a maximum value of Δ_m at temperatures
below about $0.3\,T_c$. The value for Δ_m, which measures the binding energy per
electron of a Cooper pair, is tiny, 1.4×10^{-3} eV for lead, which is typical.

Consider what must happen for a superconductor to absorb a photon: only if
the photon has energy larger than 2Δ can it break apart a Cooper pair and
promote the two electrons to higher energies. Lower energy photons will not
be absorbed: the material has an effective band gap of magnitude 2Δ, as shown
in Figure 7.26. The electrons promoted to the excited states in the "conduction"
band in the superconductor have quantum characteristics that differ from energetic electrons in an ordinary metal, and are therefore termed **quasiparticles**. For
example, the number of states available to quasiparticles at energies just above
the gap is very large. Table 7.6 lists the gap energies and transition temperatures
of a few superconductors that have been useful in astronomical detectors.

7.8.2 Light detection in an SIS junction

Two superconductors separated by a thin layer of insulator (SIS = superconductor–insulator–superconductor) constitute a **Josephson junction** if the insulator is thin enough (around 1 nm) to permit quantum-mechanical tunneling.

Table 7.6. *Some type I superconductor characteristics. Data from Kittel (2005)*

Atom	T_c, K	Band gap $2\Delta_m \times 10^4$ eV
Al^{13}	1.14	3.4
Nb^{41}	9.5	30.5
Hf^{72}	0.12	0.4
Ta^{73}	4.48	14

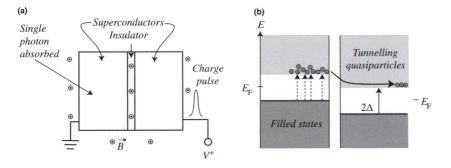

(a) Single photon absorbed — Superconductors Insulator — Charge pulse — $\vec{B}$ — V^+

(b) E — Tunnelling quasiparticles — E_F — Filled states — 2Δ — $-E_F$

Fig. 7.27 An STJ diode. (a) A cross-section of the physical device. In most practical detectors the three layers and their contacts are deposited as films on a transparent substrate, so a more accurate diagram would extend vertically several page heights. The band structure is shown in (b). Not shown is the possibility that quasiparticles on the right can tunnel back to the left to break apart additional pairs.

Figure 7.27 shows such a junction arranged as a light-detecting diode: a positive bias voltage less than $2\Delta/q$ is applied to the right-hand superconductor, and a magnetic field is applied parallel to the junction. If the junction is very cold, all excited states are empty. In a normal Josephson junction it would be possible for the Cooper pairs to tunnel from left to right, but the magnetic field suppresses that current, so the diode does not conduct.

If the left-hand superconductor absorbs a single visible light photon of wavelength λ (energy hc/λ), it receives enough energy to break apart multiple Cooper pairs, promoting a maximum of $hc/\lambda\Delta$ electrons into excited states. These quasiparticles *can* tunnel across the insulator, and those that do produce a current pulse whose total charge is inversely proportional to the wavelength of the exciting photon.

Devices of this kind, called ***superconducting tunnel junctions (STJs)***, operated with sufficient time resolution, can count individual incoming photons and determine the wavelength (from X-ray to near infrared) of each. The uncertainty of the wavelength determination depends on the wavelength. Although still very much in the development stages, a few experimental but practical multi-pixel STJ-based detectors have begun to appear at telescopes. See chapter 4 of Rieke (2003) and the references by Peacock *et al.* (1997) and Verhoeve *et al.* (2004).

Superconducting tunnel junctions promise to be the near-ideal astronomical detector: They can be fashioned into an array that produces an image yielding both spectroscopic information and high time resolution. Especially because they must operate at milli-kelvin temperatures, there are formidable engineering issues in developing them as practical and affordable astronomical tools, but there is no doubt about their superiority as detectors.

Summary

- Quantum mechanics accounts for a quantized pattern of permitted states for the energies, angular momenta, magnetic interactions, and spins of electrons bound to an isolated atom. Concepts

free state	*ground state*	*Pauli exclusion principle*
valence electron	*periodic table*	*quantum number*
fermion	*boson*	*spectroscopic notation*

- The outer (optical) electrons of an atom gain or lose energy by making transitions between permitted states. Concept:

excitation	*photo-emission*	*photo-ionization*
ground state	*photo-absorption*	*absorption edge*
thermal excitation	*Boltzmann distribution*	

- Permitted quantum states of isolated molecules are distinguished by the electronic states of their component atoms, but also by the quantized rotation and vibration states of the molecule. Concept:

  molecular absorption bands

- The energy states for electrons in solid-state crystals typically arrange themselves in continuous bands separated by forbidden band gaps. Concepts:

diamond lattice	*sp^3-hybrid orbitals*	*bonding state*
antibonding state	*valence band*	*conduction band*
holes	*semiconductor*	*intrinsic semiconductor*
electron degeneracy	*phonon*	*Fermi–Dirac statistics*
Fermi energy	*band-gap energy*	*cutoff wavelength*

- Adding small quantities of a selected impurity can produce desirable properties in a semiconductor. Concepts:

dopant	*extrinsic semiconductor*	*donor atom*
acceptor atom	*p-type*	*n-type*
impurity band		

- Photoconductors absorb a photon and create an electron–hole pair, thereby increasing the electrical conductivity of the material. Concepts:

absorption coefficient	*breakdown*	*absorption depth*

- The MOS capacitor is a block of extrinsic semiconductor separated from a metal electrode by a thin layer of insulation. With the proper voltage across the insulator, the device can store charges produced by photo-absorptions. Concepts:

  SiO_2 depletion region   potential well

  full-well capacity   saturation

- The p–n junction produces a depletion region where photo-absorptions can generate charge carriers and an electric current. Concepts:

  p–n junction   recombination current   generation current
  diode   forward bias   reverse (back) bias
  breakdown   diode equation   p-n photodiode
  avalanche photodiode   PIN photodiode   Schottky photodiode

- Electrons can leave the surface of material in a vacuum if they have energies greater that the material's work function. Photons can supply the needed energy, and thus produce an electric current in a vacuum.

  photocathode   vacuum photodiode   anode
  electron affinity   NEA photocathode

- A superconducting junction diode produces a number of conduction-band electrons that is proportional to the energy of the incoming photon. An SJD in pulse-counting mode can therefore measure both the intensity and the wavelength distribution of a source. Concepts:

  Cooper pair   Josephson junction   quasiparticles

Exercises

1. Using the nl^x notation, write down the electron configuration for the ground state, first excited state, and third excited state of iron (atomic number 26) as suggested by Table 7.2.

2. There are several exceptions to the configuration-filling scheme presented in Table 7.2. The configuration of the ground state of copper is an example. Look up a table of electron configurations in atoms and find at least five other examples.

3. Suppose a certain diatomic molecule has an energy-level diagram similar to Figure 7.3 and consider only transitions within the $\Lambda = 0$ states. Suppose that relative to the ground state, state ($J = 1$, $v = 0$) has an energy of 1 eV. Suppose also that, no matter what the rotational state is, the relative energies of the lowest vibrational states are $v(v + 1)d$, where $d = 10^{-5}$ eV and v is the vibrational quantum number. (a) Compute the wavelengths of all permitted emission lines arising between levels $J = 0$ and $J = 1$, and involving vibrational states $v = 0, 1, 2, 3, 4$. The only permitted transitions are the ones in which $\Delta v = \pm 1$. (b) Sketch the emission spectrum for these lines.

4. Compute the relative probability of finding an electron at the bottom of the conduction band relative to the probability of finding an electron at the top of the valence band in a silicon crystal at a temperature of (a) 3 K and (b) 300 K. Use Fermi–Dirac statistics. Compare your answer with the one given by the Boltzmann equation.

5. Compute the fraction of incident photons absorbed by a 100-µm-thick layer of bare silicon if the photons have wavelength (a) 500 nm and (b) 800 nm. Assume the index of refraction of silicon is 4.4 at 500 nm and 3.8 at 800 nm.

6. How does an MOS capacitor made of an n-type semiconductor work? Why do you think p-type material is usually preferred for these devices?

7. Derive a relationship between the full-well capacity of an MOS capacitor and the maximum possible relative precision that the device can produce in a brightness measurement. What is the risk in planning to achieve this precision with a single measurement?

8. Assume you have a meter that measures electric current with an uncertainty (noise) of 100 picoamps. (One picoamp $= 10^{-12}$ amp $= 10^{-12}$ coulomb s^{-1}). You employ your meter with a photodiode in a circuit like the one in Figure 7.18a. You have a 2-meter telescope at your disposal, and use a filter to limit the light received to those wavelengths at which the detector is most sensitive. Compute the magnitude of the faintest star you can detect with this system. "Detect" in this case means the signal-to-noise ratio is greater than 3. Assume the photon flux from a zero-magnitude star in the bandpass you are observing is 10^{10} photons m^{-2} s^{-1}. Your photodiode detects 45% of the photons incident, and you may ignore any background signal.

9. In response to an incoming photon, a niobium-based STJ diode detects a pulse of 500 electrons. Assume tunneling operates with 100% efficiency, and the only source of noise is counting statistics. (a) Compute the energy of the incoming photon and its uncertainty. (b) What is the wavelength of the photon and its uncertainty? (c) Compute the spectroscopic resolution ($R = \delta\lambda/\lambda$) of this device as a function of wavelength. (d) Find the equivalent expression for a device in which the superconductor is hafnium instead of niobium.

Chapter 8
Detectors

Honestly, I cannot congratulate you upon it. Detection is, or ought to be, an exact science, and should be treated in the same cold and unemotional manner. You have attempted to tinge it with romanticism, which produces much the same effect as if you worked a love-story or an elopement into the fifth proposition of Euclid.

"But romance was there," I remonstrated.

– Arthur Conan Doyle, *The Sign of the Four*, 1890

Astronomical detection, even more than the work of Sherlock Holmes, is an exact science. Watson, though, has an equally important point: no astronomer, not even the coldest and most unemotional, is immune to that pleasant, even romantic, thrill that comes when the detector *does* work, and the Universe *does* seem to be speaking.

An astronomical detector receives photons from a source and produces a corresponding *signal*. The signal characterizes the incoming photons: it may measure their rate of arrival, their energy distribution, or perhaps their wave phase or polarization. Although detecting the signal may be an exact science, its characterization of the source is rarely exact. Photons never pass directly from source to detector without some mediation. They traverse both space and the Earth's atmosphere, and in both places emissions and absorptions may modify the photon stream. A telescope and other elements of the observing system, like correcting lenses, mirrors, filters, optical fibers, and spectrograph gratings, collect and direct the photons, but also alter them. Only in the end does the detector do its work. Figure 8.1 illustrates this two-stage process of signal generation: background, atmosphere, telescope, and instruments first modify light from the source; then a detector detects.

An astronomer must understand both mediation and detection if she is to extract meaning from measurement. This chapter describes only the second step in the measurement process, detection. We first outline the qualities an astronomer will generally find important in any detector. Then we examine a few important detectors in detail: the CCD, a few photo-emissive devices, the infrared array, and the bolometer.

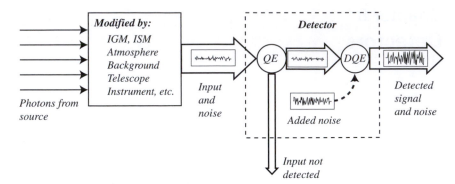

Fig. 8.1 Mediation and detection of a light signal (IGM = intergalactic medium, ISM = interstellar medium). The detection step may fail to record some of the mediated signal, and may introduce additional noise to the part of the signal that is recorded.

8.1 Detector characterization

Why does an astronomer choose one detector instead of another? Why did optical astronomers in the 1980s largely abandon photography, the then-dominant detector for imaging, in favor of solid-state arrays? Why are these same arrays useless for other purposes, such as measuring very rapid changes in brightness? Is there a *perfect* detector? We begin an answer with a list of several critical characteristics of any detector.

8.1.1 Detection modes

We can distinguish three distinct modes for detecting light.

Photon detectors produce a signal that depends on an individual photon altering the quantum-mechanical state of one or more detector electrons. For example, in the last chapter, we saw how a change in electron energy in a photoconductor or photodiode can produce a change in the macroscopic electrical properties like conductivity, voltage, or current. Other changes in quantum state might produce chemical reactions (as in photography) or a pulse of free electrons, as in vacuum photomultipliers. Photon detectors are particularly suited to shorter wavelengths (infrared and shorter), where the energies of individual photons are large compared to the thermal energies of the electrons in the detector.

Thermal detectors absorb the energy of the incoming photon stream and convert it into heat. In these devices the signal is the temperature change in the body of the detector. Although thermal detectors are in principle useful at all wavelengths, in practice, thermal detectors, especially a class called **bolometers**, have been fundamentally important in the infrared and microwave regions, as well as very useful in the gamma and X-ray regions.

Wave detectors produce signals in response to the oscillating electric or magnetic field of the incoming electromagnetic waves, usually by measuring the interference effect the incoming fields have on a wave produced by a local oscillator. In principle, these detectors, unlike photon and thermal detectors, can gauge the phase, intensity, and polarization of the detected wave. Wave detectors are especially useful in the radio and microwave parts of the spectrum.

8.1.2 Efficiency and yield

Thou shalt not waste photons.

— Anonymous, *c*. 1980

A good detector is efficient. We construct costly telescopes to gather as many photons as possible, and it seems perverse if a detector does not use a large fraction of these expensive photons to construct its signal.

Photography, for example, is relatively inefficient. The photographic detector, the emulsion, consists of a large number of tiny crystals, or **grains**, of silver halide (usually AgBr) suspended in a transparent gelatin matrix. Photons can interact with a grain to eventually turn the entire grain into elemental silver. The more silver grains present in the emulsion after it has been processed, the stronger is the signal.

Why is the process inefficient? Some photons reflect from the surface of the emulsion and are not detected. Some pass right through the emulsion, while others are absorbed in its inactive parts without contributing to the signal. Nevertheless, silver halide grains absorb something like 40–90% of the incident photons. These absorbed photons produce photoelectrons that can induce a chemical change by reducing a silver ion to a neutral atom. The corresponding neutral bromine atom (the hole produced by photo-absorption) can vanish, either combining with the gelatin or with another bromine to form a molecule that escapes the crystal. Most holes do not vanish, however, and most photo-electrons recombine with holes before they can neutralize a silver ion. Some neutral silver atoms are created, but most are re-ionized by holes before the grain can be developed. Finally, it is only after three to six silver atoms drift and clump together at a spot on the grain that the crystal becomes developable. In the end, very few of the incident photons actually have an effect in photography. The process is inefficient.

The **quantum efficiency, QE**, is a common measure of detector efficiency. It is usually defined as the fraction of photons incident on the detector that actually contribute to the signal.

$$QE = \frac{N_{\text{detect}}}{N_{\text{in}}} \tag{8.1}$$

In a perfect detector, every incident photon would be absorbed in a fashion that contributed equally to the signal, and the detector would have a QE of 100%.

Photographic emulsions have QE valuess in the range 0.5–5%.[1] Solid-state devices – like silicon photodiodes, superconducting tunnel junction (STJ) diodes, or metal-oxide-semiconductor (MOS) capacitors – have QE values in the 20%–95% range. Astronomers prefer these devices, in part, because of their high quantum efficiencies.

The quantum efficiency of a particular device is not always easy to measure, since (as in photography) the chain of events from incident photon to detection may be difficult to describe and quantify. *Absorptive quantum efficiency* is physically more straightforward, but somewhat less informative. It is defined as the photon flux absorbed in the detector divided by the total flux incident on its surface:

$$\eta = \frac{N_{abs}}{N_{in}}$$

Because absorbed photons are not necessarily detected, $QE \leq \eta$.

The *quantum yield* of a photon detector is the number of detection "events" per incident photon. For example, in silicon photoconductors, the detection event is the production of an electron–hole pair. If an incident photon has energy less than about 5 eV, it can produce at most one electron–hole pair, so the quantum yield is 1. For higher energy photons, a larger number of pairs are produced, around one e–h pair per 3.65 eV of photon energy. What happens in detail is that the first electron produced has so much kinetic energy that it can collide with the lattice to produce phonons that generate additional pairs. A 10-angstrom X-ray, therefore, will yield (on average) 34 photoelectrons. An STJ-based detector, you will recall, is particularly attractive because of its very large, wavelength-sensitive quantum yield.

8.1.3 Noise

> There are two kinds of light – the glow that illuminates, and the glare
> that obscures.
>
> – James Thurber (1894–1961)

Although efficiency in a detector is important, what really matters in evaluating a measurement is its uncertainty. The uncertainty in the output signal produced by a detector is often called the *noise*, and we are familiar with the use of the *signal-to-noise ratio*, *SNR*, as an indication of the quality of a measurement. It

---

[1] Quantum efficiency is a bit of a slippery concept in photography. For example, once a grain has formed a stable clump of three–six silver atoms, absorbed photons can make no further contribution to the signal, even though they create additional silver atoms. The entire grain is either developed or not developed depending only on the presence or absence of the minimum number of atoms. In photography, QE is thus a strong function of signal level – the highest efficiencies only apply if the density of developed grains is relatively low.

would seem that a perfect detector would produce a signal with zero noise. This is not the case.

You will recall that there is an uncertainly *inherent* in measuring the strength of any incident light ray. For a photon-counting device, this uncertainty arises from the Poisson statistics[2] of photon arrivals, and is just

$$\sigma = \sqrt{N}$$

where N is the number of photons actually counted. A perfect detector, with QE = 1, faithfully counts all incident photons and will therefore produce

$$(\text{SNR})_{\text{perfect}} = \frac{N_{\text{out}}}{\sigma_{\text{out}}} = \frac{N_{\text{in}}}{\sigma_{\text{in}}} = \sqrt{N_{\text{in}}}$$

Real detectors will differ from this perfect detector by either counting fewer photons (reducing the output noise, but also reducing both the output signal and the output SNR) or by exhibiting additional noise sources (also reducing the SNR). The ***detective quantum efficiency (DQE)*** describes this departure of a real detector from perfection. If a detector is given an input of N_{in} photons and has an output with signal-to-noise ratio $(\text{SNR})_{\text{out}}$, then the DQE is defined as a ratio:

$$\text{DQE} = \frac{(\text{SNR})_{\text{out}}^2}{(\text{SNR})_{\text{perfect}}^2} = \frac{N_{\text{out}}}{N_{\text{in}}} \tag{8.2}$$

Here N_{out} is a fictitious number of photons, the number that a perfect detector would have to count to produce a signal-to-noise ratio equal to $(\text{SNR})_{\text{out}}$. The DQE gives a much better indication of the quality of a detector than does the raw QE, since it measures how much a particular detector degrades the information content of the incoming stream of photons. For a perfect detector, DQE = QE = 1. For any detector, it should be clear from Equation (8.2) that DQE $\leq$ QE. If two detectors are identical in all other characteristics, then you should choose the detector with the higher DQE. If a parameter (wavelength of the observation, for example) affects both the input signal and the DQE, then you should choose a value that maximizes the value

$$(\text{Signal})_{\text{in}} \sqrt{(\text{DQE})} = (\text{SNR})_{\text{out}}$$

Returning to the example of the photographic emulsion, the noise in an image is experienced as ***granularity***: the microscopic structure of, say, a star image consists in an integral number of developed grains. Statistically, counting grains in an image is a Poisson process, and has an uncertainty and a SNR of $\sqrt{N_{\text{grains}}}$.

---

[2] Although we have been treating the photon-counting process as if it were perfectly described by Poisson statistics, both theory and experiment show this is not the case. Photon arrivals are not statistically independent – real photons tend to clump together slightly more than Poisson would predict. This makes little practical difference in the computation of uncertainties.

Since it takes something like 10–20 absorbed photons to produce one developed grain, the photographic process clearly degrades SNR. In addition, grains are not uniformly distributed in the emulsion, and some grains not activated by photons will nevertheless get developed to produce a background "fog." Both of these effects contribute noise, and thus reduce the DQE. A typical emulsion might have $\eta = 0.5$, QE = 0.04 and DQE = 0.02. Many solid-state detectors do not degrade the information in the input to anything like the degree that photography does, and their DQE values are close to their QE values – in the range 20–90%.

The DQE generally is a function of the input level. Suppose, for example, a certain QE = 1 detector produces a background level of 100 electrons per second. You observe two sources. The first is bright. You observe it for 1 second, long enough to collect 10,000 photoelectrons (so $SNR_{in} = 100$). For this first source, $SNR_{out} = 10,000/\sqrt{(10,100 + 10^2)} = 98$, and DQE = 0.96. The second source is 100 times fainter. You observe it for 100 seconds, and also collect 10,000 photoelectrons. For the second source, $SNR_{out} = 10,000/\sqrt{20,000 + 10,000} = 57.8$, and DQE = 0.33.

8.1.4 Spectral response and discrimination

The QE of a detector is generally a function of the wavelength of the input photons. Some wonderful detectors are useless or have low QE at some wavelengths. Silicon devices, for example, cannot respond to photons with $\lambda > 1.1$ μm since these photons have energies below the silicon band-gap energy. The precise relationship between efficiency and wavelength for a particular detector is an essential characteristic.

One can imagine an ideal detector that measures both the intensity and the wavelength distribution of the incoming beam. An STJ diode, operated in a pulse-counting mode, for example, *discriminates* among photons of different wavelength.

8.1.5 Linearity

In an ideal detector, the output signal is directly proportional to the input illumination. Departures from this strict linearity are common. Some of these are not very problematic if the functional relation between input and output is well known and well behaved. For example, in the range of useful exposures, the density of a developed photograph is directly proportional to the logarithm of the input flux. Figure 8.2 illustrates two very typical departures from linearity. At lower light levels, a detector may not respond at all – it behaves as if there were an input *threshold* below which it cannot provide meaningful information. At the other extreme, at very large inputs, a detector can *saturate*, and an upper threshold limits its maximum possible response. Further increases in input will not move the output signal above the saturation level.

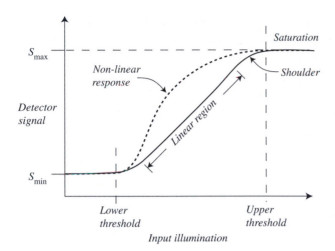

Fig. 8.2 Linear and non-linear regions in a typical detector response curve. The dashed response is completely non-linear.

8.1.6 Stability

The environment of a detector will change over time, perhaps because of variation in temperature, atmospheric conditions, or orientation with respect to gravity or to local magnetic fields. The detector itself may age because of chemical or mechanical deterioration, electrical damage, or radiation and particle exposure. Unrecognized changes can introduce systematic effects and increase uncertainties.

Two general approaches cope with detector instability. The first is to avoid or minimize anticipated changes: e.g. use thermostatic controls to maintain a constant temperature, keep the detector in a vacuum, shield it from radiation, use fiber-optic feeds so that the detector remains motionless. Basically, employ whatever strategies seem reasonable to isolate the detector from the environment. The second approach is to recognize that some changes are unavoidable and calibrate the detector to correct for the instability. For example, if the response of a detector deteriorates with age, make repeated observations of the same standard source so you can compute a correction that compensates for the deterioration.

Hysteresis is a form of detector instability in which the detector response depends on its illumination history. Human vision, for example, exhibits the phenomenum of positive and negative afterimage. Some solid-state detectors can continue to report ghost signals from bright objects long after the source has been removed.

8.1.7 Response time

How quickly can the detector make and report a measurement, then make and report the next measurement? The minimum time required is an important parameter. Readout procedures for large CCDs, for example, can limit their response time to a hundred seconds or more, while STJs and photo-emissive devices have sub-millisecond response times.

8.1.8 Dynamic range

What is the maximum range in output signal that the detector will produce in response to input? From Figure 8.2, you might surmise (correctly) that the upper and lower detection thresholds limit the dynamic range. However, other details of the detection process can influence the dynamic range. For example, if the signal is recorded digitally as a 16-bit binary integer, then the smallest possible signal is 1, and the largest is $65,535$ ($= 2^{16} - 1$). Thus, even if the range set by saturation is larger, the dynamic range is limited by data recording to $1:65,535$.

8.1.9 Physical size and pixel number

The physical size of the detector can be very important. To measure the light from a single star in the telescope focal plane, for example, it will be advantageous to match the detector size with the image size produced by the telescope: if the detector is too small, it will not intercept all the light from the source; if it is too large, it will intercept unwanted background light and probably produce a higher level of detector noise. For some detectors, physical size is related to other properties like dynamic range and response time.

A *single-channel* detector measures one signal at a time, while a *multi-channel* detector measures several at once. An astronomer might use a simple two-channel detector, for example, to simultaneously measure the brightness of a source and the brightness of the nearby background sky. A *linear array* (a string of closely packed detectors arranged in a straight line) might be a good configuration for sensing the output of a spectrograph. A *two-dimensional array* of detectors can record all parts of an astronomical image simultaneously.

Clearly, the physical size of each detector of an array determines how closely spaced its elements, or *pixels* (for *picture element*) can be. Sometimes there must be some inactive area between the sensitive parts of the pixels, sometimes not. Large arrays are more easily manufactured for some types of detectors (e.g. MOS capacitors) than for others (e.g. bolometers and wave detectors). There is an obvious advantage in field of view for detectors with a large number of pixels.

Astronomers currently employ mosaics of solid-state arrays of up to one billion pixels, with the largest individual arrays (CCDs of up to 100 megapixels in size) finding application in the X-ray through optical regions. Somewhat smaller arrays (1–4 megapixel) are in use at near-infrared (NIR) and mid-infrared (MIR) wavelengths. Focal-plane arrays of hundreds of pixels are used on some far-infrared (FIR) and sub-millimeter telescopes. Radio detectors are almost always single-pixel or few-pixel devices. At the beginning of the CCD era, photographic plates had a clear advantage in pixel number: for a very moderate cost, a photographic plate had a very large area (tens of centimeters on a side), and thus, in effect, contained up to 10^9 pixels. Mosaics of CCD arrays, although quite expensive, now match the size of medium-sized photographic plates.

8.1.10 Image degradation

Astronomers go to extremes to improve the resolution of the image produced by a telescope – minimize aberrations, launch the telescope into space, and create active and adaptive optics systems. Two-dimensional detectors like arrays should preserve that resolution, but in practice can often degrade it. *Sampling theory* was originally developed to understand electronic communications in media such as radio broadcasting and music reproduction. The Nyquist theorem states that the sampling frequency of a waveform should be greater than two times the highest frequency present in the wave. Extending this theorem to the spatial domain means that to preserve maximum detail, pixel-to-pixel spacing should be less than the *Nyquist spacing*. The Nyquist spacing is one-half the full width at half-maximum (FWHM) of the point-spread function of the telescope. If pixel spacing is larger than the Nyquist value, the resulting *under-sampling* of the image degrades resolution.

Other effects can degrade resolution. Signal can drift or bleed from its pixel of origin into a neighboring pixel, or photons can scatter within the array before they are detected.

8.2 The CCD

> One morning in October 1969, I was challenged to create a new kind of computer memory. That afternoon, I got together with George Smith and brainstormed for an hour or so. ...When we had the shops at Bell Labs make up the device, it worked exactly as expected, much to the surprise of our colleagues.
>
> — Willard Boyle, Canada Science and Technology Museum, 2008

When Boyle and Smith (1971) invented the first *charge-coupled devices* at Bell Laboratories in 1969 they quickly recognized the CCD's potential as multi-pixel light detector instead of a computer memory. By 1976, astronomers had recorded the first CCD images of celestial objects.[3] Since that time, the CCD has become a standard component in applications that include scanners, copiers, mass-market still and video cameras, surveillance and medical imagers, industrial robotics, and military weapon systems. This large market has diluted the research and development costs for astronomy. The consequent rapid evolution

---

[3] The first CCD images reported from a professional telescope were of the planets Jupiter, Saturn, and Uranus, taken in 1976 by Bradford Smith and James Janesick with the LPL 61-inch telescope outside Tucson, Arizona.

of the scientific CCD has profoundly revolutionized the practice of optical observational astronomy. This section gives a basic introduction to the principles of operation of the CDD and its characteristics as a detector.

8.2.1 General operation

Recall how an MOS capacitor stores photoelectrons in a potential well. A CCD is an array of MOS capacitors (one capacitor per pixel) equipped with circuitry to read out the charge stored in each pixel after a timed exposure. This read-out scheme (called "charge-coupling") moves charges from one pixel to a neighboring pixel; pixel-by-pixel shifting is what makes the array a CCD, rather than something else.

The basic ideas behind the array operation are simple. Imagine a matrix of MOS capacitors placed behind a shutter in the focal plane of a telescope. To take a picture, we first make sure all the capacitor wells are empty, open the shutter for the exposure time, then close the shutter. While the shutter is open, each pixel accumulates photoelectrons at a rate proportional to the rate of photon arrival on the pixel. At the end of the exposure, the array stores an electronic record of the image.

Figure 8.3 describes how the CCD changes this stored pattern of electrons into a useful form – numbers in a computer. In Figure 8.3a we show the major components of the detector. There is the light-sensitive matrix of MOS capacitors: in this case an array three columns wide by three rows tall. A column of pixels in the light-sensitive array is called a ***parallel register***, so the entire light-sensitive array is known collectively as the parallel registers. There is one additional row, called the ***serial register***, located at the lower edge of the array and shielded from light. The serial register has one pixel for each column of parallel registers (in this case, three pixels). Both the serial and parallel register structures are fabricated onto a single chip of silicon crystal.

Reading the array requires two different charge-shifting operations. The first (Figure 8.3b) shifts pixel content down the columns of the parallel registers by one pixel. In this example, electrons originally stored in row 3 shift to the serial register, electrons in row 2 move to row 3, electrons in row 1 move to row 2. Just before this first shift is initiated, the serial register is cleared of any charges that may have accumulated before or during the exposure.

The second operation now reads the newly filled serial register by shifting its contents to the right by one pixel (Figure 8.3c1). The electrons in the rightmost pixel shift into a new structure – a series of ***output amplifiers*** – that ultimately converts the charge to a voltage. This voltage is in turn converted to a binary number by the next structure, the ***analog-to-digital converter (ADC)***, and the number is then stored in some form of computer memory. The CCD continues this shift-and-read of the serial register, one pixel at a time (Figures 8.3c2 and 8.3c3) until all serial register pixels have been read.

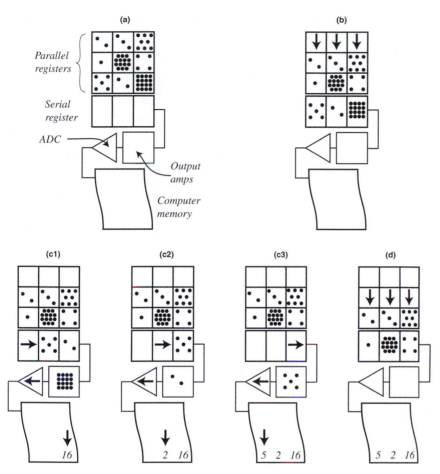

Fig. 8.3 CCD components and readout. (a) The accumulated photo-electrons in a 3×3 array of capacitors – the parallel register. (b) Shift of the bottom row into the serial register, all remaining rows shift down in the parallel register. (c) Read of the serial register one column at a time. (d) Next row shifts down into the empty parallel register.

Now the whole operation repeats for the next row: there is another shift of the parallel registers to refill the serial register with the next row (Figure 8.3d); the serial register is in turn read out to memory. The process continues (parallel shift, serial shifts, and reads) until the entire array has been read to memory. The first stage of the output amplifier is usually fabricated onto the same silicon chip as the registers. The subsequent amplifiers and the ADC are usually located in a separate electronics unit.

How does the CCD persuade the electrons stored in one capacitor to move to the neighboring capacitor? Many strategies are possible, all of which depend upon manipulating the depth and location of the potential well that stores the electrons. A parallel or serial register is like a bucket brigade. The bucket (potential well) is passed down the line of pixels, so that its contents (electrons) can be dumped out at the end. Figure 8.4 illustrates one strategy for moving the well. The depth of a potential well depends on the voltage applied to the metal, and is greatest at the Si–SiO$_2$ junction, closest to the metal layer. (See, however

Detectors

Fig. 8.4 Gate structure in
a three-phase CCD. Two
pixels are shown in cross-
section. Collection and
barrier potentials on the
gates isolate the pixels
from each other during
an exposure.
Overlapping gates
produce a gradient in the
barrier region (dashed
curve in lower figure) that
enhances collection.

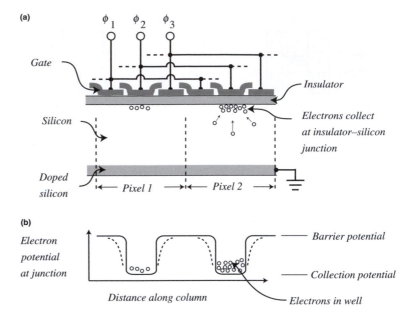

the section below on the buried-channel CCD.) The figure shows two pixels in the same register (column) of a ***three-phase CCD***. In this device, the metal electrode is separated into three ***gates***, and these are interconnected so that gate 1 of every pixel connects to gate 1 of every other pixel, and likewise for gates 2 and 3. Thus, a single pixel can simultaneously have three separate voltages or ***phases*** applied to its front side, producing a corresponding variation in the depth of the potential well, as illustrated in the figure. The interconnection of gates insures that the pattern of well depth is identical in every pixel of the register.

Setting the correct voltages on three separate gates implements both charge-shifting and pixel isolation. For example, during an exposure, phase 2, the voltage on the central metal electrode, can be set to a large positive value (say 15 V), producing what is known as the ***collection potential*** in the semi-conductor. The other two phases are set to a smaller positive voltage (say 5 V), which produces the ***barrier potential***. The barrier potential maintains the depletion region in the silicon, but prevents electrons from drifting across pixel boundaries. Photoelectrons generated in the barrier region of the silicon will diffuse into the nearest deep well under the collection phase and remain there. Each isolated pixel thus stores only charges generated within its boundaries.

To illustrate how the three gates might be used for charge shifting, assume again that the pixels are isolated during an exposure with collection under phase 2 ($\varphi_2 = +15$ V) and a barrier under the other phases ($\varphi_1 = \varphi_3 = +5$ V).

Figure 8.5 illustrates the three voltage changes that will shift charges by one pixel.

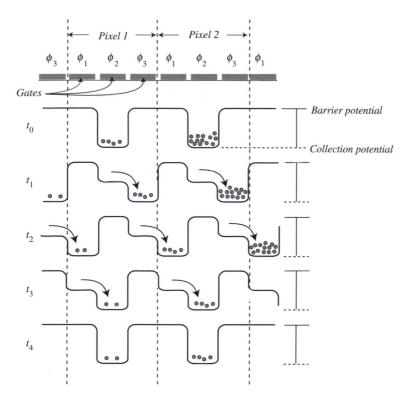

Fig. 8.5 Shifting potential wells in a three-phase CCD. See Figure 8.4 for the corresponding physical structure. Two pixels in the same register (either parallel or serial) are illustrated here. At the end of the shift, electrons stored in pixel 1 have shifted to pixel 2.

1. At time t_1, gate voltages change so that $\varphi_3 = 15$ V and $\varphi_2 = 10$ V. The electrons under φ_2 will diffuse to the right, and collect under φ_3.
2. At time t_2, after a delay that is long enough for all electrons to diffuse to the new location of the deep well, voltages change again, so that $\varphi_1 = 15$ V, $\varphi_3 = 10$ V and $\varphi_2 = 5$V. Stored electrons drain from phase 3 of the original pixel to phase 1 of the neighboring pixel.
3. A third cycling of gate voltages ($\varphi_1 = 10$ V, $\varphi_2 = 15$ V and $\varphi_3 = 5$ V) brings the electrons to the middle of the pixels at time t_3, and the one-pixel shift is complete.

The values of the barrier and collection potentials are somewhat arbitrary, but there are usually some fairly well-defined optimal values. These values, along with the properties of the insulator layer, determine required values of the **clock voltages** (the input values for φ_1, φ_2, and φ_3). An electronic system called the **CCD controller** or **CCD sequencer** sets the clock voltages and manages the very precise timing of their changes. The controller, usually built around a simple microprocessor, is generally housed in the same electronics box as the ADC and output amplifiers. Alternatively, the controller can be a program on a general-purpose computer. Besides manipulating the clock voltages, the controller also performs and coordinates several other functions, generally including:

- clearing the appropriate registers before an exposure and or a read;
- opening and closing the shutter;

- controlling the sequence of reads of the parallel and serial registers, including the patterns for special reads (see the discussions of on-chip binning and windowing below);
- controlling the parameters of the output amplifiers and the ADC (in particular, setting two constants called the **bias level** and the **CCD gain** discussed below);
- communicating with the computer that stores the data.

Two-phase and four-phase readout schemes are also sometimes used in CCDs. Most modern consumer digital cameras utilize arrays of (complementary-metal-oxide-semiconductor) **CMOS** capacitors, in which individual output amplifiers are fabricated onto the front side of each pixel. This design means that the pixels can be read out in parallel, rather than one at a time. The CMOS detectors are less expensive than CCDs of the same size, consume less power, and read out very rapidly (around 70 megapixels per second). They have not seen much use in astronomy, since they suffer from much higher read noise, dark current, pixel-to-pixel charge diffusion, and (usually) lower QE; however, they are gradually becoming more competitive with CCDs.

8.2.2 Channel stops, blooming, full well, and gain

The barrier potential prevents electrons from migrating from one pixel to another along a column in the parallel registers. What about migration along a row? In a classical CCD, shifts along a row are never needed, except in the serial register. The CCDs prevent charge migration along a row in the parallel registers by implanting (by heavily diffusing a dopant) a very highly conductive strip of silicon between columns. These **channel stops** held, say, at electrical ground, produce a permanent, extra-high barrier potential for stored electrons. Think of a pixel as a square bucket that holds water (or electrons). Two sides of the bucket, those that separate it from the adjacent columns, are maintained by the channel stop and are permanently tall and thin. The other two sides, the ones that separate it from its neighbors on the same column, are not as tall, and can be lowered or moved by "clocking" the gate voltages.

Consider what might happen if a pixel in an array fills with electrons during an exposure. As additional photoelectrons are generated in this saturated pixel, they will be able to spill over the barrier potential into the adjacent wells along their column, but cannot cross the channel stop. This spilling of charge along a column is called **blooming** (see Figure 8.6). Bloomed images are both unattractive and harmful: detection of photons in a pixel with a filled well becomes very non-linear; moreover, blooming from a bright source can ruin the images of other objects that happen to lie on the same CCD column. Nevertheless, in order to optimize the exposure of fainter sources of interest, astronomers will routinely tolerate saturated and bloomed images in the same field.

Fig. 8.6 Blooming on a CCD image: the saturated vertical columns are the bloom. The other linear spikes on the bright star image result from diffraction by the vanes supporting the telescope's secondary mirror.

There are designs for **anti-blooming CCDs**. Recent designs utilize special clocking during the exposure in a buried-channel CCD (see below) to temporarily trap excess electrons at the oxide interface.

The maximum number of electrons that can be stored in a single pixel without their energies exceeding the barrier potential is called the CCD's *full well*. The size of the full well depends on both the physical dimensions of the pixel, design of the gates, and the difference between the collecting and barrier potentials. Typical pixels in astronomical CCDs are 8–30 μm on a side and have full-well sizes in the range 25,000 to 500,000 electrons.

The final output from a scientific CCD is an array of numbers reported by the ADC to the storage computer. The number for a particular pixel is usually called its *pixel content*, and is measured in *ADU*s (analog-to-digital units). Pixel contents are proportional to the voltage the ADC receives from the output amplifier. The *gain* of the CCD is the number of electrons that need to be added to a pixel in order to increase the output contents for that pixel by one ADU.

For example, suppose a particular CCD has a full well of 200,000 electrons, and is equipped with a 16-bit ADC. The ADC is limited to digital outputs between 0 and 65,535 (= $2^{16} - 1$). A reasonable value for the gain might be 200,000/65,535 = 3.05 electrons/ADU. A smaller gain would mean that the CCD is better able to report small differences in pixel content, but would reach *digital saturation* before reaching the electronic full well. One might do this intentionally to avoid the non-linear shoulder in Figure 8.2. At a larger gain, the CCD would reach full well before the output could reach the maximum possible digital signal, so dynamic range would be reduced.

8.2.3 Readout time, read noise, and bias

To maximize DQE, the amplifier and ADC of an astronomical CCD should introduce the smallest possible noise to the output signal. A technique called *correlated double sampling* (*CDS*) is capable of very low noise operation – only a few electrons per pixel. The noise added by the CDS circuit depends crucially on how quickly it does its job – the faster, the noisier. Another consideration – the time needed for the analog-to-digital conversion – also limits the read time per pixel. Practical times correspond to a pixel sample frequency of 10 to 200 kHz, with higher frequencies producing higher noise. Except for low frequencies, noise added by the amplifier stage is proportional to the square root of the frequency.

The basis of charge-coupled readout is the one-pixel-at-a-time movement of the array contents through a single amplifier, and this is a bottleneck. A low-noise CDS stage in a scientific CCD must read out slowly, and the larger the array, the longer the read time. An important difference between scientific-grade CCDs and the commercial-grade CCDs and CMOS arrays in camcorders is the readout rate – to obtain real-motion video images, an array must read out about 30 times a second. The large read noise that results is usually not objectionable in a consumer camera because of the high input level. In contrast, the astronomical input signal is usually painfully low, and a low-noise, *slow-scan* CCD for astronomy may require many tens of seconds to read a single image.

There are some cases in astronomy where the large read noise of a rapid scan CCD is not objectionable, and in which time resolution is very important – observations of occultations of bright stars or rapid changes in solar features, for example. Also note that a rapid scan is not a problem if no data are being digitized. Thus, reading an array to clear it before an exposure can be done very quickly.

For the usual astronomical tasks, though, it is mainly lengthy readout time that puts a practical limit on the number of pixels in a CCD. (Time spent reading the detector is time wasted at the telescope!) Two strategies can speed read times. The first uses multiple amplifiers on a single array. Imagine, as in Figure 8.7a, an array with an amplifier at each corner. The CCD has two serial registers, at the top and bottom. The controller clocks the readout to split the parallel registers – they read out to both ends simultaneously – and does the same with each serial register. Each amplifier reads one quarter of the array, so the total read time is reduced by the same factor. The image can then be re-assembled in software. Multi-amplifier astronomical CCDs up to 9000×9000 pixels now (2010) exist.

A second strategy is to build a mosaic of several very closely spaced but electrically independent CCDs. Figure 8.7b shows an eight-element mosaic read by 16 amplifiers. An early device similar to this, the *Mosaic Imager*, was placed in service at the Kitt Peak National Observatory in 1998. It contained eight 2048×4096 CCDs arranged to form an 8196×8196 pixel (64 megapixel) detector that is 12 cm (5 inches) on a side. Gaps between the individual CCDs are about 0.6 mm (40 pixels). A relatively simple combination of shifted multiple exposures will fill in those parts of an image masked by the gaps on a single

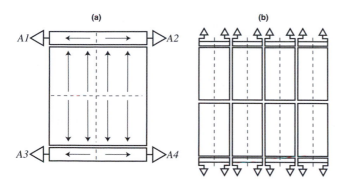

exposure. Mosaics have become so important that some modern CCDs are manufactured to be "almost-four-side-buttable" – so that the width of the gaps in a mosaic need be only to 30–100 pixels on all sides. At the present time (2009), there are several 50–120 megapixel mosaic arrays in service, and several observatories are about to introduce mosaics of over 100 devices and up to 3 gigapixels (see chapter 4 of Howell, 2006). These huge arrays expect to have not-very-objectionable readout times in the 20–60 second range. A major problem with these large-format arrays, in fact, may turn out to be simple data storage: an observer can expect to generate terabytes of image data in a few nights.

8.2.4 Dark current, cooling, and vacuum enclosures

At room temperature, a CCD is a problematic detector for astronomy. The energy of thermal agitation generates electron–hole pairs in the depletion zone and the resulting steady flow of electrons into the CCD potential wells is called *dark current*. Dark current is bad for three reasons:

1. It adds some number of electrons, N_D, to whatever photoelectrons are produced in a pixel. You must make careful calibrations to subtract N_D from the total.
2. Dark current adds not only a background *level*, N_D, but also introduces an associated uncertainty or *noise* to any signal. Since the capture of dark-current electrons into the pixel wells is a random counting process, it is governed by Poisson statistics. The noise associated with N_D dark electrons should be $\sqrt{N_D}$. This noise is more insidious than the background level, since it can never be removed. Dark current *always* degrades SNR.
3. At room temperature, the dark current can saturate a scientific CCD in seconds, which makes it impossible to record faint objects. Not good.

Lower the temperature of the CCD, and you reduce dark current. The Fermi distribution provides an estimate for the rate at which dark charges accumulate in a semiconductor pixel:

$$\frac{dN_D}{dt} = A_0 T^{\frac{3}{2}} e^{-\frac{E_G}{2kT}}$$

Here T is the temperature in kelvins, A_0 is a constant that depends on pixel size and structure, and E_G is the band-gap energy. A large fraction of dark current in a pixel arises at the Si–SiO_2 interface of the capacitor, where discontinuities in the crystal structure produce many energy states that fall within the forbidden band. Electrons in these interface states have small effective band gaps, and hence produce a large dark current.

A common method for cooling a CCD is to connect the detector to a **cryogen** – a very cold material with a large thermal mass. A very popular cryogen is a bath of **liquid nitrogen (LN2)**, a chemically inert substance that boils at 77 K = –196 °C. Since it is generally a good idea to keep the CCD at a somewhat warmer temperature (around –100 °C), the thermal link between detector and bath is often equipped with a heater and thermostat.

A cold CCD produces difficulties. The CCD and the LN_2 reservoir must be sealed in a vacuum chamber for two reasons. First, a CCD at –100 °C in open air will immediately develop a coating of frost and other volatiles. Second, the vacuum thermally insulates the LN_2 reservoir from the environment, and prevents the supply of cryogen from boiling away too rapidly. Filling the CCD chamber with an inert gas like argon is a somewhat inferior alternative. Vacuum containers, called **Dewars**, can be complicated devices (see Figure 8.8), but are quite common in observatories. At a minimum, the dewar must provide a transparent window for the input, a method for feeding electrical signals though the vacuum seal, a system for adding cryogen, and a method for periodically renewing the vacuum.

Another option for more modest cooling is dry ice (solid CO_2), which is less expensive than LN_2. Dry ice sublimates at –76 °C = 197 K.

Compact and relatively inexpensive thermoelectric (**Peltier junction**) coolers instead of cryogens require very small dewar sizes. These solid-state coolers can maintain a detector in the –30 to –50°C range, where the dark current of an ordinary CCD is still quite high, but where the dark current from an MPP CCD (see below) is acceptable for many astronomical applications. Such coolers are considerably more convenient to use than cryogens.

At the other extreme, superconducting junctions, many small band-gap detectors for the infrared, and most bolometers, require temperatures below what liquid nitrogen provides. **Liquid helium**, which boils at 4.2 K, is an expensive cryogen that is difficult to handle. Liquid ^{3}He boils at 3.2 K, but is even more difficult and expensive. To avoid the expense of evaporating helium into the air, one option is a **closed-cycle refrigerator** that compresses and expands helium fluid in a cycle. If they employ two or three stages, these systems can cool detectors to the 10–60 K range. Special closed systems using helium-3 evaporation can bring small samples to temperatures in the 0.3–3.2 K range.

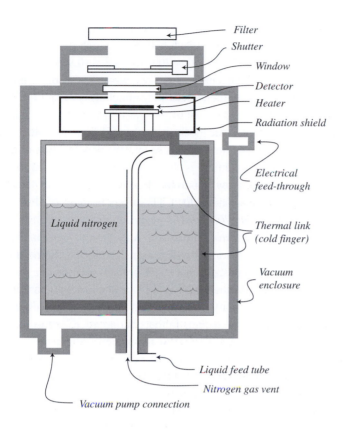

Fig. 8.8 A simple dewar for cooling a detector using liquid nitrogen. This design is common for devices that "look upward," and prevents cryogen from spilling out of the reservoir as the dewar is tilted at moderate angles.

8.2.5 Charge-transfer efficiency

The charge-coupled readout works perfectly only if all the electrons in a well shift from pixel to pixel. Disaster results if significant numbers of electrons are left behind by a shift. Images will appear streaked, and photometry becomes inaccurate. Signal loss because of charge-transfer inefficiency is greatest from the pixels furthest from the amplifier. The fraction of electrons in a pixel that are successfully moved during a one-pixel transfer is the ***charge-transfer efficiency***, or ***CTE***. Although one transfer will require three clock cycles and sub-pixel transfers in a three-phase device, CTE is always computed for a full pixel transfer. In a single-amplifier CCD, p is the actual number of full pixel transfers needed to read a particular charge packet. If the rows and columns of the parallel registers are numbered from the corner nearest the amplifier, then $p = R + C$, where R and C are the row and column numbers of the pixel in question. The fraction of the original charge packet that remains after p transfers (the total transfer efficiency, or TTE) is just

$$\text{TTE} = (\text{CTE})^p$$

The CTE needs to be very close to one. For example, suppose a 350×350 pixel array has a CTE of "three nines" (CTE = 0.999), which in this context is *not* very close to 1. Then $p = 350 + 350 = 700$, so TTE = $(0.999)700 = 0.49$; this device will lose over half the charge from the most distant pixel in the array before bringing it to the amplifier. Multi-megapixel arrays require CTE values approaching six nines.

What limits CTE? One issue is time — when CCD gate voltages change during a read, electrons need time to diffuse into the new location of the potential well. Usually, the required time is shorter than the time needed for the CDS and amplifiers to complete a low-noise read. However, at very low temperatures, electron velocities can be so small that CTE suffers because of slow diffusion, and so operation below about $-100\,°C$ is inadvisable.

Charge *traps* are a more serious limitation. A trap is any location that will not release electrons during the normal charge-transfer process. Some traps result from imperfections in the gates, channel stops, or the insulation of a pixel — flaws that deform the potentials during a read cycle to create unwanted barriers. Other traps are due to radiation damage, to unintended impurity atoms (usually metals like iron or gold), to structural defects in the silicon lattice, and to some effects not completely understood. The surface of the silicon layer in contact with the insulator will invariably have a large number of charge traps; these are such a serious problem that all modern CCDs are designed so that the potential well excludes the front surface (see the next section). Some traps affect only a few electrons each. If scattered throughout the entire body of a CCD, they produce a small decrease in the overall CTE. Other traps can render a pixel non-functional, so that it will not transfer charge in a meaningful way. This compromises the entire column upstream from the trap. Devices with a "bad column" or two are still very useful, but place additional demands on the observing technique.

Manufacturing defects can also cause a complete failure of charge transfer. The usual problems are short circuits and open circuits in the gate structure, or shorts between a gate and the semiconductor. Any of these can render a single pixel, a partial or complete column, or an entire device unreadable. The expense of a particular CCD is directly related to the ***manufacturing yield*** — if many devices in a production run need to be discarded, the cost of a single good device must rise. In the early days of CCD manufacture, yields of acceptable devices of a few percent were not uncommon.

8.2.6 The buried-channel CCD

The simple MOS/MIS (metal-insulator-semiconductor) capacitor we have been discussing up until now has its minimum electron potential (i.e. the bottom of the collection well) at the $Si-SiO_2$ interface. A CCD made of these capacitors is a ***surface-channel*** device, since charge transfer will require movement of electrons close to the interface. The high density of trapping

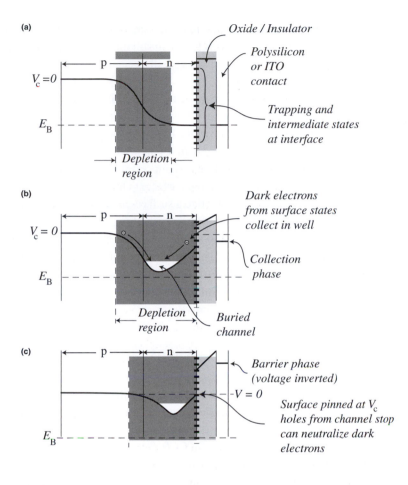

(a)

$V_c = 0$

E_B

Oxide / Insulator

Polysilicon
or ITO
contact

Trapping and
intermediate states
at interface

Depletion
region

(b)

$V_c = 0$

E_B

Dark electrons
from surface states
collect in well

Collection
phase

Depletion
region

Buried
channel

(c)

E_B

Barrier phase
(voltage inverted)

$-V = 0$

Surface pinned at V_c
holes from channel stop
can neutralize dark
electrons

Fig. 8.9 A buried channel
in a p–n junction
capacitor. (a) There is no
buried channel in the
electron potential when
the normal collection
phase voltage is applied.
If the gate voltage is
reduced, as in (b),
electrons collect away
from the interface. (c)
Inverting the voltage on
the barrier-phase
electrodes pins the
surface potential to the
channel-stop value and
allows a current of holes
to flow to neutralize dark-
current electrons.

states at the interface makes it impossible to achieve acceptable charge-transfer efficiency in a surface-channel CCD. (Values of only 0.99 are typical.) All modern scientific CCDs are designed so that the transfer channel is located several hundred nanometers below the interface. In these **buried-channel CCDs (BCCDs)**, all electrons collect in a region safely removed from the surface traps, and all charge transfers take place within the unperturbed interior of the semiconductor lattice.

Manufacturers can produce a buried channel by constructing a p–n junction near the semiconductor surface. Figure 8.9 illustrates the basic principle. Figure 8.9a shows the potential energy for electrons in an MOS or MIS device in which the semiconductor consists of a thin n-type region (perhaps 300–800 nm thick) layered on top of a much thicker p-type region. Within the semiconductor, the potential exhibits the basic pattern for a junction diode – there is a high-resistivity region depleted of majority charge carriers near the junction, and a potential difference, E_B, across the depletion zone. In Figure 8.9a we connect the p side to electrical ground, and set the gate voltage a relatively large positive voltage near

E_B. In this state, photoelectrons created in the depletion zone will be swept into the broad channel in the n region, where they can still interact with surface traps. Making the gate voltage even more positive will deepen the well and create a surface channel.

To create the buried channel, the voltage on the gate is made *more negative*. In Figure 8.9b, the gate voltage has been lowered so that electrons are repelled from the surface. This alters the shape of the potential and produces a minimum in the n-region, which is called the **collection potential**. (The required voltage on the gate is the **collection phase**.) Note two important features: First, electrons that collect in the well do not contact the surface. (This is good.) Second, the capacity of the well is reduced compared to a surface-channel device made from the same material. (This is not so good).

Figure 8.9c illustrates the electron potential under the barrier phase. Here the gate voltage is even more negative. The potential minimum in the semiconductor, although somewhat closer to the surface, is still buried. As a result, electrons generated under the barrier phase also avoid the surface traps as they move internally to the nearest collection potential.

8.2.7 Alternative CCD readout designs

You should be aware of several alternative methods for reading out the CCD that offer some specialized advantages. Consult Howell (2006) or the manufacturers and observatory websites (e.g. pan-STARRS, e2v, Kodak; see Appendix I) for further details.

The **orthogonal-transfer CCD**, or **OTCCD** (see Tonry *et al.*, 1997) has a gate structure that permits charge-coupled shifting of pixel contents either along the row or along the column, on either the entire array or on subsections. Orthogonal-transfer CCDs can make small image shifts to compensate for tip—tilt seeing-disk motion during an exposure, and are being used for the 1 gigapixel mosaic of the pan-STARRS project.

Frame-transfer CCDs permit a very short time interval between successive frames. They recognize that it is the amplifier stage that limits the readout rate of a scientific CCD, so rapidly read an acquired frame into an inactive (shielded) set of parallel registers. The device then reads the shielded frame slowly through the amplifier while the next frame is being acquired.

Low-light-level CCDs or **L3CCDs** have additional extra-large, deep-well MOS capacitors in a "charge multiplication" extension of the serial register. The device clocks charges from the serial register into these capacitors at a very high voltage, so that the energy of a transferred electron can produce an additional electron–hole pair when it enters a multiplication capacitor. Several hundred multiplication transfers typically produce multiplication gains of 100–1000 before amplification, so read noise is insignificant, permitting rapid readout (1–10 MHz) and true photon-counting at low light levels.

8.2.8 The MPP CCD

Interface states at the Si–SiO$_2$ junction remain the major source of dark current in a simple BCCD. Thermal electrons can reach the conduction band by "hopping" from one interface state to another across the forbidden gap. You can eliminate this electron hopping by *pinning* a phase, as in Figure 8.9c. To pin the phase, you set the voltage on the gate to so negative a value that the potential at the interface *inverts*, that is, it reaches the same potential as the back side of the p region, which is also the same, V_c, as the potential of the conductive channel stops. Any further reduction in the gate voltage has little effect on the interface potential, since the surface is now held at ground by holes that flood in from the channel stops. The abundance of holes means that thermal electrons are neutralized before they can hop through the interface states. Dark current in a pinned phase is reduced by several orders of magnitude.

A *partially inverted* three-phase CCD operates with one non-inverted phase (the collection phase, as in Figure 8.9b), and with the other two phases pinned and serving as the barrier phases, as in Figure 8.9c. Dark current in such a device is about one third of what it would be in a completely non-inverted mode. If all three phases are pinned, the CCD is a *multi-pinned-phase* (*MPP*) device, and dark current less than 1% the rate in non-inverted mode. The obvious difficulty with MPP operation is that there is no collection phase – the buried channel runs the entire length of a column. Multi-pinned-phase devices therefore require additional doping under one of the phases to make a permanent collection potential. This is possible because the value of E_B in Figure 8.9 depends on the density of dopants in the semiconductor. In an MPP device, for example, the surface under phase 2 might invert with the collection phase set at –5 V, while the other two (barrier) phases require –7 V for inversion.

With their remarkably low dark currents, MPP CCDs can operate at room temperature for several minutes without saturation. In recent designs, dark rates below 0.1 electron per second are routine at –40 °C, a temperature attainable with inexpensive thermoelectric coolers. An MPP CCD controlled by a standard personal computer is a formidable and inexpensive astronomical detector within the financial means of many small observatories, both professional and amateur. As a result, modern observers using telescope apertures below 1 meter are making quantitative astronomical measurements of a kind that would have been impossible at the very best observatories in the world in 1975.

The full-well capacity of an MPP device is a factor of two or three less than a partially inverted BCCD. Modern MPP devices nevertheless have respectable full wells. Appendix I gives the specifications for a few devices currently on the market. If the larger full well is more important than the reduced dark current, the proper selection of clock voltages makes it possible to run a device designed for MPP operation in a partially inverted mode.

8.2.9 Surface issues

We need to address the very practical question of getting light into the depletion region of the CCD pixels.

Frontside options. The most direct approach sends light through the metal gates. Since even very thin layers of most metals like copper or aluminum are poor transmitters, the "metal" layer of the CCD is usually made of highly doped *polysilicon*: silicon in a glass-like, amorphous state – a random jumble of microscopic crystals. A thin (about 0.5 micron) layer of doped polysilicon is both relatively transparent as well as a good electrical conductor, but it does, however, absorb green, blue and (especially) ultraviolet light. Other conductive materials, like doped *indium tin oxide* (*ITO*) have better transparency properties than polysilicon; ITO electrodes are becoming common, but are somewhat harder to fabricate.

There are two general strategies for further improving the short wavelength QE of a front-illuminated CCD. The first is somehow to make the gate structure more transparent. The second is to change the wavelength of the incoming light to one at which the gates are more transparent.

Open-electrode architecture improves transparency with a gate structure that leaves part of the pixel uncovered. For example, the collection-phase electrode might be oversized and shaped like a hollow rectangle. It is even possible to fabricate pixel-sized *microlenses* over the frontside to redirect much of the incoming light to the uncovered area of each pixel.

A related approach is the *virtual-phase CCD*, where a single gate covers half of the pixel, and a four-step potential profile is constructed by implanting dopants in the semiconductor. Changing the voltage on the single gate can produce pixel-to-pixel charge transfer similar to a four-phase CCD. Virtual-phase CCDs have even better blue QEs than open-electrode devices, especially if equipped with microlenses, but are more difficult to fabricate and generally have relatively poor CTE values.

A different strategy applies a thin coating of *phosphor* on top of the gates. The useful phosphors are organic molecules that absorb a short-wavelength photon to move to an excited state, then de-excite by emitting one or more longer-wavelength photons. Lumigen (or lumogen), for example, is a commercial compound that absorbs light shortward of 420 nm, and is otherwise transparent. Upon de-excitation, it emits photons at around 530 nm, which can easily penetrate polysilicon gates. Since phosphors emit in all directions, they will slightly degrade image resolution at short wavelengths. Another drawback is that some phosphors tend to evaporate in a vacuum, especially at high temperatures.

Backthinning. A completely different solution sends the light in through the back (from the bottom of Figure 8.10) of the device, avoiding the gates completely. This *backside illumination* has the advantage that green, blue, and ultraviolet, which would be absorbed by a polysilicon or ITO layer, will pass

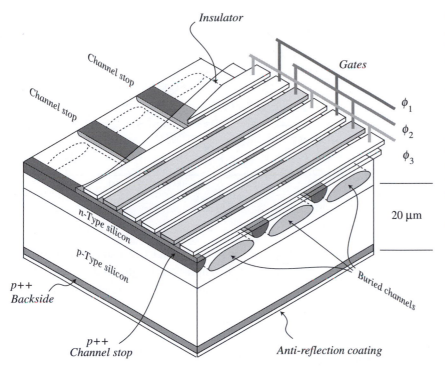

Fig. 8.10 Schematic of a thinned, three-phase CCD. In a conventional CCD, insulated gate electrodes usually overlap, while in an open architecture, gaps more closely follow the pixel pattern. This drawing is of a backthinned device. A front-illuminated device would have a much thicker silicon layer, with the AR coating above the gates.

directly into the silicon. Since these photons have a short absorption depth, they create photoelectrons mainly near the back face of the device. This is a serious problem. In order for the electrons to be able to diffuse from the back face into the depletion zone without recombining, the semiconductor layer needs to be very thin (10–20 μm). "Thinning" the silicon will in turn reduce its ability to absorb NIR photons, which have a large absorption depth. The final geometry needs to be something of a compromise. Nevertheless, astronomers have generally embraced backthinned CCDs, since they detect a considerably larger fraction of incident photons of all wavelengths than does any frontside-illuminated device (see Figure 8.11). Their main drawback is that they are difficult to manufacture and therefore expensive, if available at all.

If red and near-infrared QE is very important, the ***deep-depleted CCD*** offers some improvement over the normal backthinned device. Because the depth of the light-sensitive depletion zone is inversely proportional to the dopant concentration, use of a lightly doped (high resistivity) silicon layer means that the total layer thickness of the CCD can be increased to about 50 μm. The thicker detector has greater long-wavelength sensitivity, and is mechanically easier to fabricate. However, achieving the required resistivity can be difficult, and cosmetic quality to date has been inferior to thin devices.

Anti-reflection coatings. An anti-reflection (AR) coating is most effective for light of a particular wavelength, so a CCD designer must choose the coating

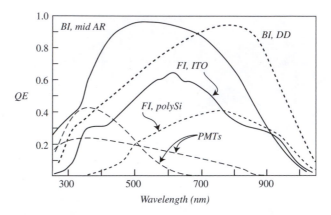

with the intended use of the detector in mind. Often CCD manufacturers offer a choice of coatings to enhance either the short-wavelength, mid-wavelength or NIR response. Figure 8.11 shows a selection of the QE characteristics of a few different CCD designs.

8.3 Photo-emissive devices

Researchers have developed the simple vacuum photodiode described in the last chapter from a detector of very limited capability (with poor QE in the red, very low signal levels, mechanical fragility, and single-channel operation) into devices that compete with or enhance CCDs in special circumstances. In this section we examine three astronomical detectors that depend upon the vacuum photoelectric effect.

8.3.1 The photomultiplier tube

One disadvantage of the simple vacuum photodiode described in the last chapter (Figure 7.24) is low signal level. The ***photomultiplier tube*** (***PMT***) is a vacuum device that increases this signal by several orders of magnitude. Figure 8.12 illustrates its operation. In the figure, a voltage supply holds a semi-transparent photocathode on the inside of the entrance window at large negative voltage, usually around one or two kilovolts. A photon hits the cathode and ejects a single electron. In the vacuum, this electron accelerates towards the more positive potential of a nearby electrode called a ***dynode***, which is coated with a material (e.g. Cs_3Sb, $CsKSb$, BeO, GaP) that can easily release electrons to the vacuum if hit by an energetic particle. Because the original photoelectron impacts the dynode with 100 eV or so of kinetic energy, it usually ejects several secondary electrons. The number of secondary electrons is a statistical quantity whose mean value, δ, usually lies between 2 and 10. The group of electrons

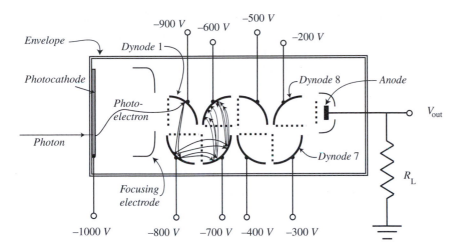

Fig. 8.12 A simple photomultiplier tube. The potential of the first dynode accelerates a single photoelectron emitted from the cathode. Its impact releases several secondary electrons, which accelerate and hit dynode 2, releasing another generation of secondaries. After (in this case) eight stages of amplification, a large pulse of electrons flows through the anode and load resistor to ground.

ejected from the first dynode then accelerates to the second dynode, where each first-dynode electron produces δ second-dynode electrons. The process continues through n dynodes, until the greatly multiplied pulse of electrons lands on the anode of the PMT. If each dynode is equivalent, the total number of electrons in a pulse generated by a single photoelectron is

$$N = A\delta^n$$

where the factor A accounts for inefficiencies in redirecting and collecting primary and secondary electrons.

In the figure, the signal is the average DC voltage measured across a load resistor. However, for weak sources, the large pulses of electrons that arrive at the anode are easily counted electronically, and the PMT can operate in a ***pulse-counting mode***: each pulse is generated by the arrival of a *single* photon at the cathode. In this mode, the QE of the PMT depends on the QE of the photocathode, which can be as high as 40–50% for some materials (see Figure 8.11).

The single-channel PMT was the detector of choice for precise astronomical brightness measurements from 1945 until the advent of CCDs in the early 1980s. The spectral responses of the available PMT photocathode materials defined, in part, some of the now-standard photometric band-passes (the U, B, and V bands in Table 1.2, for example). Since photomultipliers have few advantages over CCDs, they have become rare at observatories. One important advantage of the PMT, however, is response time. The temporal spread of a single pulse at the anode limits the shortest interval over which a PMT can sense a meaningful change in signal. Pulse widths are so narrow (5–10 nanosecond) for many PMTs that they can, in principle, detect signal changes as rapid as a few milliseconds. The response time of a CCD, in contrast, is several tens of seconds for a standard slow-scan device, with quicker response possible with increased noise. Arrays of STJs, although still in the development phase, have the potential for response times similar to PMTs.

8.3.2 The microchannel plate

The upper part of Figure 8.13 shows an important variation on the PMT. Take a glass capillary with a diameter between 5 and 25 μm, and a length around 40 times its diameter. Coat the inside surface of this tube with a semiconductor that has good secondary electron-emitting properties, and connect the ends of the channel coating to the voltages as shown. You have created a ***microchannel***. Place this microchannel assembly in an evacuated chamber between a photo-cathode and an anode, and it can serve in place of the dynode chain of a PMT. A

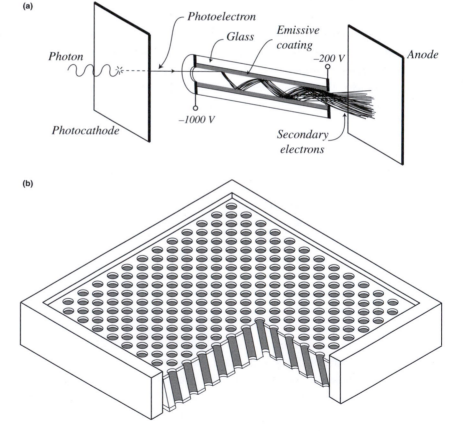

Fig. 8.13 (a) A single microchannel – a small glass tube whose interior is coated with dynode-type material. A large potential drop along the tube insures that an electron impact at one end will produce a burst of secondary electrons at the other. (b) A closely packed array of channels forming an MCP.

photoelectron from the cathode will accelerate towards the upper end of the channel, where it strikes the wall and generates a spray of secondary electrons. These secondary electrons will in turn strike the channel wall further down, multiplying their numbers. After several multiplications, a large pulse of electrons emerges from the end of the microchannel and accelerates to the anode.

A *microchannel plate* (MCP), as illustrated in Figure 8.13b, consists of an array of up to several million microchannels closely packed to form a plate or disk several millimeters in diameter and less than a millimeter thick. The electrical contact that coats the front surface can be made of a metal that has some secondary-electron emission capabilities, so that photoelectrons that do not strike the inside of a channel might still be detected via emission from the contact. You can make a high gain but very compact PMT by sandwiching several MCPs between a photocathode and anode in a vacuum enclosure, as in Figure 8.14. Such *MCP PMTs*, operated as single-channel devices, have an advantage in size, power consumption, response time and stability in magnetic fields compared to dynode-based devices.

The MCP, however, is most valuable as a component in a two-dimensional detector. Various anode configurations or electron-detection devices can generate an output image that faithfully reproduces the input on the cathode of an MCP. The *multi-anode microchannel array detector* (*MAMA*) is an example. In the MAMA, the anode of the MCP PMT is replaced with two planes of parallel wires that form an x–y grid. A pulse of electrons emerging from a

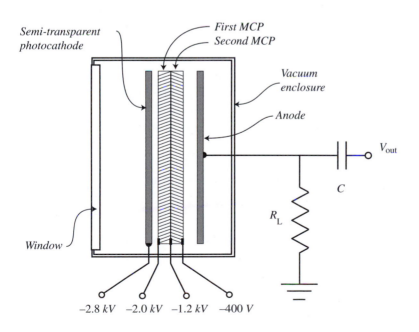

Fig. 8.14 A schematic MCP PMT. The dynodes of the photomultiplier are replaced by two microchannel plates. The capacitor at the output means the device is used in pulse-counting mode. Compare with Figure 8.11.

particular microchannel will impact with maximum intensity on one x-wire and one y-wire. Special circuitry then increments the signal count at the corresponding x–y address in the output image.

The MAMA detectors are especially useful at short wavelengths where the DQE of the device can be very high if it is equipped with a "solar-blind" photocathode insensitive to visual and infrared photons. Space astronomy has employed MAMA detectors to great advantage in for the detection of X-rays and far-ultraviolet light. Although silicon CCDs are also sensitive at these wavelengths, they suffer from high sky background levels from starlight and scattered sunlight that cannot be completely removed by filtering.

8.3.3 Image intensifiers and the ICCD

An ***image intensifier*** is not a detector, but a vacuum device that amplifies the brightness of an image. Because military interest in night vision drives the development of intensifiers, the military terminology (Generation I, II, III[4] etc.) for different designs has become standard. Figure 8.15 shows a Generation II intensifier coupled by optic fibers to a CCD. The intensifier resembles a MCP PMT, but it has a phosphor screen instead of an anode. A photoelectron leaving the cathode produces a pulse of high-energy electrons that excites multiple molecules in the phosphor. These then de-excite by photo-emission. The location of the phosphor emission maps the location of the original photo-absorption on the cathode. A single input photon can generate 10^4 to 10^7 phosphor photons.

As shown in Figure 8.15, an intensifier can be a useful first stage for an array detector like a CCD. It is important to understand, however, that although an intensifier will vastly increase signal strength and decrease exposure times, it will always decrease the *input* SNR for the CCD.

For example, consider a source that produces N_i photons at the photo-cathode of an image intensifier during an integration. If the input is dominated by photon noise (assume background is negligible) then the uncertainty in the input signal is just $\sqrt{N_i}$. The intensifier output at the phosphor is

$$N_{out} = gN_i$$

where g is the gain factor of the intensification. The variance of N_{out} is therefore

$$\sigma_{out}^2 = g^2 \sigma_{in}^2 + \sigma_g^2 N_i^2 = \left(g^2 + \sigma_g^2 N_i\right) N_i$$

Here σ_g is the uncertainty in the gain. Thus, the SNR at the input and output are

---

[4] Generation I devices (now obsolete) used electric or magnetic fields to accelerate photoelectrons from a cathode and then re-focus them directly onto the phosphor. Generation II and III devices use an MCP to form the image as described in the text. Generation III devices have advanced photocathodes sensitive in the NIR.

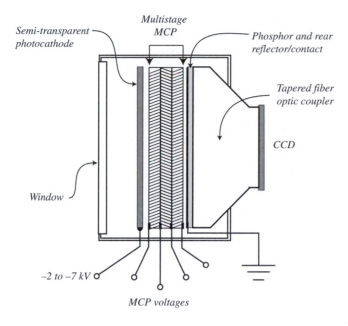

Semi-transparent photocathode

Multistage MCP

Phosphor and rear reflector/contact

Tapered fiber optic coupler

CCD

Window

-2 to -7 kV

MCP voltages

$$SNR_{in} = \sqrt{N_i}$$

$$SNR_{out} = gN_i \left[\left(g^2 + \sigma_g^2 N_i \right) N_i \right]^{-\frac{1}{2}} = \sqrt{N_i} \bigg/ \left[1 + \frac{\sigma_g^2 N}{g^2} \right]^{\frac{1}{2}} \le SNR_{in}$$

So long as intensifier gain is uncertain, intensification will degrade the SNR.

Intensified CCDs (ICCDs) are thus useful in situations where the primary noise source is NOT photon noise in the signal, and/or where rapid signal changes need to be monitored. In such cases (e.g. read noise or dark-current noise dominant), using an intensifier can improve the DQE of the entire device by decreasing the required exposure times.

A related use for the image intensifier is as a *signal conditioner* for a CCD – the image on the phosphor is not only brighter than the one that arrives at its photo-cathode, it also emits photons of a different wavelength. You can select a cathode sensitive to ultraviolet light, for example, and a phosphor that emits near the CCD QE peak in the red. The ICCD thus detects sources that it would find absolutely invisible without the intensifier.

8.4 Infrared arrays

Although modern CCDs in some respects approach perfection as astronomical detectors, the large size of the forbidden band gap of silicon means that they are blind to all light with wavelength longer than 1.1 µm. The development of infrared-sensitive arrays has faced great technical difficulties, but advances have come quickly. These advances have had an even greater impact on infrared astronomy than the CCD has had in the optical. This is because prior to the CCD, optical astronomers had an excellent, although inefficient, multi-pixel detector – the photographic plate. Prior to infrared arrays, infrared astronomers had only single-pixel devices. Different wavelength regions in the infrared place different demands on detector technology. We first make a brief survey of these differences, then examine the general method of infrared detector fabrication. Chapter 6 of Glass (1999) gives a general qualitative discussion of infrared technology, and chapter 11 of McLean (2008) gives a more technical treatment.

8.4.1 Detectors at different wavelengths

In the near infrared (1–5 μm; J, H, K, L, and M bands) practical arrays for astronomy first appeared at observatories in 1986. Although a number of materials were tried, the most successful eventually proved to be arrays of junction diodes made of indium antimodide (*InSb* – often pronounced "ins-bee") or HgCdTe (mercury–cadmium telluride, or *MCT*). The initial arrays contained only a few pixels but, by 2009, manufacturers were producing 2048×2048 pixel buttable arrays, and astronomers were assembling infrared-sensitive mosaics that were only an order of magnitude smaller (70 megapixels) in size than contemporary CCD-based devices. Modern NIR arrays have QE values less than 70%, read noise levels less than 10 electrons, and dark currents below 0.1 electrons per second.

In the MIR (5–28 μm) progress has been more modest, partly because high background levels limit ground-based observing at these wavelengths. At even the best sites in the MIR, the atmosphere is marginally transparent only in the N band (8–13 μm) and, to a lesser extent, the Q band (17–28 μm) so that any useful observations require a very large telescope or an orbiting observatory like Spitzer. At present, the most advanced MIR arrays are of blocked-impurity-band (BIB) photoconductors, usually fabricated from silicon doped with antimony (Si:Sb) or arsenic (Si:As). As of 2009, arrays in a 1024×1024 pixel format have begun to appear at the world's largest telescopes. Arrays of Schottky photodiodes (with PtSi as the metal layer) are easier to fabricate and have seen some use, but are limited by poor quantum efficiency.

In the far infrared (25–350 μm) the Earth's atmosphere is completely opaque (there is a weak and erratic window at 40 mm at high-altitude sites). Far-infrared detectors, therefore, must be flown in spacecraft or very high-altitude aircraft. In general, extrinsic detector arrays have been the most useful in this region. Doped silicon with the smallest band gap, Si:Sb, has a cutoff wavelength at around 30 μm, so for longer wavelengths, extrinsic germanium has been used. The Spitzer Space Telescope, for example, carries 32×32 pixel arrays of Ge:Ga (cutoff near

115 μm) and a 2×20 stressed[5] Ge:Ga array (cutoff near 190 μm). Mosaics of Spitzer-sized devices are under construction. For even longer wavelengths, observers have used small arrays of bolometers (discussed in a later section).

8.4.2 Infrared detector construction

Building an infrared array of photon detectors of any of the types discussed above is different from building a CCD in an important way. Charge-coupled devices are based on a mature technology. Buoyed by the ballooning market in computers and consumer electronics over the past forty years, manufacturers have refined their skill in the fabrication of electronic components in intrinsic silicon and devices based on p–n junctions in silicon. Expertise with more difficult materials like InSb, MCT, and extrinsic silicon and germanium is limited in comparison. That infrared arrays exist at all is due in large part to their applicability to battlefield imaging, surveillance, and remote sensing. Because building electronics is so much easier in silicon, almost all modern infrared arrays are built as two-layer **hybrids**: one layer is composed of the infrared-sensitive material, the other, made of silicon, provides the electronics for reading the signal.

Figure 8.16 sketches some of the details in the construction of two pixels of a NIR hybrid array. The light-sensitive elements are junction photodiodes made of MCT – that is, the alloy $Hg_{(1-x)}Cd_xTe$. The MCT has an adjustable cutoff wavelength, and although it has appeared primarily in NIR detectors, it is a potentially useful material at longer wavelengths.[6] 256×256 Arrays of this design were installed in the NIR camera and multi-object spectrograph (NIC-MOS) of the Hubble Space Telescope. Successors to the NICMOS array (first the PICNIC and then the HAWAII arrays manufactured by Rockwell Scientific Corporation) have grown to 2048×2048 size.

The figure illustrates a cross-section of two pixels. The top layer is the silicon readout array, which contains several CMOS field-effect transistors (MOS-FETs) at each pixel. The lower layer contains the infrared-sensitive material – in this case a p–n photodiode at each pixel. The total thickness of the MCT is

---

[5] Creation of a majority carrier in p-type material requires breaking an atomic bond and remaking it elsewhere (movement of an electron from the valence band to an acceptor state). It is easier to do the bond breaking (it takes less energy) if the crystal is already under mechanical stress. Thus, the cutoff wavelength of a stressed p-type crystal is longer than for an unstressed crystal. Maintaining the proper uniform stress without fracturing the material is a delicate operation.

[6] The pure form of HgTe behaves like a metal, while pure CdTe has a band gap of 1.6 eV. The band gap of the $Hg_{(1-x)}(CdTe)_x$ alloy depends on x, the cadmium telluride fraction, and is 0.31 eV (cutoff at 4μm) at $x = 0.35$ and 0.1 eV (cutoff at 100μm) at $x = 0.17$. So far, cutoff wavelengths longward of 20μm have been difficult to achieve, but new techniques such as lattice fabrication by means of molecular beams may permit manufacturers to make low-x MCT of the required uniformity.

Fig. 8.16 Cross-section of two pixels of a hybrid array of MCT photodiodes. Each diode connects to the silicon readout circuits through an indium bump conductor.

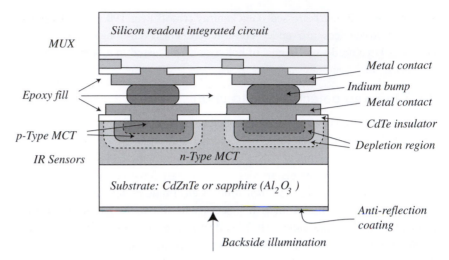

Silicon readout integrated circuit

MUX

Metal contact

Indium bump

Metal contact

Epoxy fill

CdTe insulator

p-Type MCT

Depletion region

IR Sensors   n-Type MCT

Substrate: CdZnTe or sapphire (Al$_2$O$_3$)

Anti-reflection coating

Backside illumination

quite small, and it is grown or deposited on a transparent substrate, like sapphire, to provide mechanical strength. Initially, the two layers are manufactured as separate arrays. A small bump of the soft metal indium is deposited on the output electrode of each photodiode. A matching bump of indium is deposited on the corresponding input electrode of each silicon readout circuit. The silicon and the infrared-sensitive arrays are then matched pixel-to-pixel and pressed together, so the indium bumps compress-weld against their mates to make good electrical contact. The spaces between bumps can then be filled with epoxy to secure the bond.

There are obvious and not-so-obvious pitfalls in making arrays using this "bump-bonding" approach, but the technique is becoming mature. Nevertheless, NIR arrays remain considerably more expensive than CCDs of the same pixel dimensions.

Reading an infrared array differs fundamentally from reading a CCD. There is no pixel-to-pixel charge transfer: each pixel sends output to its individual readout integrated circuit (***ROIC***) in the silicon layer. Since one of the tasks of the silicon layer is to organize the multiple signals from all pixels into a single stream of data from the amplifier, the layer is often called the ***multiplexer*** or ***MUX***. Many multiplexers, especially in large arrays, read to several (usually two or four, but sometimes many more) data lines simultaneously. Important differences from CCDs include:

- Since a pixel does not have to (nor is it able to) pass charge to and from its neighbors, a "dead" pixel (caused, for example, by a failure in the bump bond) will not kill the entire upstream column, as it might in a CCD. Although saturation occurs, the "blooming" penalty present in CCDs is not a feature of infrared arrays.
- Since readout is separate from sensing, reads can be non-destructive, and the same image read several times. Moreover, the array can be read out while the infrared layer is still responding to light.

- Very high background levels invariably hamper infrared observations from the ground. This forces very short (0.1–10 seconds) integration times to avoid saturation. To cope with the resulting data rate, controllers often co-add (average) many of the short exposure images and save only that result.
- Many infrared sensors are somewhat non-linear, so calibration for linearity is a much greater concern with an infrared array than it is with a CCD.
- Because of the smaller band gaps involved, dark currents in infrared arrays can be a severe problem, and these detectors must operate at low temperatures. Although some NIR arrays work well with liquid-nitrogen cooling, MIR arrays require refrigeration to much lower temperatures.
- Any detector sensitive to wavelengths longer than about 5 μm requires a cold enclosure to shield it from the infrared light flooding in from its warm (and therefore glowing) surroundings. These hot surroundings include the telescope structure and optics, so space telescopes that can be kept cold are superior infrared observing platforms. For similar reasons, the secondary mirrors of ground-based infrared telescopes are designed to be as small as possible. In the FIR, even the readout circuits are heat sources that need to be isolated from detectors.

8.5 Thermal detectors

Thermal detectors do not depend upon photons to move charge carriers directly from one band to another. They work, rather, as two-element devices: (1) a thermometer, which senses the temperature increase produced in (2) an absorber, when the latter is exposed to an incoming light beam. Figure 8.17 sketches a generalized thermal detector. In the figure, a heat sink at temperature T_0 encloses the absorber and thermometer in an evacuated cavity. A strip of material with conductance G connects the absorber and heat sink. To make an observation, the shutter is opened and the incoming light is allowed to deposit energy in the absorber at rate P_{in}. After a time, the temperature of the absorber will increase by amount ΔT, a quantity that therefore measures P_{in}.

For example, if the absorber were allowed to reach equilibrium with the shutter open, the following condition will apply:

$$P_{in} = P_{out} = P_{conduct} + P_{radiate}$$

$$\approx G\Delta T + \sigma A\left[(T_0 + \Delta T)^4 - T_0^4\right]$$
$$= \Delta T\left\{G + \sigma A\left[4T_0^3 + 6T_0^2\Delta T + 4T_0(\Delta T)^2 + (\Delta T)^3\right]\right\} \qquad (8.3)$$
$$= \Delta T\{G + \sigma A R(T_0)\} = \Delta T G'$$

In this equation, A is the total surface area of the absorber, and σ is Stefan's constant. We assume the solid angle subtended at the absorber by the shutter opening is small. To maximize the sensitivity, which is just $\Delta T/P_{in}$, the term in braces in the final expression must be kept as small as possible. To this end, it is

Fig. 8.17 General design
for a thermal detector. A
thermometer records the
increase in the
temperature of a light-
absorber after it is
exposed to a source. A
strip of conducting
material links the
absorber to a large heat
sink. Because it may
require power to operate,
the thermometer may
contribute to P_{in}.

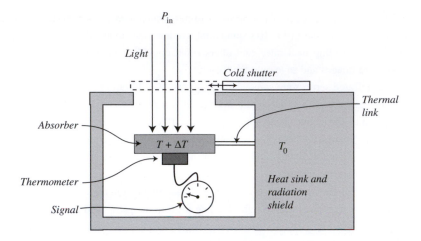

Fig. 8.17 General design for a thermal detector. A thermometer records the increase in the temperature of a light-absorber after it is exposed to a source. A strip of conducting material links the absorber to a large heat sink. Because it may require power to operate, the thermometer may contribute to P_{in}.

clear that both the conductance of the link and the area of the absorber should be small. Likewise, especially since it appears to the third power, the temperature of the sink, T_0, needs to be kept very low. An additional benefit of low temperature is that if the conductance term is large relative to the radiative term, the effective conductance, G', in W K^{-1}, will be nearly constant. Although it is not addressed in the equilibrium expression above, keeping the heat capacity of the absorber small will mean that the **response time** of the detector is short. The time dependence will be given by

$$\Delta T(t) = T_0 + \frac{P_{in}}{G'}[1 - \exp(-tG'/C)] \tag{8.4}$$

Here C is the heat capacity of the detector in J K^{-1} and the time constant is C/G'. In general, a thermal detector will employ an absorber that is black, with a small heat capacity. Physical size should be as small as possible, matching the size of the focal-plane image but not approaching the scale of the wavelength being observed. The absorber will be linked to a temperature sink that is kept as cold as possible (often, for example, with liquid helium).

In practice, the thermal detectors used in astronomy have almost always been **bolometers**, defined as thermal detectors in which the temperature sensor is a **thermistor** (a contraction of the words *thermal* and *resistor*), a small mass of semiconductor or metal whose electrical conductivity is a strong function of temperature. The small signal levels usually characteristic of astronomical measurements have restricted instruments to just a few thermistor materials: extrinsic silicon has been used, but n-type extrinsic germanium (doped with gallium) is the most common choice. The gap between the donor states and the conduction band in Ge:Ga is around 0.01 eV, so a modest increase in temperature will excite donor electrons to the conduction band and decrease the resistance of the thermistor. As in any semiconductor with a band gap, the resistance as a function of temperature will be given by an equation of the form

$$R(T) = R_0 T^{-\frac{3}{2}} e^{\frac{A}{kT}} \qquad (8.5)$$

where A is a constant that depends on the band-gap energy and R_0 depends on both the band gap and the doping level.

The situation is somewhat different at temperatures below 5 K, where the mechanism of thermal excitation of electrons across the gap between donor and conduction states becomes unimportant. The practice there is to use very highly doped thermistors, in which conductivity is due to hopping within the impurity band. Equation (8.5) does not apply in this case, but other semi-empirical expressions relate a decrease in resistance to an increase in temperature.

Thermistor resistance is usually monitored in a simple circuit in which the voltage drop across the sensitive material is observed with the thermistor placed in series with a stable reference resistor, as illustrated in Figure 8.18a. The reference resistor must be large relative to the bolometer resistance; otherwise a positive heating feedback can run away in the bolometer.

One simple bolometer design, illustrated in Figure 8.18b, has an elegant feature – the absorber and the temperature sensor are one and the same element. The wires that connect this sensor to the conductivity-measuring circuit also provide the thermal link to the heat sink. Figure 8.18c shows a compound bolometer, in which the thermistor is fastened to a radiation-absorbing plate. Both these designs can be easily incorporated in a one-dimensional array. Modified geometries have led to two-dimensional arrays of a few hundred pixels.

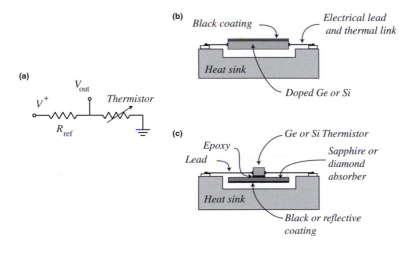

Fig. 8.18 Bolometers. (a) Electrical characteristics: a thermistor, whose electrical resistance depends on temperature, is connected in series with a reference resistor. Voltage across the reference changes with the temperature of the detector. (b) A bolometer in which the thermistor is also the absorber. Here the electrical lead is also the thermal link to the heat sink. Alternatively, in (c), the thermistor can be mechanically and thermally bonded to the absorber.

At very low temperatures, a bolometer can employ a type of superconducting thermistor called a ***transition edge sensor*** (***TES***) to attain very high sensitivity. See Rieke (2003), chapter 9, for a more detailed discussion of the TES and of bolometers in general.

Summary

- An important measure of detector quality is the detective quantum efficiency:

$$\text{DQE} = \frac{(\text{SNR})_{\text{out}}^2}{(\text{SNR})_{\text{perfect}}^2}$$

- Characteristics of a detector include its mode (photon, wave, or thermal). Important detector concepts:

signal	*noise*	*quantum efficiency (QE)*
absorptive QE	*quantum yield*	*spectral response*
spectral resolution	*linearity*	*saturation*
hysteresis	*stability*	*response time*
dynamic range	*physical size*	*array dimensions*
Nyquist spacing	*image sampling*	

- The charge-coupled device, or CCD, is usually the preferred astronomical detector at visible wavelengths. Concepts:

parallel registers	serial register	gates
clock voltages	barrier potential	ADU
output amplifier	collection potential	CCD gain
blooming	channel stop	CMOS capacitor
full well	digital saturation	read noise
corrrelated double sampling	multi-amplifier arrays	dark current
orthogonal-transfer CCD	cryogen	Dewar
buried-channel CCD (BCCD)	CTE	traps
multi-pinned phase (MPP)	inverted gate	L3CCD
frontside illumination	backthinned CCD	ITO
deep-depleted CCD	open electrode	microlens
virtual-phase CCD		

- Several important astronomical detectors depend on the vacuum photoelectric effect. Concepts:

photomultiplier (PMT)	dynode	pulse-counting
microchannel plate	MCP PMT	MAMA
image intensifier	signal conditioner	ICCD

- Observational techniques and device performance with infrared arrays is highly dependent on the wavelength region observed. Concepts:

near-, mid- and far-infrared	InSb	MCT
BIB detectors	*Si:Sb*	*NICMOS*
HAWAII	*hybrid array*	*ROIC*
MUX	*indium bump bond*	

- Semiconductor thermistor resistance:

$$R(T) = R_0 T^{-\frac{3}{2}} e^{\frac{A}{kT}}$$

- Thermal detector concepts:

heat sink	*bolometer*
thermistor	*time constant*

Exercises

1. A photodiode has an overall quantum efficiency of 40% in the wavelength band 500–600 nm. The reflectivity (fraction of photons reflected) at the illuminated face of the detector in this band is measured to be 30%. If this face is treated with AR coatings, its reflectivity can be reduced to 5%. Compute the QE of the same device with the AR coating in place.

2. A certain detector measures the intensity of the light from a stable laboratory blackbody source. The signal in three identical trials is 113, 120 and 115 mV. From the blackbody temperature, the experimenter estimates that 10^4 photons were incident on the detector in each trial. Compute an estimate for the DQE of the detector.

3. A photon detector has a QE of q and a quantum yield of y. The uncertainty in y is $\sigma(y)$. Show that DQE $= q$ if $\sigma(y) = 0$, but that DQE $< q$ otherwise.

4. A CCD has pixels whose read noise is 3 electrons and whose dark current is 1 electron per second. The QE of the detector is 0.9. Compute the DQE of a single pixel if 1000 photons are incident in a 1-second exposure. Compute the DQE for the same pixel if the same number of photons is incident in a 400-second exposure.

5. An MOS capacitor observes two sources in the band 400–600 nm. Source A has a spectrum such that the distribution of photons in the 400–600 nm band is given by $n_A(\lambda) = A\lambda^3$. Source B has a distribution of photons given by $n_B(\lambda) = B\lambda^{-2}$ in the same band. If the two sources generate photoelectrons at exactly the same rate, compute their (energy) brightness ratio. You may assume the detector's QE is not a function of wavelength.

6. Construction of a monolithic 8192 × 8192 pixel CCD array is technologically possible. How long would it take to read this array through a single amplifier at a pixel frequency of 25 kHz?

7. The gate structure for four pixels of a certain orthogonal transfer CCD is sketched at below. Propose a pattern for (a) assigning gate voltages during collection, (b) a method for clocking voltages for a one-pixel shift to the right, and (c) a method for clocking voltages for a one-pixel shift downwards. Gates with the same numbers are wired together.

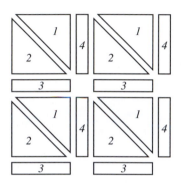

8. At an operating temperature of 300 K, a certain CCD exhibits a dark current of 10^5 electrons per second. (a) Estimate the dark rate, in electrons per second, if this CCD is operated at –40 °C (233 K). (b) Compute the operating temperature at which the dark current will be 10 electrons per second.

9. A CCD has a CTE of "three nines" (i.e. CTE = 0.9990). What fraction of the charge stored in the pixel most distant from the amplifier actually reaches the amplifier if the array is (a) 128 pixels on a side or (b) 2048 on a side?

10. A rapid-scan CCD has a read noise of 200 electrons per pixel. You observe a source that produces 400 photoelectrons spread over 25 pixels. Dark current and background are negligible. (a) Compute the SNR for this measurement. (b) Suppose an image intensifier is available with a gain of 10^4 and a gain uncertainty of $\pm5\%$. Repeat the SNR computation for the intensified CCD. Should you use the bare or the intensified CCD for this measurement?

11. Consider the general situation in which a bare CCD would record N photoelectrons with a total read noise of R electrons in a given exposure time. An intensifier stage has a gain of g and a gain uncertainty of σ_g. If $g \gg 1$, show that the intensifier will improve the overall DQE in the same exposure time if $R^2 g^2 > \sigma_g N$.

12. A single-element bolometer operates with a heat sink at 12 K. The thermal link has a conductance of $G' = 5 \times 10^{-7}\,\mathrm{W\,K^{-1}}$ and a heat capacity of $C = 3 \times 10^{-8}\,\mathrm{J\,K^{-1}}$. (a) Compute the time constant and temperature change after 2 seconds of exposure to a source that deposits 10^{-10} W in the bolometer. (b) If the bolometer is a doped germanium thermistor with a resistance of R_0 ohm at 12 K and effective energy gap of A = 0.02 eV, compute the fractional change in resistance due to the exposure in (a).

Chapter 9
Digital images from arrays

All the pictures which science now draws of nature and which alone seem
capable of according with observational fact are mathematical pictures.

— Sir James Jeans, *The Mysterious Universe*, 1930

Astronomers normally present the output of a sensor array in the form of a
digital image, a picture, but a mathematical picture. One appealing characteristic of a digital image is that the astronomer can readily subject it to mathematical manipulation, both for purposes of improving the image itself, as well as
for purposes of extracting information.

Accordingly, the chapter will proceed by first presenting some general
thoughts about array data, and some general algorithms for image manipulation.
Because they are so useful in astronomy, we next examine some procedures for
removing image flaws introduced by the observing system, as well as some
operations that can combine multiple images into a single image. Finally, we
look at one important method for extracting information: *digital photometry*,
and derive the *CCD equation*, an expression that describes the quality you can
expect from a digital photometric measurement.

9.1 Arrays

Astronomers usually use ***panoramic detectors*** to record two-dimensional
images and, at optical wavelengths, they most often use a charge-coupled
device (CCD). Unlike a photographic plate (until the 1980s, the panoramic
detector of choice), a CCD is an ***array*** – a grid of spatially discrete but
identical light-detecting elements. Although this chapter discusses the CCD
specifically, most of its ideas are relevant to images from other kinds of arrays.
These include broadband superconducting tunnel junctions (STJs), hybrid
semiconductor arrays used in the infrared, ultraviolet-sensitive devices like
microchannel plates, and bolometer arrays used in the far infrared and at other
wavelengths.

9.1.1 Pixels and response

A telescope gathers light from some astronomical scene and forms an image in its focal plane. At each point (x', y') in the focal plane, the image has brightness, $B(x', y')$, measured in W m^{-2}. The function B is an imperfect representation of the original scene. Every telescope has limited resolving power and optical aberrations. Often a telescope will transmit some parts of the scene more efficiently than other parts, perhaps because of dust on a filter or some structural obstruction. In addition, some contributions to the brightness of the image do not originate from the remote source: the background glow from the atmosphere or (in the infrared) from the telescope, for example.

We introduce a panoramic detector (or *focal-plane array*) to record this imperfect image. It is invariably a *rectangular* array, with elements arranged in N_X columns and N_Y rows. We denote the location of an individual detector element in this array as $[x, y]$, where it will be convenient to restrict x and y to integer values (running from 1 to N_X and 1 to N_Y, respectively). Instead of the phrase "individual detector element," we use the word *pixel* (from "picture element").

Figure 9.1 shows a few pixels of some array. The sensitive area of a single pixel is a rectangle of dimensions δ_{px} by δ_{py}, and the pixels are separated by distances d_{px} horizontally and d_{py} vertically. For most direct-imaging devices, pixels are square ($d_{px} = d_{py} = d_p$) and have sizes in the 5–50 μm range. Linear arrays ($N_x \gg N_y$), sometimes used in spectroscopy, are more likely to employ oblong pixel shapes.

If $d_p > \delta_p$ in either direction, each pixel has an insensitive region whose relative importance can be measured by the geometric *fill factor*,

$$\frac{\delta_{px}\,\delta_{py}}{d_{px}\,d_{py}}$$

For many CCDs $\delta_p = d_p$, and the fill factor is unity.

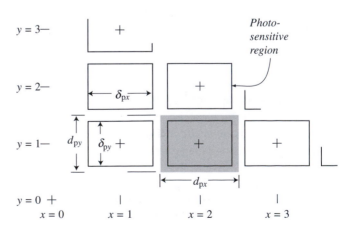

Fig. 9.1 Pixels near one corner of a detector array. The shaded region indicates the pixel at [2, 1], which consists of a photosensitive region surrounded by an insensitive border.

Our detector lies in the focal plane of the telescope, with the x (for the detector) and x' (for the function B) axes aligned. We are free to choose the origin of the primed coordinate system, so can make the center of a pixel with coordinates $[x, y]$ have primed coordinates:

$$x' = x \cdot d_\mathrm{p}$$
$$y' = y \cdot d_\mathrm{p}$$

The light falling on the pixel $[x, y]$ will have a total power, in watts, of

$$P[x,y] = \int_{\left(y-\frac{1}{2}\right)\delta_\mathrm{p}}^{\left(y+\frac{1}{2}\right)\delta_\mathrm{p}} \int_{\left(x-\frac{1}{2}\right)\delta_\mathrm{p}}^{\left(x+\frac{1}{2}\right)\delta_\mathrm{p}} B(x',y')\mathrm{d}x'\mathrm{d}y' \qquad (9.1)$$

In Equation (9.1), we use square brackets on the left-hand side as a reminder that the detector pixel takes a discrete sample of the continuous image $B(x', y')$ and that x and y can only take on integer values. This **pixelization** or **sampling** produces a loss of image detail if the pixel spacing, d_p, is less than half the resolution of the original image. Such **undersampling** is usually undesirable.

We expose the pixel to power $P[x, y]$ for a time interval, t. It responds by producing (in the case of the CCD) a number of photoelectrons. We call this the photo-response, $r_0[x, y]$. Note that for many photon detectors, including the CCD, $r_0[x, y]$ depends on the *number of incident photons*, not on the energy, $tP[x, y]$. To complicate matters, the photo-response signal usually mixes indistinguishably with that produced by other mechanisms (thermal excitation, light leaks, cosmic-ray impacts, radioactivity, etc. The pixel gives us, not $r_0[x,y]$, but $r[x,y]$, a total response to all elements of its environment, including $P[x, y]$; see Figure 9.2.

Although it is convenient to think of the CCD response on the microscopic level of individual electrons, this may not be the case for other devices. In some

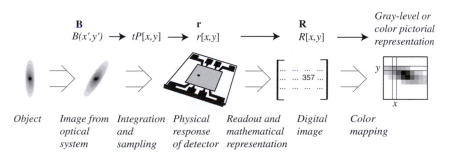

B
$B(x',y') \longrightarrow tP[x,y] \longrightarrow r[x,y] \longrightarrow R[x,y] \longrightarrow$ *Gray-level or color pictorial representation*

| Object | Image from optical system | Integration and sampling | Physical response of detector | Readout and mathematical representation | Digital image | Color mapping |

Fig. 9.2 Production of an astronomical digital image

arrays, it will be better to regard $r[x,y]$ as an analog macroscopic property like a change in temperature or conductivity.

9.1.2 Digital Images

Our instrument must communicate a quantification of $r[x,y]$ to the outside world. In the case of the CCD, the clock circuits transfer charge carriers through the parallel and serial registers, and one or more amplifiers convert each charge packet to a voltage (the *video signal*). Another circuit, the analog-to-digital converter (ADC), converts the analog video signal to an electronic representation of an integer number, primarily because integers are much easier to store in a computer. We symbolize the integer output for pixel $[x, y]$ as $R[x, y]$, its *pixel value*.

The entire collection of all $N_x \times N_y$ integers, arranged as a mathematical array to echo the column–row structure of the detector, is **R**, a *digital image*. Sometimes we call a digital image a *frame*, or an *exposure*. We use boldface symbols for an entire array (or image), as in **R**, and the subscripts in square brackets to indicate one element of an array (a single pixel value), as in $R[x, y]$. The digital image, **R**, is the digital representation of the detector response, **r**. The relation between **R** and **r** may not be simple.

Digital images are simply collections of numbers interpreted as images, and they can be produced in a variety of ways – perhaps by scanning and digitizing an analog image, by an artistic effort with a computer "paint" program or by any method that assembles an ordered array of numbers. Often, to help interpret the array, we *map* the numbers onto a gray-scale or color-scale and form a pictorial representation.

For example, the "picture" of the nearby galaxy M51 in Figure 9.3 is a representation of a digital image in which a grid of squares is colored according to the corresponding pixel values. Squares colored with 50% gray, for example, correspond to pixel values between 2010 and 2205, while completely black squares correspond to pixel values above 4330. A mapping like Figure 9.3 usually cannot show all the digital information present, since pixel values are

Fig. 9.3 A CCD image of the galaxy M51.

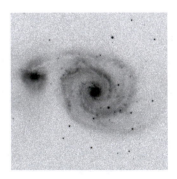

often 16-bit integers,[1] while human vision only distinguishes at most a few
hundred gray levels (which code as 7- or 8-bit integers).

9.1.3 CCD Gain

We use some special terminology in the case where image **R** represents the
response of an astronomical array. Quantifying detector response usually means
measuring a voltage or current (i.e. an *analog* quantity) and subsequently
expressing this as a *digital* quantity. Hence, each pixel of **R** is said give a count
of how many ***analog-to-digital units*** (ADUs) were read from the detector. Each
pixel value, $R[x, y]$, has "units" of ***ADU***. The terms ***data number*** (DN) and
counts are sometimes used instead of ADU.

The differential change in $r[x, y]$ that produces a change of one ADU in $R[x, y]$
is called the ***gain***

$$g[x,y] = \text{gain} = \frac{dr[x,y]}{dR[x,y]}$$

In the general case, gain will differ from pixel to pixel, and may even depend on
the signal level itself. In the case of the CCD, gain is set primarily by the output
amplifier and the ADC, and the astronomer might even set the gain with the
controlling software. We expect approximately identical gain for all pixels.
Moreover, CCD amplifiers are generally linear, so we usually assume $g[x, y]$
is independent of $r[x, y]$. The CCD gain has units of electrons per ADU:

$$g = \text{CCD gain} = \frac{dr[x,y]}{dR[x,y]} \text{ [electrons per ADU], independent of } r, x \text{ and } y$$

Gain may differ (by a small amount, one hopes) for each amplifier on a multi-
amplifier CCD chip, or for the components in a mosaic.

9.1.4 Pictures lie

> The world today doesn't make sense, so why should I paint pictures that do?
>
> — Pablo Picasso (1881–1973)

Figure 9.3, the gray-scale map of a CCD image of the galaxy M51, imperfectly
represents **R**, the underlying digital image. But even the underlying image is a
lie. There are interstellar, atmospheric, and telescopic effects that mask, distort,
and destroy information as light travels from M51 to the detector, as well as
additions and transformations introduced by the detector itself — all information

---

[1] The number of bits (binary digits), n_B, in a computer memory location determines the value of the
largest integer that can be stored there. (It is $2^{n_B} - 1$.) Thus, a 16-bit integer can have any value
between 0 and 65,535, while an 8-bit integer can have values between 0 and 255.

Fig. 9.4 Additive (ADD), multiplicative (MUL) and non-linear (Non-Lin) effects produce imperfections in detector output. Alterations by optics include intentional restrictions by elements like filters. The local environment may add signal by introducing photons (e.g. light leaks) or by other means (e.g. thermal dark current, electronic interference, cosmic rays).

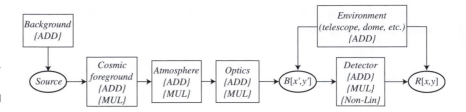

gains and losses that we would rather not have. Figure 9.4 schematically represents the most obvious elements that might influence the raw digital image.

For the moment, imagine a "perfect" digital image, $\mathbf{R}^*$. In $\mathbf{R}^*$, the number of ADUs in a pixel is directly proportional either to the energy or number of photons arriving from the source located at the corresponding direction in the sky. The image $\mathbf{R}^*$ is not influenced by any of the elements represented in Figure 9.4. Mathematically, three kinds of processes can cause $\mathbf{R}$, the raw image, to differ from $\mathbf{R}^*$, the perfect image:

Additive effects contribute or remove ADUs from a pixel in a way that is independent of the magnitude of $R^*[x, y]$. Examples include:

- background radiation emitted by the telescope, the Earth's atmosphere, foreground or background stars, or any other objects visible to the pixel;
- impacts of cosmic rays and other energetic particles;
- the ambient thermal energy of the pixel;
- a voltage intentionally added to the video signal to guarantee amplifier linearity.

Multiplicative imperfections change $R^*[x, y]$ to a value proportional to its magnitude. Examples include:

- spatial or temporal variations in quantum efficiency or in gain;
- absorption by the Earth's atmosphere;
- absorption, reflection, or interference effects by optical elements like filters, windows, mirrors and lenses, as well as dirt on any of these

Non-linear imperfections change $R^*[x, y]$ to a value that depends on a quadratic or higher power of its magnitude. An example would be a quantum efficiency or gain that depends on the magnitude of $R^*[x, y]$. *Saturation*, a decrease in detector sensitivity at high signal levels, is a common non-linear imperfection.

All these imperfections are least troublesome if they are *flat*, that is, if they have the same effect on every pixel. Subtracting a spatially uniform background is relatively easy. In contrast, if the imperfection has detail, removing it requires more work. Subtracting the foreground stars from an image of a galaxy, for example, is relatively difficult. Not every imperfection can be removed, and every removal scheme inevitably adds uncertainty. No image ever tells the complete truth.

9.2 Digital image manipulation

> If a man's wit be wandering, let him study the mathematics.
>
> — Francis Bacon (1561–1626)

One of the great benefits of observing with modern arrays is that data take the form of digital images – numbers. Astronomers can employ powerful and sophisticated computing tools to manipulate these numbers to answer questions about objects. We usually first find numerical answers, but eventually construct a narrative answer, some sort of story about the object. Our concern in the remainder of this chapter is to describe some of the computational and observational schemes that can remove the imperfections in astronomical images, and some schemes that can *reduce* those images to concise measurements of astronomically interesting properties. We begin with some simple rules.

9.2.1 Basic image arithmetic

First, some conventions. As before, boldface letters will symbolize complete digital images: **A**, **B**, **C**, and **D**, for example, are all digital images. Plain-faced letters, like h and k, represent single-valued constants or variables. As introduced earlier, indices in square brackets specify the location of a single pixel, and $A\,[2,75]$ is the pixel value of the element in column 2, row 75, of image **A**.

If $\{op\}$ is some arithmetic operation, like addition or multiplication, then the notations

$$\mathbf{A} = \mathbf{B}\ \{op\}\ \mathbf{C}$$
$$\mathbf{A} = k\ \{op\}\ \mathbf{D}$$

mean that

$$A[x,y] = B[x,y]\ \{op\}\ C[x,y], \text{ and}$$
$$A[x,y] = k\ \{op\}\ D[x,y],$$
$$\text{for all indices, } 1 \leq x \leq N_X \text{ and } 1 \leq y \leq N_Y$$

That is, the indicated operation is carried out on a pixel-by pixel basis over the entire image. Clearly, all images in an equation must have the same size and shape for this to work. For example, suppose you take an image of Mars, but, by mistake, leave a lamp on inside the telescope dome. The image actually recorded, as depicted in Figure 9.5, will be

$$\mathbf{A} = \mathbf{M} + \mathbf{L}$$

where **M** is the image due to Mars alone and **L** is the image due to the lamp and everything else. You might be able to obtain a good approximation to **L** by

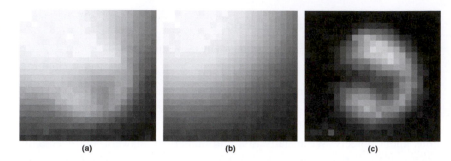

(a)   (b)   (c)

Fig. 9.5 Image subtraction. The image in (a) is of the planet Mars plus background and foreground illumination $(\mathbf{M} + \mathbf{L})$. The image $\mathbf{L}$ in (b) is of the blank sky at the same altitude and azimuth, obtained after sidereal motion took Mars out of the field. The image in (c) is the difference between the two.

leaving the lamp on and taking an image of the blank sky. If so, as shown in the figure, you can computationally recover $\mathbf{M}$:

$$\mathbf{M} = \mathbf{A} - \mathbf{L}$$

9.2.2 Image dimensions and color

We find it natural to think of digital images as two-dimensional objects – brightness arrayed in rows and columns. But a digital image is just a way to interpret a string of numbers, and there are many cases in which it makes sense to think of images with three or more dimensions. For example, you take a series of 250 images of the same star field to search for the period of a suspected variable. Each image has 512 rows and 512 columns. It makes sense to think of your data as a three-dimensional stack, with dimensions $512 \times 512 \times 250$. You will therefore encounter terms like ***data cube*** in the astronomical literature. Another common example would be the output of an array of STJ detectors, where spectral distribution would run along the third dimension. Higher dimensions also make sense. Suppose you take 250 images of the field in each of five filters – you then could have a four-dimensional "data-hypercube." We will not discuss any special operations for these higher-dimensional objects. How they are treated will depend on what they represent, and often will come down to a series of two-dimensional operations.

Color images are a special case. Digital color images pervade modern culture, and there are several methods for encoding them, most conforming to the device intended to display the image. For example, each pixel of a color computer monitor contains three light sources: red (R), green (G), and blue (B). The ***RGB color model*** represents an image as a three-dimensional stack, one two-dimensional digital image for each color. Each pixel value codes how bright the corresponding colored light source should be in that one pixel. The RGB is an additive color model: increasing pixel values increases image brightness.

Subtractive color models are more suited to printing images with ink on a white background. The most common, the ***CMYK model***, uses a stack of four two-dimensional images to represent amounts of cyan, magenta, yellow, and black ink in each pixel. In a subtractive model, larger pixel values imply a darker color.

Astronomers almost never detect color images directly, but will frequently construct *false color* images as a way of displaying data. For example, you might create an RGB image in which the R channel was set by the pixel values of a K-band (i.e. infrared) image, the G channel was set by the pixel values of a V-band (i.e. visual) image and the B channel was set by the pixel values of a far-ultraviolet image. The resulting image would give a sense of the "color" of the object, but at mostly invisible wavelengths.

Astronomers also use *color mapping* to represent the brightness in a simple digital image. In a color mapping, the computer uses the pixel value to reference a color look-up table, and then displays the corresponding color instead of some gray level. Since the eye is better at distinguishing colors than it is at distinguishing levels of gray, a color map can emphasize subtle effects in an image.

9.2.3 Image functions

We expand our notation to include functions of an image. In the following examples, each pixel in image $\mathbf{A}$ is computed from the pixels with the same location in the images in the right side of the equation:

$$\mathbf{A} = -2.5\log(\mathbf{C})$$
$$\mathbf{A} = h(\mathbf{B})^2 + k\sqrt{\mathbf{C}}$$
$$\mathbf{A} = \max(\mathbf{B}, \mathbf{C}, \mathbf{D})$$
$$\mathbf{A} = \text{median}(\mathbf{B}, \mathbf{C}, \mathbf{D})$$

The "max" function in the third example would select the largest value from the three pixels at each x, y location:

$$A[x, y] = \text{the largest of} \{B[x, y], C[x, y], D[x, y]\},$$

for each x and y in $\mathbf{A}$.

Likewise, the fourth example would compute the median of the three indicated values at each pixel location. We can think of many more examples. We also introduce the idea of a function that operates on an entire image and returns a *single* value. For example, the functions maxP and medianP:

$$a = \text{maxP}(\mathbf{A})$$
$$b = \text{medianP}(\mathbf{A})$$

will treat the pixels of image $\mathbf{A}$ as a list of numbers, and pick out the largest value in the whole image and the median value of the whole image, respectively. Again, you can think of a number of other examples of functions of this sort.

9.2.4 Image convolution and filtering

The concept of *digital filtration* is a bit more complex. Image *convolution* is an elementary type of digital filtration. Consider a small image, $\mathbf{K}$, which measures

$2V + 1$ rows by $2W + 1$ columns (i.e. the number of rows and columns are both odd integers). We define the convolution of $\mathbf{K}$ on $\mathbf{A}$ to be a new image, $\mathbf{C}$,

$$\mathbf{C} = \mathrm{conv}(\mathbf{K}, \mathbf{A}) = \mathbf{K} \otimes \mathbf{A}$$

Image $\mathbf{C}$ has the same dimensions as $\mathbf{A}$, and its pixels have values

$$C[x, y] = \sum_{i=1}^{(2V+1)} \sum_{j=1}^{(2W+1)} K[i, j] A[(x - V - 1 + i), (y - W - 1 + j)] \qquad (9.2)$$

The array $\mathbf{K}$ is sometimes called the **kernel** of the convolution. For example, consider the kernel for the 3×3 boxcar filter:

$$\mathbf{B} = \begin{bmatrix} \frac{1}{9} & \frac{1}{9} & \frac{1}{9} \\ \frac{1}{9} & \frac{1}{9} & \frac{1}{9} \\ \frac{1}{9} & \frac{1}{9} & \frac{1}{9} \end{bmatrix} = \frac{1}{9} \begin{bmatrix} 1 & 1 & 1 \\ 1 & 1 & 1 \\ 1 & 1 & 1 \end{bmatrix}$$

Figure 9.6 suggests the relationship between the kernel, the original image, and the result. (1) The center of the kernel is aligned over pixel $[x, y]$ in the original image. (2) The value in each pixel of kernel is multiplied by the value in the image pixel beneath it. (3) The sum of the nine products is stored in pixel $[x, y]$ of the filtered image. (4) Steps (1)–(3) are repeated for all valid values of x and y.

Figure 9.7 shows an image before and after convolution with a boxcar filter. What happens in the convolution is that every pixel in the original image gets replaced with the average value of the nine pixels in the 3×3 square centered on itself. You should verify for yourself that this is what Equation (9.2) specifies. The boxcar is a filter that blurs detail – that is, it reduces the high spatial frequency components of an image. Figure 9.7c shows that a larger-sized boxcar kernel, 7×7, has an even greater blurring effect.

Note that in convolution, there is a potential problem at the image edges, because Equation (9.2) refers to non-existent pixels in the original image $\mathbf{A}$. The usual remedy is artificially to extend the edges of $\mathbf{A}$ to contain the required

Fig. 9.6 Image convolution operation. The kernel is aligned over a set of pixels in the original centered on position *x, y*. The result is the sum of the products of each kernel pixel with the image pixel directly beneath it. The result is stored in pixel *x, y* of the filtered image.

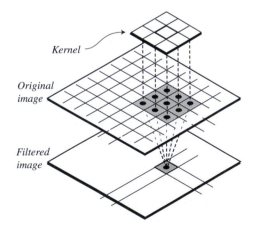

pixels, typically setting the value of each fictitious pixel to that of the nearest actual pixel.

Convolutions that blur an image are called *low-pass filters*, and different kernels will blur an image in different ways — a *Gaussian kernel* (whose values are set by a two-dimensional Gaussian function) can simulate some atmospheric seeing effects, for example. Other kernels are *high-pass filters*, and emphasize image detail while suppressing large-scale (low spatial frequency) features. Representative of these is the *Laplacian kernel*. The 3×3 Laplacian is

$$\begin{bmatrix} -1 & -1 & -1 \\ -1 & 8 & -1 \\ -1 & -1 & -1 \end{bmatrix}$$

It essentially computes the average value of the second derivative of the intensity map — enhancing pixels that differ from the local trend. Figure 9.8 shows an example. Other filter kernels can produce image sharpening without loss of large-scale features, edge detection, gradient detection, and embossing effects.

A particularly useful filtering process is *unsharp masking*. The filtered image is the original image minus an "unsharp-mask" image — this mask is a low-pass filtered version of the original. The unsharp mask enhances the high-frequency components and reduces the low-frequency components of the image, emphasizing detail at all brightness levels. Since convolution is distributive, unsharp masking can be accomplished by convolution with a single kernel. For example, convolution with the 5×5 identity kernel

$$\mathbf{I} = \begin{bmatrix} 0 & 0 & 0 & 0 & 0 \\ 0 & 0 & 0 & 0 & 0 \\ 0 & 0 & 1 & 0 & 0 \\ 0 & 0 & 0 & 0 & 0 \\ 0 & 0 & 0 & 0 & 0 \end{bmatrix}$$

leaves the image unchanged. Convolution with a 5×5 Gaussian ($\sigma = 1.25$ pixels)

$$\mathbf{G} = \frac{1}{3.58} \begin{bmatrix} 0.03 & 0.08 & 0.11 & 0.08 & 0.03 \\ 0.08 & 0.21 & 0.29 & 0.21 & 0.08 \\ 0.11 & 0.29 & 0.38 & 0.29 & 0.11 \\ 0.08 & 0.21 & 0.29 & 0.21 & 0.08 \\ 0.03 & 0.08 & 0.11 & 0.08 & 0.03 \end{bmatrix}$$

creates a blurred mask. An unsharp masking filtration operation would be

$$\mathbf{C} = 2(\mathbf{I} \otimes \mathbf{A}) - \mathbf{G} \otimes \mathbf{A}$$

or, using the distributive properties of the convolution operation,

$$\mathbf{C} = (2\mathbf{I} - \mathbf{G}) \otimes \mathbf{A} = \mathbf{F} \otimes \mathbf{A}$$

In this example,

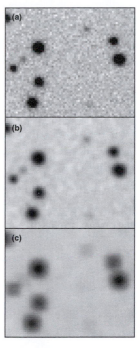

Fig. 9.7 Boxcar filter. (a) A CCD image of a few stars in the cluster M67, displayed as a negative gray-scale (stars are black). (b) The previous image convolved with a 3×3 boxcar. The smoothing effect is most obvious in the sky background. (c) The image in (a) convolved with a 7×7 boxcar.

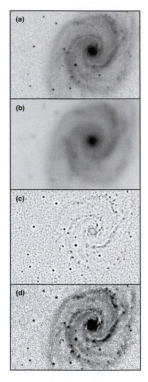

Fig. 9.8 Digital filtration. (a) Image of M51. Full width at half-maximum (FWHM) of the star images is about 2 pixels. (b) After application of a 7-pixel FWHM Gaussian filter to (a). (c) The original after application of a 5 × 5 Laplacian filter, which emphasizes features like stars and the higher-contrast spiral features. Note that sky noise is also enhanced. (d) After application of an unsharp mask based on the Gaussian in (b).

$$\mathbf{F} = 2\mathbf{I} - \mathbf{G} = \frac{1}{3.58}\begin{bmatrix} -0.03 & -0.08 & -0.11 & -0.08 & -0.03 \\ -0.08 & -0.21 & -0.29 & -0.21 & -0.08 \\ -0.11 & -0.29 & -6.78 & -0.29 & -0.11 \\ -0.08 & -0.21 & -0.29 & -0.21 & -0.08 \\ -0.03 & -0.08 & -0.11 & -0.08 & -0.03 \end{bmatrix}$$

Other forms of filtration are not convolutions as defined by Equation (9.2), but do utilize the idea illustrated in Figure 9.6 – the value of a pixel in the filtered image is determined by applying some algorithm to the neighboring pixels as described by a kernel. For example, a 3 × 3 *local-median filter* sets the filtered pixel value equal to the median of the unfiltered pixel and its eight neighbors. Other examples use conditional operators: you might map the location of suspicious pixels by computing the statistics, include the sample standard deviation, s, inside a 9 × 9 square surrounding a pixel, then apply the operation:

> if the original pixel differs from the local mean by more than 3s,
>
> then:
>
> set the value of the filtered pixel to zero,
>
> otherwise:
>
> set the value of the filtered pixel to one.

Any digital filtration destroys information, so use it with caution.

9.3 Preprocessing array data: bias, linearity, dark, flat, and fringe

When astronomers speak of *data reduction*, they are thinking of discarding and combining data to reduce their volume as well as the amount of information they contain. A single CCD frame might be stored as a few million numbers – a lot of information. An astronomer usually discards most of this. For example, he may only care about the brightness or position of a single object in the frame – information represented by just one or two numbers. Ultimately, he might reduce several hundred of these brightness or position measurements to determine the period, amplitude, and phase of a variable star (just three numbers and their uncertainties) or the parameters of a planet's orbit (six numbers and six uncertainties).

Few astronomers enjoy reducing data, and most of us wish for some automaton that accepts what we produce at the telescope (raw images, for example) and gives back measurements of our objects (magnitudes, colors, positions, chemical compositions). In practice, a great deal of automation is possible, and one characteristic of productive astronomy is a quick, smooth path from telescope to final measurement. The smooth path is invariably paved with one or more computer programs working with little human intervention. Before using or writing such a program, the astronomer must get to know his data, understand their imperfections, and have a clear idea of what the data can or cannot reveal.

Eventually, data reduction permits ***data analysis*** and ***interpretation*** – for example, what kind of variable star is this, what does that tell us about how stars evolve? Properly, the boundaries between reduction, analysis, and interpretation are fuzzy, but each step towards the interpretation stage should become less automatic and more dependent on imagination and creativity.

The first and most automatic steps remove the most obvious imperfections. Data should characterize the astronomical source under investigation, not the detector, telescope, terrestrial atmosphere, scattered light, or any other perturbing element. This section examines the very first steps in reducing array data, and explains reductions that must be made to all CCD data (and most other array data), no matter what final measurements are needed. Other authors sometimes refer to these steps as the ***calibration*** of the image. I prefer to separate these steps into the ***preprocessing*** and then ***de-fringing*** of the image.

Consider, then, a raw image, **R**. Of the many imperfections in **R**, preprocessing attempts to correct for:

- **Bias**. If a detector is exposed to no light at all, and is given no time to respond to anything else in its environment, it may nonetheless produce positive values for a particular pixel, $R[x, y]$, when it is read out. In other words, even when $r[x, y]$, the response of the detector, is zero, $R[x, y]$, *is not*. This positive output from a zero-time exposure is called the bias level, or the ***zero level***, and will be present in every frame as a quantity added to the output.

- **Dark response**. If a detector is not exposed to a signal from the telescope, but simply sits in the dark for time t, it will in general respond to its dark environment so that $r[x, y]$ is not zero. In most detectors, this dark response is the result of thermal effects. In a CCD, electron–hole pairs are created from the energy present in lattice vibrations at a rate proportional to $T^{3/2} \exp(-a/kT)$, where a is a constant that depends on the size of the band gap. Like the bias, the dark response adds ADUs to the readout of every frame. Unlike the bias, dark response will depend on exposure time.

- **Linearity**. The response of a *linear* detector is directly proportional to incoming signal. All practical detectors are either completely non-linear or have a limited range of linearity. One of the appealing characteristics of CCDs is the large range of signal over which their response is linear. Even CCDs, however, ***saturate*** at large signal levels, and eventually cease to respond to incoming photons.

- **Flat field response**. Identical signals generally do not produce identical responses in every pixel of a detector array. Not all pixels in the array respond to light with equal efficiency. This defect can arise because of structural quantum-efficiency differences intrinsic to the array. It can also arise because of vignetting or other imperfections in the optical system like dust, fingerprints, and wildlife (insects turn up in unexpected locations) on filters or windows.

The observer wants to remove these instrument-dependent characteristics from her images in preprocessing. To do so, she must make some reference observations and appropriate image manipulations. We consider each of the four preprocessing operations in turn. The books by Howell (2006) and by Martinez and Klotz (1998) treat CCD data reduction in greater detail.

9.3.1 Bias frames and overscans

If the observer simply reads her array with zero integration time (actually, the CCD first clears, then immediately reads out), never exposing it to light, she has obtained a ***bias frame***. The bias frame represents the electronic background present in every frame, no matter how short the integration time. The idea, of course, is that this is uninteresting information, and the astronomer needs to subtract the bias frame from every other frame she plans to use. In practice, one bias frame may well differ from another. For larger CCD arrays readout time may be long enough for much to happen, including cosmic-ray hits, local radio-activity, and electronic interference.

It is good practice to obtain many bias frames. For one thing, properly combining several frames will reduce uncertainty about the average level of the bias, as well as minimize the influence of cosmic-ray events. For another, the careful observer should monitor the bias level during an observing run, to guard against any drift in the average level, and to make sure any two-dimensional pattern in the bias is stationary.

Assume for the moment that the bias does not change with time, and that the astronomer takes N bias frames during the run. Call these $\mathbf{z}_1, \mathbf{z}_2, \ldots, \mathbf{z}_N$. How should he combine these frames to compute $\mathbf{Z}$, the one representative bias image he will subtract from all the other frames? Here are some possibilities:

(1) **Mean**. Set $\mathbf{Z} = \text{mean}(\mathbf{z}_1, \mathbf{z}_2, \ldots, \mathbf{z}_N)$

This is a bad strategy if there are any cosmic-ray hits. Computationally easy, it will dilute the effects of cosmic rays, but not remove them.

(2) **Median**. Set $\mathbf{Z} = \text{median}(\mathbf{z}_1, \mathbf{z}_2, \ldots, \mathbf{z}_N)$

This works well, since the median is relatively insensitive to statistical outliers like the large pixel values generated by cosmic rays. It has the disadvantage that the median is a less robust and stable measure of central value than the mean, and is thus somewhat inferior for those pixel locations not struck by cosmic rays.

(3) **Indiscriminant rejection**. At each $[x, y]$, reject the largest pixel value, then use (1) or (2) on the remaining $(N-1)$ values.

This removes cosmic rays, but is possibly too drastic, since it skews the central values towards smaller numbers. An alternative is to reject both the largest and the smallest values at each location. This discards two entire images worth of data, and skews cosmic-ray pixels to slightly larger numbers.

(4) **Selective rejection**. At each [x, y], reject only those pixels *significantly* larger than the mean, then apply (1) or (2) on the remaining values. To decide whether or not a pixel value is so large that it should be rejected, use a criterion like:

$$z_i[x,y] > \mu[x,y] + k\sigma[x,y]$$

where μ and σ are the mean and standard deviation of the pixel values, (a) at x, y, or (b) over a segment of the image near x, y, or (c) over the entire image. The value of the constant k determines how selective the rejection will be. For a normal distribution, $k = 3$ will reject 14 legitimate (non-cosmic-ray) pixels out of 10,000.

This is an excellent strategy, but is computationally intensive. Strategy 4b or 4c makes it possible to produce a "clean" **Z** from a single frame by replacing the rejected pixel value with the mean or median value of its neighbors.

You will undoubtedly think of other advantages or disadvantages to all these strategies, and also be able to compose alternatives. The exact strategy to use depends on circumstance, and we will use the notation

$$\mathbf{Z} = \text{combine}(\mathbf{z}_1, \mathbf{z}_2, \ldots, \mathbf{z}_N)$$

to indicate some appropriate combination algorithm.

What if the bias changes over time? The astronomer might compute different **Z** values for different segments of the run, but only if the changes are gradual. A common alternative strategy for CCDs is to use an **overscan**. You produce overscan data by commanding the clocks on the CCD so that each time the serial register is read, the read continues for several pixels *after* the last physical column has been read out.[2] This produces extra columns in the final image, and these contain the responses of "empty," unexposed pixels, elements of the serial register that have not been filled with charge carriers from the parallel registers. These extra columns are the overscan region of the image and record the bias level during the read. The usual practice is to read only a few extra columns, and to use the median pixel values in those columns to correct the level of the full two-dimensional **Z** image. If ω_i is the overscan portion of image i, and Ω_Z is the overscan portion of the combined **Z**, then the bias frame to apply to image i is

$$\mathbf{Z}_i = \mathbf{Z} + (\text{medianP}(\omega_i - \Omega_Z)) \tag{9.4}$$

Figure 9.9 shows a slightly more complicated application of an overscan. Here the zero level has changed during the read, and shows up in the image most clearly as a change in background in the vertical direction. To correctly remove

[2] It is also possible to continue to read beyond the last exposed *row*. This means the overscan of extra rows will include the dark charges generated during the full read time. For arrays operating with significant dark current, this may be significant. Some manufacturers intentionally add extra physical pixels to the serial register to provide overscan data.

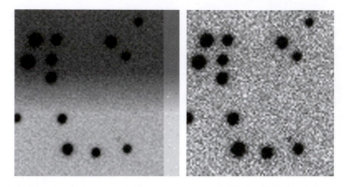

Fig. 9.9 Overscan correction. (a) This frame has a 10-column overscan region on its right edge. The frame in (b) results after the bias frame, corrected for the overscan, is subtracted and the overscan section trimmed from the image. Note that the frame in (b) is displayed with a different gray-scale mapping than in (a).

the bias, the astronomer fitted a one-dimensional function (in the y-direction) to the difference $(\omega_i - \Omega_Z)$, and added that function to $\mathbf{Z}$.

9.3.2 Dark current

Even in the absence of illumination, a detector will generate a response during integration time t. This is called the **dark response**. The rate at which the dark response accumulates is the **dark current**. Although primarily a thermal effect, dark current will not be the same for every pixel because of inhomogeneities in fabrication. Some pixels, called "hot" pixels, differ from their neighbors not in temperature, but in efficiency at thermal production of charge carriers.

To calibrate for dark current, a long exposure is taken with the shutter closed – this is called a **dark frame**, $\mathbf{d}$. In view of the earlier discussion about cosmic-ray hits and uncertainties, it is best to combine several individual dark frames ($\mathbf{d}_1, \mathbf{d}_2, \ldots, \mathbf{d}_M$) to produce one representative frame:

$$\mathbf{d} = \mathrm{combine}(\mathbf{d}_1, \mathbf{d}_2, \ldots, \mathbf{d}_M)$$

The dark frames should be obtained in circumstances (temperature, magnetic environment) as similar as possible to those prevailing for the data frames. If $\mathbf{d}$ has exposure time t, then you may compute the **dark rate** image as

$$\mathbf{D} = \frac{\mathbf{d} - \mathbf{Z}}{t}$$

or

$$\mathbf{D} = \frac{1}{t} \mathrm{combine}([\mathbf{d}_1 - \mathbf{Z}_1], [\mathbf{d}_2 - \mathbf{Z}_2], \ldots, [\mathbf{d}_M - \mathbf{Z}_M])$$

The second form applies if you are using an overscan correction for each dark frame as in Equation (9.4). You may then correct for dark current and bias on every data frame by subtraction of the image $t\mathbf{D} + \mathbf{Z}$. The units for $\mathbf{D}$ in the above equations are ADUs per second. However, dark current for CCDs is almost always quoted in units of electrons per second as $\mathbf{D}_e = \mathbf{gD}$, where g is the detector gain in electrons per ADU.

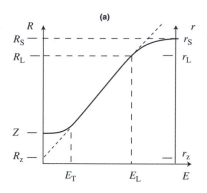
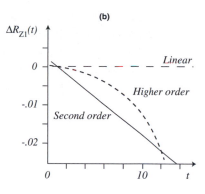

Fig. 9.10 Linearity: (a) A schematic of the output R, in ADU, and the response, r, in electrons, of a single pixel in a detector that is linear over a restricted input range. The sloped dashed line is Equation (9.5). (b) An experimental determination of a correction for non-linearity as explained in the text.

Observers routinely cool detectors to reduce dark current and its associated noise. In some cases (e.g. a CCD at $-90\,°C$) the dark rate may be so low that you can omit the correction. A careful observer, though, will always take dark frames periodically, even if only to verify that they are not needed. It also appears that the dark current in some multi-pinned-phase (MPP) CCDs is somewhat non-linear, which means you must either model the non-linearity or take dark frames whose exposure times match those of the data frames.

9.3.3 Detector linearity

All practical devices depart from linearity. If a pixel receives photons at rate $P[x, y]$ for time t, it has a **linear output** if

$$R[x,y] = R_z[x,y] + t(D[x,y] + Q[x,y]P[x,y]) = R_z[x,y] + bE[x,y] \qquad (9.5)$$

where $R_z[x,y]$, $D[x,y]$ and $Q[x,y]$ are pixels in time-independent arrays, respectively a zero level, the dark rate, and the efficiency in ADUs per photon. A similar equation applies to the response in electrons, r. The CCD response curves resemble Figure 9.10a, where the horizontal variable, E, the exposure, is a quantity proportional to total input (photon count plus dark current). The detector in the figure is linear between a threshold exposure, E_T, and an upper limit, E_L. The output labeled Z is the **bias** of the pixel.

The pixel **saturates** at response r_S and output R_S. Recall that saturation in a CCD pixel results if its potential well fills and the MOS capacitor can store no additional charge carriers. Saturated pixels are insensitive to light and to dark current, but in many devices can **bloom** and spill charge carriers into neighboring pixels.

Charge-coupled devices have remarkably good linearity over several orders of magnitude. Equation (9.5) typically holds to better than 1% over a range $E_L \approx E_s \approx 10^5 E_T$. Moreover, the threshold effect for a CCD is very small, so that $R_z \approx Z$. Recall also that ADCs may be set so that pixels reach digital saturation at $r < r_L$.

Many other devices have significant non-linear behavior well short of their saturation levels, and the observer must remove this non-linearity. Figure 9.10b illustrates one method for empirically measuring a correction.

Suppose R_1 is the response of one pixel to a one-second exposure. If this pixel is a linear detector, satisfying Equation (9.5), then different integration times with a constant light source should give response $R_L = R_z + t(R_1 - R_z)$. However, the actual detector has the non-linear response, R. A series of exposures of different lengths can generate a plot of the quantity

$$\Delta R_{z1}(t) = \frac{1}{t}(R - R_L) = \frac{R - R_z}{t} - (R_1 - R_z) \tag{9.6}$$

as a function of exposure time t. If a linear fit to this has non-zero slope (e.g. the solid line in Figure 9.10b) that fit indicates a quadratic equation for the corrected linear response in the form

$$\mathbf{R_L} = \lin(\mathbf{R}) = a + b\mathbf{R} + c\mathbf{R}^2 \tag{9.7}$$

Strictly speaking, the constants a, b, and c could be different for different pixels in an array. In most cases, uncertainties in the pixel-to-pixel variation justify using average values for the entire array.

9.3.4 Flat field

Correcting for pixel-to-pixel variations in device sensitivity is both the most important and the most difficult preprocessing step. Conceptually, the correction procedure is very simple. The astronomer takes an image of a perfectly uniform (or "flat") target with the complete observing system: detector, telescope, and any elements like filters or obstructions that influence the focal-plane image. If the observing system is equally sensitive everywhere, every pixel in this *flat-field image*, after correction for the bias and dark, should produce an identical output. Any departure from uniformity in the corrected flat-field image will map the sensitivity of the system, in the sense that pixels registering higher counts are more sensitive. Figure 9.11 shows a raw CCD image, a flat-field image, and the original image after the flat-field correction.

At least three practical difficulties hamper the kind of correction illustrated. First, it is difficult to produce a sufficiently (i.e. 0.5% or better) uniform target. Second, sensitivity variations are in general a function of wavelength. Therefore, the spectrum of the target should match that of the astronomical sources of interest. Spectrum matching becomes especially troublesome with multiple sources of very different colors (stars and background sky, for example) in the same frame. Finally, it is difficult to guarantee that the "complete observing system" remains unchanged between the acquisition of the flat field and acquisition of the data frames.

In common practice, observers employ three different objects as the flat-field target: (1) the bright twilight sky, (2) the dark night sky, and (3) a nearby

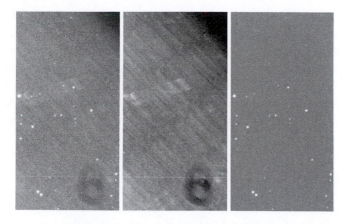

Fig. 9.11 (a) A section of an unprocessed CCD image of a star field. (b) A combined twilight flat for the same section. The two dark rings in the lower right are the shadows of two dust particles on the CCD window. The dark region in the upper right results from vignetting by the filter holder. The section is shown after preprocessing in (c).

object – usually an illuminated surface inside the observatory dome. Images of these sources are usually termed twilight, dark sky, and dome flats, respectively. Each has advantages and disadvantages.

Twilight flats

The clear twilight sky is not uniform: it is brighter all the way around the horizon than it is near the zenith, and, of course, brighter in the direction of the rising or recently set Sun. By pointing towards the zenith (the exact location of the "flat" spot – usually 5–10 degrees anti-solar from the zenith – is slightly unpredictable), the observer finds a target uniform to about 1% over a one-degree field. It is rare to do better than this. For narrow fields of view, this is acceptable. Clouds usually prohibit good flats.

The advantages of the twilight-sky target are that, for a brief interval, it is the right brightness, and relatively uniform. Moreover, observing in twilight means flat-field calibrations do not consume valuable nighttime hours. The disadvantages are:

- Large-scale uniformity is limited by the natural gradient in the twilight sky, and small-scale uniformity is limited by the gradual appearance of star images as twilight fades.
- The twilight sky has a spectrum that is quite different from that of most astronomical sources, as well as that of the night sky.
- Twilight brightness and spectrum both change rapidly. The duration of useable twilight is short, and with large arrays (long readout times), or with many filters, it becomes difficult to accumulate sufficient numbers of images.
- Scattered skylight near the zenith has a strong linear polarization, and the flat field of some systems may be polarization sensitive.

Dark-sky flats

The emission from the dark (moonless!) night sky is a tempting source for flat fields. Uniformity is perfect at the zenith and degrades to about two percent per degree at a zenith angle near 70 degrees. Moreover, the spectrum of the night sky is identical to one source of interest: the background that will usually be subtracted from all data frames, an especially important advantage if measuring sources fainter than the background sky. High sky brightness is the rule in the ground-based infrared, where dark-sky flats are the rule.

Offsetting these attractive characteristics are some potent negatives for dark-sky targets. First, stars are everywhere. Any dark-sky flat will inevitably contain many star images, marring the target's uniformity. The observer can remove star images and construct a good flat with the **shift-and-stare** or **dither** method. The astronomer takes many deep exposures of the dark sky, taking care always to "dither" or shift the telescope pointing between exposures by at least many stellar image diameters. He then combines these in a way that rejects the stars. For example, take five dithered images of a dark field. If the density of stars is low, chances are that at any $[x, y]$ location, at most one frame will contain a star image; so computing the median image will produce a flat without stars. More sophisticated combination algorithms can produce an even better rejection of stellar images. The shift-and-stare method should also be employed for twilight flats, since (1) they will usually contain star images and (2) telescope pointing should be shifted back to the flat region near the zenith for each new exposure anyway.

Understand the limitations of shift-and-stare: the scattered-light halos of bright stars can be many tens of seconds of arc in radius and still be no fainter than one percent of the background. Removing such halos, or extended objects like galaxies, can require large shifts and a very large number of exposures.

A second difficulty is that the dark sky is – well – dark. In the visible bands, one typically requires 10^2 to 10^6 times as long to count the same number of photons on a dark-sky frame as on a twilight frame. Sometimes, particularly in broad bands with a fast focal-ratio telescope, this is not a serious drawback, but for most work, it is crucial. Each pixel should accumulate at least 10^4 electrons to guarantee one percent Poisson uncertainty; so dark-sky flats will typically require long exposure times. They are consequently very costly, since time spent looking at blank sky might otherwise be spent observing objects of greater interest.

A modification of shift-and-stare can sometimes help here. If the objects of interest occupy only a small fraction of the frame, then it should be possible to dither and collect many unaligned data frames. The median of these unaligned frames is the dark-sky flat, and no time has been "wasted" observing blank sky, since the flat frames also contain the science.

Dome flats

A source inside the dome is an attractive flat-field target, since the astronomer in principle controls both the spectrum and the intensity of the illumination, and observations can be taken during daylight. With very small apertures, it is possible to mount a diffusing light box at the top of the telescope tube, but most telescopes are simply pointed at a white screen on the inside of the dome. In practice, in a crowded dome, it is often difficult to set up a projection system that guarantees uniform illumination, the shadow of a secondary may become important in the extrafocal image, and there is an increased possibility of introducing unwanted light sources from leaks or reflections. Nevertheless, dome flats are a very important flat-field calibration technique.

Computing simple flats

Assume you have collected N flat-field images, all taken through a single filter, using one of the targets discussed above. If s_i is one if these raw images, then the first step in creating the calibration frame is to remove its bias, dark, and non-linearities:

$$\mathbf{f}_i' = \lim(\mathbf{s}_i) - \mathbf{Z}_i - t_i \mathbf{D}$$

As before, $\mathbf{D}$ is the dark rate, t_i is the exposure time, and $\mathbf{Z}_i$ is the overscan-corrected bias. Next, to simplify combining frames each, frame should be normalized so that the median pixel has a value of 1.0 ADU:

$$\mathbf{f}_i = \mathbf{f}_i'/\mathrm{medianP}(\mathbf{f}_i')$$

Finally, all normalized frames should be combined to improve statistics, as well as to remove any stars or cosmic-ray events:

$$\mathbf{F}_C = \mathrm{combine}(\mathbf{f}_1, \mathbf{f}_2, ..., \mathbf{f}_N)$$

A different calibration frame must be produced for each observing configuration. Thus, there must be a different flat for each filter used, and a different set of flats whenever the observing system changes (e.g. the detector window is cleaned, or the camera rotated).

Compound flats

Given the imperfections of all three flat-fielding techniques, the best strategy sometimes combines more than one technique, applying each where its strengths are greatest. Thus, one uses a dome flat or twilight flat to establish the response of the system on a small spatial scale (i.e. the relative sensitivity of a pixel compared with those of its immediate neighbors.) Then, one uses a smoothed version of a dark-sky flat to establish the large-scale calibration (e.g. the response of the lower half of the detector relative to the upper half). The idea is to take advantage of both the absence of small-scale non-uniformities (stars) in the dome or twilight target as well as the absence of large-scale non-uniformities (brightness gradients) in dark-sky targets. To create the compound flat-field calibration, assume that $\mathbf{F}_S$ and $\mathbf{F}_L$ are calibration frames computed as described in the previous section. Frame $\mathbf{F}_S$ is from a target with good small-scale uniformity, and $\mathbf{F}_L$ from one with good large-scale uniformity. Now compute the ratio image and smooth it:

$$c = \frac{F_L}{F_S}$$
$$C = \text{conv}\{b, c\}$$

The kernel in the convolution, b, should be chosen to remove all small-scale features from image c. Image C is sometimes called an ***illumination correction***. The corrected compound flat is just

$$F = F_S \cdot C$$

9.3.5 Preprocessing data frames

Suppose a CCD has output R_i in response to some astronomical scene. Preprocessing corrects this image for non-linearity, bias, dark, and flat field:

$$R_{pi} = \frac{\text{lin}(R_i) - Z_i - t_i D}{F} \tag{9.8}$$

Preprocessing non-CCD array data can differ slightly from the above procedures. For infrared arrays read with double-correlated sampling, the output is the difference between reads at the beginning and the end of an exposure, so bias values cancel and Z is numerically zero. Also in the infrared, emission from the variable background often dominates the images, so much so that raw images may not even show the location of sources before sky subtraction. A common observing practice then is to "chop" telescope pointing between the object investigated and the nearby (one hopes, blank) sky to track its variations. Many infrared-optimized telescopes employ ***chopping secondary mirrors*** that efficiently implement rapid on-source/off-source switching. ***Chopping*** is in this context different from ***nodding*** – manually moving the telescope in the shift-and-stare technique.

In the infrared, then, these high-signal sky frames are usually combined to form the flat-field image. A typical preprocessing plan might go like this: if $s_1, s_2, \ldots, s_n$ are the sky exposures and F is the flat, then

$$S = \text{combine}(s_1, s_2, \ldots s_n)$$
$$f' = S - d$$
$$F = f'/\text{medianP}(f') \tag{9.9}$$
$$S_i = a_i f'$$
$$R_{pi} = \frac{R_i - S_i - d}{F}$$

We assume that both R_i and S are first corrected to remove non-linearity. In the fourth equation, a_i is a scaling factor that matches the medianP or modeP of S to the sky level of the data frame, which might be computed from the adjacent (in time) sky frames.

9.3.6 Fringing

Monochromatic light can produce brightness patterns in a CCD image due to reflection and interference within the thin layers of the device. Fringing is usually due to narrow night-sky emission lines, and if present means that the image of the background sky (only) contains the superimposed fringe pattern. It tends to occur in very narrow band images, or in images in the far red where night-sky upper-atmospheric OH emission is bright. The fringe pattern is an instrumental artifact, and should be removed.

The fringe pattern depends on the wavelengths of the sky emission lines, but its amplitude varies with the ratio of line to continuum intensity in the sky spectrum, which can change, sometimes rapidly, during a night. Fringes will not appear on twilight or dome flats, but will show up on a dark-sky flat produced by the shift-and-stare method.

If fringing is present, you should *not* use the dark-sky image to create a flat (use twilight), but use the dark-sky image to create a fringe calibration. If $\mathbf{S}$ is the dark-sky image as in Equation (9.9) and $\mathbf{B}$ is the processed dark sky image, we have

$$\mathbf{B} = \frac{\mathbf{S} - \mathbf{d}}{\mathbf{F}} = \mathbf{B}_\mathrm{c} + \mathbf{B}_\mathrm{f} = \mathbf{B}_\mathrm{c} + A\mathbf{b}_\mathrm{f}$$

Here, $\mathbf{B}_\mathrm{c}$ is the part of $\mathbf{B}$ due to the continuum, and $\mathbf{B}_\mathrm{f}$ the part due to fringes. We treat $\mathbf{B}_\mathrm{f}$ as the product of an amplitude, A, and normalized pattern, $\mathbf{b}_\mathrm{f}$. If $\mathbf{B}'$ is a slightly smoothed version of $\mathbf{B}$, then:

$$B_\mathrm{c}[x, y] = \mathrm{minP}(\mathbf{B}')$$
$$\mathbf{B}_\mathrm{f} = \mathbf{B} - \mathbf{B}_\mathrm{c}$$
$$A = \mathrm{maxP}(\mathbf{B}') - \mathrm{minP}(\mathbf{B}')$$

Removing the fringes from a processed science image $\mathbf{R}_\mathrm{pi}$ is then simply a matter of measuring the fringe amplitude on the science image, A_i, and subtracting the calibration fringe pattern scaled to match:

$$\mathbf{R}_\mathrm{pfi} = \mathbf{R}_\mathrm{pi} - \frac{A_\mathrm{i}}{A}\mathbf{b}_\mathrm{f}$$

9.4 Combining images

After preprocessing, astronomers often combine the resulting images. You might, for example, have acquired a dozen images of an extremely fascinating galaxy, and reason (correctly) that adding all of them together digitally will produce a single image with superior signal to noise. The combined image should show features in the galaxy, especially faint features, more clearly than do any of the individual frames. In another example, you may be trying to observe a nebula whose angular size is greater than the field of view of your CCD. You would like to assemble a complete image of the nebula by combining many of your small CCD frames into a large mosaic. Combining images is a tricky business, and this section provides only a brief introduction.

9.4.1 Where is it? The centroid

Suppose you want to combine images **A** and **B**. An obvious requirement is that the pixel location of a source in **A** must be the same as its location in **B**. But what exactly *is* the location of a source? We can compute the location by a two-step process, preferably through instructions to a computer:

1. Decide which pixels belong to the source.
2. Compute an appropriate centroid of those pixels.

It is not totally obvious how to complete step 1. In the case of point sources like stars, you can get a good idea of their approximate locations by applying a Laplacian filter (whose size matches the point-spread function – see the next section) to a digital frame and noting the maxima of the filtered image. To decide which pixels around these locations are part of a star image and which are not requires some thought. For example, if you ask which pixels in a typical CCD image receive light from a bright star in the center of the frame, the answer, for a typical ground-based point spread function, is: "all of them." A better question is: "which pixels near the suspected star image receive a signal that is (a) larger than (say) 3σ above the background noise and (b) contiguous with other pixels that pass the same test?"

Figure 9.12 illustrates this approach (there are others) – the bar heights indicate pixel values in a small section of a CCD frame. Although most of the pixels in the area probably registered at least one photon from the star in

Fig. 9.12 Bar heights represent pixel values near a faint star image. Darker bars are high enough above the background to qualify as image pixels.

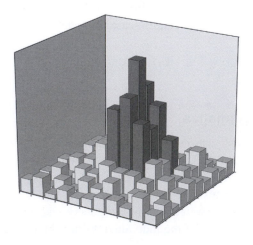

the center, only those colored dark gray stand out from the background according to the 3σ contiguous criterion.

With the "star pixels" identified, you can then compute their **centroid**. Typically, you consider only that part of the dark gray volume in Figure 9.12 that is above the background level, and compute the (x, y) coordinates (fractional values permitted) of its center of mass. If $R[x, y]$ is a pixel value and if B is the local background level, then the centroid coordinates are:

$$x_{\text{cen}} = \frac{\sum_x \sum_y x(R[x,y] - B)}{\sum_x \sum_y (R[x,y] - B)}, \quad y_{\text{cen}} = \frac{\sum_x \sum_y y(R[x,y] - B)}{\sum_x \sum_y (R[x,y] - B)} \tag{9.10}$$

The sums include only star pixels. Depending upon the signal-to-noise ratio (SNR) in the sums in Equations (9.10), the centroid can locate the image to within a small fraction of a pixel.

9.4.2 Where is it, again? PSF fitting

Finding the centroid of an image is computationally simple, but works well only in cases where images are cleanly isolated. If images blend together the centroid finds the center of the blended object. Even if there is no confusion of images, one object may asymmetrically perturb the background level of another (a galaxy near a star, for example).

In situations like this, you can use knowledge of the **point-spread function** (*PSF*) to disentangle blended and biased images. The procedure is to fit each of the *stellar* (only) images on the frame with a two-dimensional PSF, adjusting fits to account for all the flux present. The actual algorithm may be quite complex, and special complications arise if there are non-stellar objects present or if the shape of the PSF varies from place to place due to optical aberrations or to anisoplanatism in adaptive systems. Despite the difficulties, PSF fitting is nevertheless essential for astrometry and photometry in crowded fields.

9.4.3 Aligning images: shift, canvas and trim

Figure 9.13 shows two CCD frames, **A** and **B**, of M33 at different telescope pointings. Each frame has dimensions $x_{\text{max}} = 256 \times y_{\text{max}} = 256$. We consider the problem of **aligning** the two images by applying a **geometric transformation** to each – a geometric transformation changes the pixel coordinates of image data elements. In this example, we make the transformation by first measuring the $[x, y]$ coordinates for three stars in the area common to both frames. Suppose that on average, we find for these objects that $x_B - x_A = \Delta x_B = -115$ and that $y_B - y_A = \Delta y_B = 160$. (Assume for now that coordinates are restricted to integers.) There are two possible goals in making the transformation.

First, we might wish to make a new image that contains data from *both* **A** *and* **B**, perhaps to improve the SNR. Do this by creating **A′** and **B′**, two small images that contain only the overlap area from each frame:

$$A'[\xi, \eta] = A[\xi + \Delta x_B, \eta]$$
$$B'[\xi, \eta] = B[\xi, \eta + \Delta y_B] \tag{9.11}$$

The values stored in the pixels of **A′** and **B′** and are the same as the values in **A** and **B**, but they have different coordinates. The ***translation*** operation executed by Equation (9.11) simply slides **B** and **A** until coordinates match. An important step in making the new images discards or ***trims*** any pixels that fall outside the overlap region. Specifically, we trim all pixels except those with coordinates $1 \leq x_{max} - |\Delta x_B|$ and $1 \leq y_{max} - |\Delta y_B|$. Both trimmed images thus have the same size, which means we can combine them (add, average, etc.) using image arithmetic. For example:

$$\mathbf{C}_{AND} = \mathbf{A'} + \mathbf{B'}$$

More complicated combination algorithms might be appropriate, especially with large numbers of images.

A second mode of image combination arises when we note that the large galaxy in Figure 9.13 does not fit in a single frame, and we wish to combine the two frames to make a wide-angle view. We want an image that includes *every* valid pixel value from *either* **A** *or* **B**. The procedure is simple: we make two ***canvases***, **C$_A$** and **C$_B$**, each with dimensions $x_{max} + |\Delta x_B|$ by $y_{max} + |\Delta y_B|$, large enough to include all pixels. Then, we "paste" each image in the appropriate section of its canvas, and then combine the large canvases into one large final image. In our example, the canvases have coordinates ξ', η', and the operations that paste the images onto their canvases are:

Fig. 9.13 Aligning and combining two images. Alignment and transformation are based on the coordinates of the three marked stars in the overlap region. See text for details.

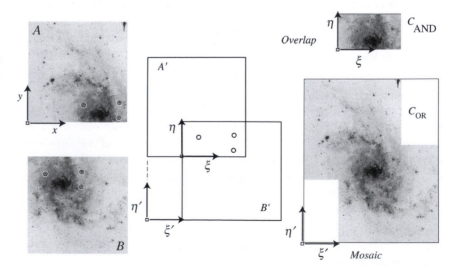

$$C_A(\xi',\eta') = \begin{cases} A'(\xi',\eta'-\Delta y_B) = A(\xi',\eta'-\Delta y_B), & 1 \leq \xi' \leq x_{max}, 1 \leq \eta' \leq y_{max} + \Delta y_B \\ -10,000 & \text{otherwise} \end{cases}$$

$$C_B(\xi',\eta') = \begin{cases} B'(\xi',\eta'-\Delta y_B) = B(\xi'+\Delta x_B,\eta'), & 1 \leq \xi' \leq x_{max} + \Delta x_B, 1 \leq \eta' \leq y_{max} \\ -10,000 & \text{otherwise} \end{cases}$$

The arbitrary large negative value of −10,000 simply flags those pixels for which there are no data. Any value that cannot be confused with genuine data can serve as a flag. We can combine the two canvases

$$\mathbf{C}_{OR} = \text{combine}(\mathbf{C}_A, \mathbf{C}_B)$$

with some appropriate algorithm. For example, the pseudo-code:

$$\text{IF } \{C_A[x,y] \neq -10,000 \quad \text{AND} \quad C_B[x,y] \neq -10,000\}$$

$$\text{THEN } C_{OR}[x,y] = \frac{1}{2}[C_A[x,y] + C_B[x,y]]$$

$$\text{ELSE } C_{OR}[x,y] = \max[C_A[x,y], C_B[x,y]]$$

will compute values for mosaic pixels for which there are some data, and put a flag (−10,000) in those where there is no data.

9.4.4 Aligning images: geometric transformations

Translations are only one of several kinds of geometric transformation. Suppose, for example, you wish to combine images from two different instruments. The instruments have different pixel scales[3] (in seconds of arc per pixel); so one set of images requires a scale change, or **magnification**. The transformation is

$$x = \xi/M_x$$
$$y = \eta/M_y$$

Again, $[\xi, \eta]$ are the coordinates in the new image, and the equations allow for stretching by different amounts in the x- and y-directions.

Small **rotations** of one image with respect to another might occur if a camera is taken off and remounted on the telescope, or if images from different telescopes need to be combined, or even as the normal result of the telescope mounting (e.g. imperfect polar alignment in an equatorial, an imperfect image rotator in an alt-azimuth, or certain pointing schemes for a space telescope). If

---

[3] Scale differences can have subtle causes: the same CCD–telescope combination can have slightly different scales because of focal-length changes caused by thermal effects on mirrors or chromatic effects in lenses.

A' is the image produced when $\mathbf{A}$ is rotated about its origin counterclockwise through angle θ, then $A'[\xi, \eta]$ has the same pixel value as $A[x, y]$ if

$$x = \xi \cos \theta + \eta \sin \theta$$
$$y = \eta \cos \theta - \xi \sin \theta$$

For wide fields, optical **distortions** can become significant (e.g. the Seidel pincushion or barrel distortion aberration, which increases with the cube of the field size). These require relatively complicated transformations.

In creating mosaics from images with different telescope pointings, **projection effects** due to the curvature of the celestial sphere also need to be considered. Such effects have long been an issue in photographic astrometry, and chapter 11 of Birney *et al.* (2006) outlines a simple treatment of the problem.

To derive any geometric transformation, the general approach is to rely on the locations of objects in the field. In the final transformed or combined image we require that a number of reference objects $(1, 2, 3, \ldots, N)$ have pixel coordinates $(\xi_1, \eta_1), (\xi_2, \eta_2), \ldots, (\xi_N, \eta_N)$. We can call these the **standardized coordinates** – they might be coordinates derived from the known right ascension (RA) and declination (Dec) of the reference objects, or might be taken from the actual pixel coordinates on a single image. Now, suppose one of the images you wish to transform, image $\mathbf{B}$, contains some or all of the reference objects, and these have coordinates

$$(x_{B1}, y_{B1}), (x_{B2}, y_{B2}), \ldots, (x_{BM}, y_{BM}) \quad M \leq N$$

Your task is to find the transformations

$$x = f_{\mathrm{B}}[\xi, \eta]$$
$$y = g_{\mathrm{B}}[\xi, \eta]$$

that will tell you the pixel values in $\mathbf{B}$ that correspond to every pair of standardized coordinates. You specify the forms for the functions from your knowledge of how the images are related. You might, for example, expect that narrow-field images from the same instrument would require just a simple translation, while wide-field images from different instruments might need additional correction for magnification, rotation, distortion, or projection. For a given functional form, the usual approach is to use a least-squares technique to find the best values for the required constants $\Delta x_A, \Delta y_A, \theta, M_x$, etc. Note that some geometric transformations may not **conserve flux** (see the next section).

Reducing data from digital arrays very commonly involves a two-step **align and combine** procedure:

(a) apply geometric transforms on a group of images to produce a new set aligned in a common system of coordinates, correcting for flux changes if necessary, then

(b) combine the aligned images with an appropriate algorithm.

This procedure is often termed **shift and add**. Basic observational issues make shift and add an indispensable technique, and we already discussed some of these in the context of the **shift-and-stare** observing technique for flat-field calibration images. (You do shift and *stare* at the telescope, shift and *add* the data-reduction computer.) To shift and stare, or **dither**, the observer takes several exposures of the same scene, shifting the telescope pointing slightly between exposures. The aim is to produce a number of equivalent exposures, no two of which are perfectly aligned.

There are many reasons to take several short exposures rather than one long one. For one thing, all arrays saturate, so there may well be an exposure time limit set by the detector. Second, the only way to distinguish a pixel illuminated by a cosmic-ray strike from one illuminated by an astronomical object is to take multiple images of the scene. Astronomical objects are present in every image at the same standardized coordinate location; cosmic rays (and meteor trails and Earth satellites) are not. Similarly, bad pixels, bad columns, and the insensitive regions in array mosaics always have the same pre-transformation coordinate, but different standardized coordinates. When images are aligned, the bad values due to these features in one frame can be filled in with the good values from the others.

9.4.5 Interpolation

Geometric transforms set the values of the pixel at standardized coordinates $[\xi_j, \eta_j]$ in a new image to those at pixel at (x_j, y_j) in the original image; see Figure 9.14. Now, ξ_j and η_j must be integers, but x_j and y_j generally contain fractional parts. Therefore, we use round brackets (non-integers permitted) to write, symbolically

$$B'[\xi_j, \eta_j] = B\left(f_B[\xi_j, \eta_j], g_B[\xi_j, \eta_j]\right) = B(x_j, y_j)$$

Since we only know the pixel values for the image **B** at locations where x and y are integers, we must use the pixel values at nearby integer coordinates to *estimate* $B(x_j, y_j)$ – the value a pixel *would* have if it were centered precisely at the non-integer location (x_j, y_j).

We could, for example, ignore any image changes at the sub-pixel level, and simply round x_j and y_j up or down to the nearest integers, and set $B(x_j, y_j)$ equal to the value of the **nearest pixel**. This is simple, and largely preserves detail, but will limit the astrometric accuracy of the new image.

Bilinear interpolation often gives a more accurate positional estimate. Figure 9.14 shows the point (x_j, y_j) relative to the centers of actual pixels in the original image: we compute x_0 and y_0, the values of x_j and y_j *rounded down* to the next lowest integers. Thus, the values of the four pixels nearest the fractional location (x_j, y_j) are

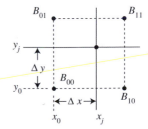

Fig. 9.14 Bilinear interpolation. The methods finds the value of the image intensity at point (x_j, y_j), given the nearest pixel values.

$$B_{00} = B[x_0, y_0], \qquad B_{10} = B[x_0 + 1, y_0]$$
$$B_{01} = B[x_0, y_0 + 1], \qquad B_{11} = B[x_0 + 1, y_0 + 1]$$

If we assume that **B** changes linearly along the axes, we can make two independent estimates for the value of $B(x_j, y_j)$. These average to the bilinear interpolated value:

$$B(x_j, y_j) \approx (1 - \Delta x)(1 - \Delta y)B_{00} + (\Delta x)(1 - \Delta y)B_{10} + (1 - \Delta x)(\Delta y)B_{01}$$
$$+ (\Delta x)(\Delta y)B_{11} \tag{9.12}$$

where

$$\Delta x = x_j - x_0, \quad \Delta y = y_j - y_0$$

Bilinear interpolation preserves astrometric precision and affects photometry in predictable ways. (Any geometric transformation in which the output grid does not sample the input grid uniformly will change the photometric content of the transformed image.) As you can see from Equation (9.12), the procedure essentially takes a weighted average of four pixels – as such, it *smoothes* the image. Bilinear interpolation chops off peaks and fills in valleys, so an interpolated image is never as sharp as the original; see Figure 9.15. Furthermore, the smoothing effect artificially reduces image noise.

If resolution is of great concern, it is possible to fit the pixels of the original image with a higher-order function that may preserve peaks and valleys. The danger here is that higher-order surfaces may also produce artifacts and photometric uncertainties, especially for noisy images. Nevertheless, it is not unusual for astronomers to use higher-order fitting techniques like bicubic interpolation or B-spline surfaces.

Fig. 9.15 The original pixel values in the upper left are shifted by 0.5 pixels to the right. In the lower left, the shift and linear interpolation smoothes the original, removing peak P, which may be due to noise or some real feature. At the lower right, 3 × resampling preserves more detail after shift and interpolation.

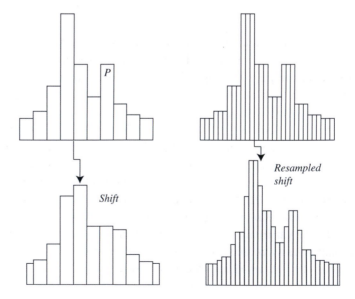

9.4.6 Resolution: resampling, interlace, and drizzle

Geometric transformations are essential for combining images with the shift-and-add image technique. Transformations, however, require either interpolation (which degrades resolution) or the "nearest-pixel" approximation, which degrades positional information. **Resampling** the original image at higher magnification circumvents some of the image degradation that accompanies interpolation, and in some cases can actually *improve* the resolution of the combined image over that of the originals.

The idea is to make the pixels of the output, or transformed, image smaller (in seconds of arc), and thus more closely spaced and numerous, than the pixels of the input image. In other words, the scale (in arcsec per pixel) of the standardized coordinates is larger than the scale of the original input coordinates. We discuss three resampling strategies.

The first is just a modification of the shift-and-add (and interpolate) algorithm. All that is done is to resample each input image by an integral number (e.g. each original pixel becomes nine pixels in the resampled version). After shifting or other transformations, resampling mitigates the smoothing effect produced by interpolation, since this smoothing effect is on the scale of the output pixels. Figure 9.15 shows a one-dimensional example. An image is to be shifted 0.5 pixels to the right from its position in the original. The left-hand column shows the result of the shift and linear interpolation without resampling, and the right column shows the same result if the output pixels are one-third of the size of the input. Linear interpolation in each case produces some smoothing, but the smoothing is less pronounced with the finer grid. Compared to using the original pixel sizes, aligning multiple images on the finer output grid will of course improve the resolution of their combined image.

The second method is usually called **interlace**, and is in some ways analogous to the nearest-pixel approach described earlier. The interlace algorithm examines each input pixel (i.e. $B[x, y]$ at only integer coordinates), locates its transformed center in a particular output pixel in a finer grid (but again, only integer coordinates), and copies the input value to that single output pixel. There is no adjustment for fractional coordinates, nor for the fact that the input pixel may overlap several output pixels. Figure 9.16a gives an example of a shifted and rotated input grid placed on an output grid with smaller pixels. The center of each input pixel is marked with a black dot. Interlacing this single input places values in the output pixels (i.e. the dark-colored pixels), "hit" by the dots, and "no value" or "zero-weight" flags in the other pixels.

Interlace for a single image is a flawed approach. First, it creates a discontinuous image, since only some fraction of the output pixels will score a "hit", and the remainder will have zero weight. Second, we have introduced positional errors because we ignore any fractional coordinates.

Fig. 9.16 Resampling an input grid. The interlace technique (a) regards values in the input grid as if concentrated at points. Grayed pixels on the output copy the values from the input points, white-colored output pixels have no value. The drizzle method (b) assumes values are spread over a square "drop" smaller than an input pixel. Most output pixels overlap one or more input drops, although some, as illustrated, may overlap none.

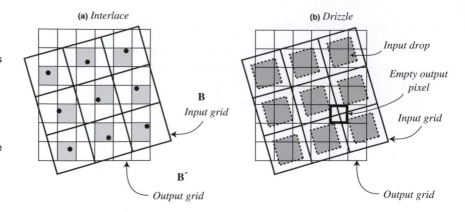

Both problems become less significant as more images of the same scene are added to the output. If each addition has a slightly different placement on the output grid, a few additions could well fill in most output pixels with at least one valid value. Moreover, positional information improves as the interlace fills and averaging reduces uncertainty in the brightness distribution.

The combined image is a weighted mean of all the shifted frames, with the weight, $w_i[\xi, \eta]$, of a particular pixel either one (if it is a hit) or zero (if no hit *or* if we decide the hit is by a cosmic ray or by a bad pixel). Thus, the combined image **C** is

$$C[\xi, \eta] = \frac{1}{\sum\limits_{i=1}^{N} w_i[\xi, \eta]} \left(w_1[\xi, \eta]B'_1[\xi, \eta] + w_2[\xi, \eta]B'_2[\xi, \eta] + \ldots + w_N[\xi, \eta]B'_N[\xi, \eta] \right)$$

(9.13)

We cannot use Equation (9.13) for any pixel in **C** with a combined weight of zero. In this case, the pixel has no valid value. It is possible to interpolate such a missing value from the surrounding output pixels, but this will cause photometric errors unless the "no-value" status is due to masking cosmic rays or bad pixels.

Interlacing shifted images has the potential for actually improving image resolution in the case where the camera resolution is limited by the detector pixel size rather than by the telescopic image itself. Figure 9.17, shows the interlaced result for a one-dimensional example: a double source with a separation of 1.3 input pixels, with each source FWHM = 0.8 pixels. Three dithered input images are shown, none of which shows the double nature of the source, as well as the interlaced combination with 1/3-size output pixels. The combined image resolves the two components.

The interlace technique is powerful, but unfortunately difficult to execute observationally. Suppose, for example, a detector has 0.8 arcsec pixels. To effectively interlace images on an output grid of pixels half that size, the

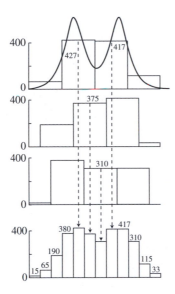

Fig. 9.17 The interlace
method in a one-
dimensional example.
The actual brightness
distribution of a double
source is sampled with
large pixels. In three
samples (upper plots)
displaced from one
another by 1/3 of a pixel,
no indication of the
double nature of the
source appears, yet the
combined and interlaced
image (bottom plot) does
resolve the source.

astronomer must observe four frames, displaced orthogonally from one another by an odd multiple of 0.4 arcsec. Some observers can achieve the placement needed for an efficient interlace, but the limited precision of many actual telescope controls usually produces a set of exposures whose grids are dithered randomly at the sub-pixel level.

The ***variable-pixel linear reconstruction*** method, more commonly known as ***drizzle***, can be much more forgiving about input grid placement. Drizzle assumes that the flux in a square input pixel of size (length) d is not spread over the pixel, but is uniformly concentrated in a smaller concentric square, called a "drop," whose sides have length fd; see Figure 8.16b, where the drops are the shaded squares. The fractional size of the drops, i.e. the value of f, can be varied to accommodate a particular set of images. As $f \to 0$ the drizzle method approaches the interlace method, and as $f \to 1$ drizzle, drizzle approaches resampled shift and add.

We introduce a parameter, s, to measure the relative scale of the output pixels: for input pixels of length d, output pixels have length sd. The drizzle algorithm then runs as follows: Input pixel $B_i[x, y]$ in frame i will contribute to output pixel $B'_i[\xi, \eta]$ if any part of the input *drop* overlaps the output pixel. If the area of overlap is $a_i[x, y, \xi, \eta](fd)^2$, then the contribution will be

$$B_i[x, y]W_i[x, y]a_i[x, y, \xi, \eta]s^2$$

The factor s^2 conserves surface brightness in the final image, and the weighting factor $W_i[x, y]$ accounts for bad pixels and other effects (e.g. exposure time) in the input frame. Adding all contributions from the input image (up to four input drops can overlap a single output pixel), we assign the output value and weight as

$$B'_i[\xi, \eta] = s^2 \sum_{x,y} B_i[x, y] W_i[x, y] a_i[x, y, \xi, \eta]$$

$$w_i[\xi, \eta] = \sum_{x,y} W_i[x, y] a_i[x, y, \xi, \eta]$$

We make the final combination of images by computing the weighted mean of all the input frame contributions to each pixel as in Equation (9.13).

$$C[\xi, \eta] = \frac{\sum\limits_{i=1}^{N} B'_i[\xi, \eta]}{\sum\limits_{i=1}^{N} w_i[\xi, \eta]}$$

9.4.7 Cleaning images

Images inevitably have defects caused by bad detector pixels or by unwanted radiation events like cosmic-ray impacts or radioactive decays in or near the detector. Most methods for removing such defects require multiple dithered images of the same scene.

One familiar prescription works quite well. Start with $N > 2$ dithered images $\{\mathbf{R}_1, \mathbf{R}_2, \ldots, \mathbf{R}_N\}$ whose intensities are scaled to the same exposure time. Align them (i.e. use a geometric transform to make all astronomical sources coincide):

$$\mathbf{R}'_i = \text{GXform}(\mathbf{R}_i), \ i = 1, \ldots, N$$

Now combine the transformed images to form the median image:

$$\mathbf{C} = \text{median}(\mathbf{R}'_1, \mathbf{R}'_2, \ldots, \mathbf{R}'_N)$$

The median is relatively insensitive to pixel values (like many radiation events or bad pixels) that differ greatly from the central value, so it produces a "clean" version of the image. Although simple to execute, the median becomes less graceful with images of differing weights and does have some shortcomings:

1. At a location where all pixel values are good, the median is not as good an estimator of the central value as is the mean.
2. The median is not *completely* insensitive to deviant values: e.g. the median will be slightly biased towards higher values at the location of cosmic-ray hits.
3. The median will perform very poorly in special cases (e.g. if multiple values at the same location are bad).

A more sophisticated cleaning method is to *flag* the defects in the original images, sometimes by assigning the affected pixels a special value (a large negative number, for example) or by assigning them a weight of zero. In one technique of this sort, the astronomer generates a special companion image, the *mask*, for each $\mathbf{R}_i$. The mask values (usually either one or zero) indicate whether the corresponding image pixel is to be included or excluded in any subsequent operations; see Figure 9.18.

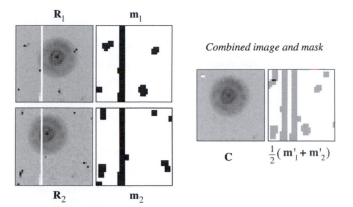

Combined image and mask

$$\mathbf{C} \qquad \tfrac{1}{2}(\,\mathbf{m}'_1 + \mathbf{m}'_2\,)$$

Fig. 9.18 Pixel masks. Two offset images of the planetary nebula NGC 2392 (The Eskimo) are marred by an insensitive column and many cosmic ray strikes. The mask next to each raw image on the left blocks (black pixels = 0, white = 1) every bad pixel and its immediately adjacent neighbor. The right-hand images show the combined image and mask after alignment. Since there are only two images, the combined image shows noticeably different noise levels in masked and unmasked regions. Two pixels in the upper left are masked in both images and have zero weight. They show as black in the right-hand image of the combined masks.

How can you generate a mask for a particular image? Usually, bad detector pixels or columns are well documented or are easily discovered on flat-field exposures. You can identify radiation events, which occur at random locations and can mimic images of astronomical objects, with the median-image method described at the start of this section. Once the complete mask is generated for an input image, a conservative approach might be to mask all pixels that are adjacent to bad pixels as well, since radiation events tend to spill over. At the end of this process, there will be a separate mask for each input image.

You then geometrically transform all input images, along with their masks, so that all are aligned. The final combination of these aligned images is a weighted mean in which all defective pixels are ignored. That is, if $\mathbf{m}_i$ is the mask for input image i, w_i is the image weight, and $\mathbf{m}'_i$ is the transformed mask:

$$C[\zeta, \eta] = \frac{\displaystyle\sum_{i=1}^{N} w_i m'_i[\zeta, \eta] R'_i[\zeta, \eta]}{\displaystyle\sum_{i=1}^{N} w_i m'_i[\zeta, \eta]}$$

Figure 9.18 illustrates a simple combination of two small images using masks.

9.5 Digital aperture photometry

We have discussed the preprocessing of individual images (the linearity, dark, bias, flat, and fringe corrections) and the combination of multiple frames to produce a deeper and possibly wider image. As a reminder, we summarize those steps here:

$$\mathbf{R}_{\text{pf}i} = \frac{\text{lin}(\mathbf{R}_i) - \mathbf{Z}_i - t_i \mathbf{D}}{\mathbf{F}} - \frac{A_i}{A}\mathbf{b}_f$$

$$\mathbf{R}'_i = \text{GXform}\left(\mathbf{R}_{\text{pf}i}\right)$$
$$\mathbf{C} = \text{combine}\left(\mathbf{R}'_1, \mathbf{R}'_2, ..., \mathbf{R}'_N\right)$$

Here we understand that the combination will be something like a median image or weighted mean, perhaps utilizing masks and a drizzle.

The next task in the reduction procedure is often measurement of the brightness of one or more objects. Measuring brightness is at heart a simple task — we did it in the exercises in Chapters 1 and 2. Start with the preprocessed image — an individual frame, $\mathbf{R}_{\text{pf}i}$, or an aligned/combined accumulation of such frames, $\mathbf{C}$. Then just add up the emission from the object of interest, which usually is spread over many pixels. In doing so, remember to remove the background, which contains positive contributions made by sources both behind and in front of the object of interest. The latter include scattered light from other astronomical objects as well as the glow of the atmosphere and (especially in the thermal infrared) of the telescope. We will use the terms **sky** and **background** interchangeably for all this unwanted light, no matter where it originates. Once we have isolated the signal attributable to the source alone, we will need to quantify the uncertainty of the result.

Finally, the signal measured will only be meaningful if it is calibrated — expressed in units like magnitudes or watts per square meter. We consider the calibration process in the next chapter, and confine ourselves here to the tasks of separating signal from background and of estimating the uncertainty of the result.

9.5.1 Digital apertures and PSF fits

Consider a very common situation: from a digital image, you want to determine the brightness of a *point* source — a star, quasar, or small object in the Solar System. Define a circular area, the ***digital aperture***[4], that is centered on the centroid of the object (see Figure 9.19). The radius of the digital aperture should include a substantial fraction of the emission from the star. Now make three simple computations:

1. Add up all the pixel values inside the aperture. This sum represents the total emission from the aperture — the light from the star plus the light from the background. To deal with fractional pixels (see Figure 9.19) at the edges, multiply every value by $A[x, y]$, the fraction of the pixel's area that is inside the aperture.

$$\text{Total} = \sum_{x,y} A[x,y] R_{\text{p}}[x,y]$$
$$n_{\text{pix}} = \sum_{x,y} A[x,y]$$

---

[4] Yes, aperture means "opening". The terminology recalls the days of photoelectric photometry, when it was necessary to place an opaque plate with one small clear aperture in the focal plane. This passed only the light from the star and very nearby sky through to the photocathode, and blocked all other sources.

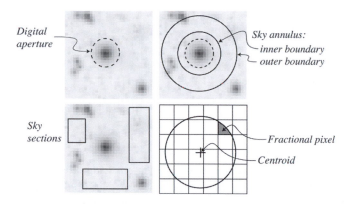

Fig. 9.19 Digital apertures. The upper left-hand image shows a circular aperture centered on a point source; lower left, rectangular apertures for sampling background emission. The image in the upper right shows an annular aperture for sampling sky emission near a point source. All curved apertures will require some strategy for dealing with pixels that contain some segment of the boundary, as in the image at the bottom right.

The sums are understood to extend over the entire x–y extent of the aperture. The number n_{pix} is just the area of the aperture in pixels.

2. Estimate $\bar{B}$, the value of the sky emission per pixel. Usually, you must estimate $\bar{B}$ from a source-free region near the object of interest (see the next section for details). Compute that part of the emission in the aperture that is due to the sky:

$$\text{sky} = n_{\text{pix}}\bar{B}$$

3. Subtract the sky emission from the total, and the remainder is the detector response attributable to the source alone; this is the signal in ADUs:

$$S_{\text{ADU}} = \text{total} - \text{sky}$$
$$S_{\text{ADU}} = \sum_{x,y} A[x,y]R_{\text{p}}[x,y] - n_{\text{pix}}\bar{B} \qquad (9.14)$$

In situations in which star images seriously overlap, digital aperture photometry fails, because it is impossible easily to estimate the polluting star's contribution to the background of the object of interest. We have already discussed (Section 9.4.2) the idea of fitting a **PSF** to each star image on a frame. Point-spread-function fitting is required in crowded-field photometry since (at the cost of considerable computational complexity) it can separate the contributions of individual overlapping images from one another and from the diffuse background. Once all overlapping images are accounted for, integration of the PSF fit of the image of interest gives the *signal* term in Equation (9.14).

The question of *aperture size* is important. For PSF fitting, the aperture size is often technically infinite, e.g. a typical PSF for a ground-based image is Gaussian-like – see Appendix J. You can determine the appropriate PSF *shape* by examining high-SNR star images and comparing their profiles (especially the central regions with good signal) to expected shapes or by constructing an empirical shape. For PSFs, well over 90% (usually much more) of emission is within a diameter of three times the FWHM.

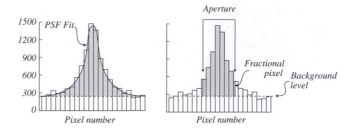

If he does not use PSF fitting, the astronomer must choose the digital aperture size; see Figure 9.20. There are two conflicting considerations: he wants a *large* aperture because it includes as much light as possible from the star, yet he wants a *small* aperture because it excludes background light and, especially, its associated noise. An aperture that includes too much sky will decrease the SNR of the final measurement, as will an aperture that includes too little of the source. The optimum size varies with the relative brightness of the star. Since point-source photometry requires the same aperture size for all stars, this generally means the astronomer chooses the aperture size based on the faintest star observed. The choice is implemented in software, so it is easy to try a range of apertures (diameters somewhere between 0.75 and 4 times the FWHM of the image profile) and identify the aperture (usually, a diameter near 1.5–2 times the profile FWHM) that yields the best SNR.

Finally, note that a digital aperture need not be circular. Indeed, many objects have decidedly non-circular shapes, and invite equally non-circular apertures. Photometry via Equation (9.14) applies as well to such shapes.

9.5.2 Measuring the sky

Both PSF fitting and digital aperture photometry demand an accurate measure of the sky emission per pixel over of the source of interest. This, of course, is one area where it is impossible to measure the sky brightness, so we measure the sky *near* the source, and hope that sky brightness does not change with location. There are some cases where this hope is forlorn. A notorious example is the photometry of supernovae in other galaxies: the background due to the host galaxy changes drastically on the scale of a digital aperture size, so any "nearby" sky measurement is guaranteed to introduce uncertainty. (Fortunately, supernovae are temporary. An image of the galaxy obtained with the same instrument after the supernova has faded can provide the needed background measurement.)

A smooth background near the source should ameliorate many difficulties. In this case, the nearest possible sample should be the most accurate, and a sample symmetrically positioned around the source stands a chance of averaging out any trends. Figure 9.19 shows a digital aperture and a ***sky annulus***. The annulus is a region between two circles centered on the source. The smaller circle (the inner boundary of the annulus) is as small as possible, but still large enough to exclude any appreciable emission from the source. The outer radius of the annulus is less strictly determined, but should be large enough to include a statistically significant number of pixels. If the outer radius is too large, it may sample sky that differs from the sky within the aperture.

The best estimate of the sky value in the annulus is clearly *not* the mean pixel value: the annulus is bound to contain images or halos of other stars. These bias the mean towards larger values. The median is less sensitive to the influence of this kind of pollution, and the mode is even better: the most common value in the annulus certainly sounds like the optimum measurement of the sky. Practical computation of the mode usually requires the construction of a smoothed histogram, with the sky value computed as the mean of the values in the most populous bin of the histogram.

Figure 9.19 illustrates a second approach to measuring the sky value. An astonomer selects one or more relatively star-free sections of the image, and computes the median or modal value. The advantage of this method is that it avoids the influence of nearby sources on background estimates, and if the field near the source of interest is crowded, this is the only alternative. The disadvantage is that the sky sections may be relatively far from the point of interest, and they may not sample uniformly enough to minimize the effects of large-scale trends in background brightness. As explained earlier, in the infrared, one generally obtains sky levels from separate (chopped) exposures.

9.5.3 Signal and noise in an aperture

Knowing the uncertainty of a digital photometric measurement is nearly as important as discovering its value. In this section, we develop an equation for the SNR in aperture photometry with a CCD. The general approach, if not the exact equation, will apply for photometry with all digital arrays.

For simplicity, we consider only the case of a single exposure, corrected for non-linearity, dark, bias, and flat. Recall the digital aperture photometry operation given in Equation (9.14):

$$S_{\text{ADU}} = \left\{ \sum_{x,y} A[x,y] R_p[x,y] \right\} - n_{\text{pix}} \bar{B} \qquad (9.15)$$

Here $A[x,y]$ is the fraction of the pixel inside the digital aperture, and $\bar{B}$ is the estimated average background emission per pixel. The pixel values are in ADUs (analog-to-digital units), values that we can convert to the number of electrons read out from the pixel by multiplying by g, the CCD gain factor. In terms of electrons, then, Equation (9.15) becomes

$$\text{Signal} = N_* = g[S_{\text{ADU}}] = \left\{ \sum_{x,y} A[x,y] r_p[x,y] \right\} - n_{\text{pix}} b_e$$

Here the signal is N_*, the total number of electrons produced by the source in the aperture. The values $r_p[x, y]$ and b_e are the preprocessed pixel value and the estimated background value in electrons. The noise, or uncertainty in N_*, follows from an application of Equation (2.17) to (9.15). Although it is not always safe to do so, we assume uncertainties in pixel values are not correlated:

$$\sigma_N^2 = \left\{ \sum_{x,y} \{A[x,y]\}^2 \sigma_{r,p}^2[x,y] \right\} + n_{\text{pix}}^2 \sigma_{b,e}^2 \tag{9.16}$$

To evaluate $\sigma_{r,p}^2[x, y]$, the uncertainty in a preprocessed pixel value, we write out the preprocessing operation for a single pixel as described for a CCD in Equation (9.8):

$$r_p[x, y] = \frac{1}{f[x, y]} \{L(r[x, y])r[x, y] - d_e[x, y] - \zeta_e[x, y]\} \tag{9.17}$$

Here:

$f[x, y]$ = the normalized flat field response,

$d_e[x, y]$ = the estimated dark count in the pixel in electrons,

$\zeta_e[x, y]$ = the estimated bias level in the pixel, in electrons, and

$L(r[x, y])$ = the linearity correction for the pixel, expressed as multiplicative factor.

We will again assume that the uncertainties in each of the variables in Equation (9.17) are not correlated, so that we can apply Equation (2.17) to compute the variance of processed pixel value:

$$\sigma_{r,p}^2[x, y] = \sigma_r^2 \frac{L^2}{f^2} + \sigma_{d,e}^2 \frac{1}{f^2} + \sigma_{\zeta,e}^2 \frac{1}{f^2} + \sigma_L^2 \frac{r^2}{f^2} + \sigma_f^2 \left\{ \frac{Lr - d_e - \zeta_e}{f^2} \right\}^2$$

To simplify the notation, we have omitted the $[x, y]$ coordinate references for all the terms on the right-hand side. We can clean up this expression further by noting that $f \approx 1$ and $L \approx 1$:

$$\sigma_{r,p}^2[x, y] = \sigma_r^2 + \sigma_{\zeta,e}^2 + \sigma_{d,e}^2 + \sigma_L^2 r^2 + \sigma_f^2 (r - d_e - \zeta_e)^2 \tag{9.18}$$

We will examine each of the terms on the right-hand side in turn. The first term in Equation (9.18) is the square of the uncertainty in the raw pixel value itself. The unprocessed pixel value is just

$$r[x, y] = r'[x, y] + \zeta[x, y] = n[x, y] + b[x, y] + d[x, y] + \zeta[x, y]$$

where the four quantities on the extreme right are, respectively, the actual single-pixel values of the signal, background, dark, and bias expressed in electrons. The variance of the raw pixel value must be

$$\sigma_r^2[x, y] = \sigma_{r'}^2 + \sigma_\zeta^2 = r' + \rho^2$$

The variance of the pixel response, $r'[x, y]$, is equal to its mean because the response is Poisson-distributed. Noise present on every readout of the device, independent of exposure time or illumination, produces the variance of the bias level, ρ^2. For a CCD, we distinguish two components:

$$\rho^2 = \sigma_{\text{read}}^2 + \sigma_{\text{digit}}^2$$

The first is the **read noise** – the uncertainty in the zero-level signal from the output amplifier. The second is the **digitization noise** – the uncertainty that results when the ADC circuit rounds the analog signal to an integer. If the analog values are uniformly distributed, the digitization noise is $g/\sqrt{12}$, where g is the CCD gain. Usually the CCD gain is adjusted so that the digitization component is smaller than σ_{read}^2, but we will include both in the parameter ρ, the "zero-level uncertainty" or "digital read noise".

The actual values for the background $b[x, y]$ and dark $d[x, y]$ levels in a particular pixel are unknown, and we will simply use estimated values: the dark is estimated from dark frames (or assumed to be zero if the detector is sufficiently cold) and the background is estimated from nearby "sky" pixels. Thus

$$\sigma_r^2[x, y] \approx n + b_e + d_e + \rho^2 \tag{9.19}$$

The second term in Equation (9.18) is the squared uncertainty in the "estimated" bias level (different from the read noise in a single pixel!). This estimate is usually computed by averaging a number of calibration frames. If the bias drifts, then $\sigma_{\zeta,e}$ might be large. If we obtain p_z bias frames, the minimum variance of the mean of the p_z values for the bias at pixel $[x, y]$ is given by

$$\sigma_{\zeta,e}^2 = \frac{\rho^2}{p_z} \tag{9.20}$$

where ρ is the digital read noise. If the bias is obtained from an overscan, and the base bias pattern is very well determined, then p_z is the number of columns in the overscan.

Likewise, the third term is the variance in the estimated dark count. If the estimate is an average of p_d dark frames, each of the same exposure time as the data frame, then the variance of the mean is

$$\sigma_{d,e}^2 = \frac{1}{p_d}\left\{ d_e + \left(1 + \frac{1}{p_z}\right)\rho^2 \right\} \tag{9.21}$$

The second term in the braces is there because dark frames are themselves processed by subtracting an estimated bias level. Note that we assume the same number of number bias frames, p_z, are used for the dark as for the data.

The fourth term in Equation (9.18) is $\sigma_L^2 r^2$, the variance in the linearity correction scaled by the square of the pixel value. Uncertainty in the linearity correction should not be of concern with most CCDs except near saturation (usually no correction is made), but it can be an issue in infrared arrays where linearity properties may vary from pixel to pixel. The value of σ_L^2 can be measured by examining linearity calibration frames.

The fifth term in Equation (9.18), $\sigma_f^2(n+b)^2$ (where $n+b = r - d_e - \zeta_e$), arises from the uncertainty in the normalized flat field. In the ideal case, σ_f^2 should approach $(n_f[x,y])^{-1}$, where $n_f[x,y]$ is the total number of photoelectrons counted at the pixel location in all flat-field calibration exposures. One should thus be able to reduce this uncertainty to insignificance just by accumulating enough calibration frames. Real observational situations seldom approach the ideal, and one can investigate uncertainties in the flat by, say, comparing combined flats taken on two different nights, or by varying the color of the flat-field target.

Substituting Equations (9.19), (9.20), and (9.21) into (9.18) gives the variance in the value of a single preprocessed pixel.

$$\sigma_{r,p}^2[x,y] = n + b_e + a_d(d_e + a_z\rho^2) + \sigma_L^2 r^2 + \sigma_f^2(n + b_e)^2 \tag{9.22}$$

where

$$a_d = 1 + \frac{1}{p_d}, \qquad a_z = 1 + \frac{1}{p_z}$$

Now return to Equation (9.16). We require a value for the uncertainty in the estimated background. We usually estimate the background by averaging $r_p[x,y]$ in a region of p_b pixels (e.g. the sky annulus) in which $n[x,y]$ is zero. That is,

$$b_e = \frac{1}{p_b} \sum_{x,y}^{sky\ section} r_p[x,y]$$

The variance, then, is

$$\sigma_{b,e}^2 = \frac{1}{p_b^2} \sum \sigma_{r,p}^2[x,y] = \frac{1}{p_b}\left\{\bar{\sigma}_{r,p}^2\right\}$$

But we have just worked out $\sigma_{r,p}^2[x,y]$, the variance of a *single* preprocessed pixel. Substituting for the average variance $\bar{\sigma}_{r,p}^2$ from Equation (9.22) for the case $n[x,y] = 0$:

$$\sigma_{b,e}^2 = \frac{1}{p_b}\left\{b_e + a_d(d_e + a_z\rho^2) + \bar{\sigma}_L^2 \bar{r}_b^2 + \bar{\sigma}_f^2(b_e)^2\right\} \tag{9.23}$$

Now we can turn to Equation (9.16) one last time:

$$\sigma_N^2 = \left\{\sum_{x,y}^{aperture} \{A[x,y]\}^2 \sigma_{r,p}^2[x,y]\right\} + n_{pix}^2 \sigma_{b,e}^2$$

We know all terms on the right-hand side, so substituting

$$\sigma_{\rm N}^2 = \left(\sum \{A[x,y]\}^2 n[x,y]\right) + \left(P^2 + \frac{n_{\rm pix}^2}{p_{\rm b}}\right)(b_{\rm e} + a_{\rm d}(d_{\rm e} + a_z \rho^2)) + s_{\rm L}^2 + s_{\rm f}^2 \quad (9.24)$$

where

$$P^2 = \sum_{x,y} (A[x,y])^2 \le n_{\rm pix}^2 = \left(\sum_{x,y} A[x,y]\right)^2$$

The parameter P is the properly weighted pixel count in the aperture for computing the uncertainty of a uniform signal. The two terms on the extreme right of Equation (9.24) are the contribution to the variance due to uncertainty in the linearity and the flat-field corrections. We can write formal expressions for these, but they can only yield values if one can examine the repeatability of flat and linearity calibrations; see Problems 9.4 and 9.5.

Equation (9.24) does not include some sources of uncertainty that could be important in a specific array, like uncertainties in corrections for charge-transfer inefficiency. If such effects can be well modeled, one could in principle represent them with additional terms.

As a tool for evaluating photometric uncertainty, the most serious problem with Equation (9.24) is its failure to account for systematic effects like those due to the non-uniformity of a flat-field target, color differences between sky, star, and flat, or variations in atmospheric transparency. You should *not* use this equation to evaluate the uncertainty in your digital photometry. *As always, the primary information about the uncertainty of your photometry comes from the scatter in repeated observations and the disagreement of your results with those of others*.

But Equation (9.24) is far from useless. It gives you a way to compare the expected random error with the actual scatter in your data – if you get something unexpected, think hard to understand why. The equation is also a very important tool for *planning* observations, for answering questions like: "how many minutes at the telescope will I need if I want to measure the brightness of my $V = 22.5$ quasar with a precision of 1%?"

9.5.3 The CCD equation

We will use Equation (9.24) for the not-so-special case of aperture photometry on a star. For most reasonable apertures, the counts due to the star are very small at the edge of the aperture where the partial pixels are located. In that case, we will not be far off in making the approximation (which applies exactly if partial pixels are not employed):

$$\sum_{x,y} A^2[x,y] n[x,y] \approx \sum_{x,y} A[x,y] n[x,y] = N_*$$

The SNR then implied by Equation (9.24) is

$$\text{SNR} = \frac{N_*}{\left\{ N_* + (P + a_\mathrm{b})(b_\mathrm{e} + a_\mathrm{d}(d_\mathrm{e} + a_\mathrm{z}\rho^2)) + s_\mathrm{L}^2 + s_\mathrm{f}^2 \right\}^{\frac{1}{2}}} \qquad (9.25)$$

This equation, in various approximations, is known as the **CCD equation**. The usual approach is to simplify Equation (9.25) by assuming good preprocessing practices as well as good fortune, so that the system will remain stable and the observer will collect a very large number of bias and (if needed) dark frames, the flat-field and linearity corrections will not contribute significant errors, and that $P = n_\mathrm{pix}$. In other words, $a_\mathrm{d} = a_\mathrm{z} = 1$, and $s_\mathrm{L} = s_\mathrm{f} = 0$, so that the CCD equation becomes

$$\text{SNR} = \frac{N_*}{\left\{ N_* + n_\mathrm{pix}\left(1 + \frac{n_\mathrm{pix}}{p_\mathrm{b}}\right)(b_\mathrm{e} + d_\mathrm{e} + \rho^2) \right\}^{\frac{1}{2}}} \qquad (9.26)$$

Since the CCD equation is often used to estimate the required exposure time, t, we rewrite this as

$$\text{SNR} = \frac{\dot{N}_* t}{\left\{ \left[\dot{N}_* + n_\mathrm{pix}\left(1 + \frac{n_\mathrm{pix}}{p_\mathrm{b}}\right)(\dot{b}_\mathrm{e} + \dot{d}_\mathrm{e}) \right] t + n_\mathrm{pix}\left(1 + \frac{n_\mathrm{pix}}{p_\mathrm{b}}\right)\rho^2 \right\}^{\frac{1}{2}}} \qquad (9.27)$$

The dotted quantities give the electron rates for source photons, background photons, and dark current. The read-noise term is independent of the exposure time. Solving for the exposure time:

$$t = \frac{B + (B^2 + 4AC)^{\frac{1}{2}}}{2A} \qquad (9.28)$$

where

$$P_\mathrm{b} = n_\mathrm{pix}\left(1 + {}^{n_\mathrm{pix}}\!/\!p_\mathrm{b}\right)$$

$$A = \frac{\dot{N}_*^2}{(\text{SNR})^2}$$

$$B = \dot{N}_* + P_\mathrm{b}(\dot{b}_\mathrm{e} + \dot{d}_\mathrm{e})$$

$$C = P_\mathrm{b}\rho^2$$

Figure 9.21 illustrates predictions based on the CCD equation in three different situations.

The **bright-star** or **photon-noise-limited** case, where the counting rate from the source, $\dot{N}_*$, exceeds all other terms in the denominator of (9.27). This case approaches the Poisson result:

$$\text{SNR} = \sqrt{N_*} = \sqrt{\dot{N}_*}\sqrt{t}$$

So in the bright-star case, the SNR improves as the square root of the exposure time, and the observer willing to devote sufficient time can produce

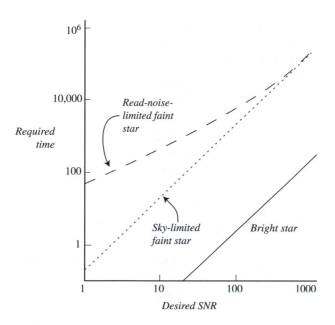

Fig. 9.21 The CCD equation. The plot shows the required time to reach a specified signal-to-noise ratio in the three limiting cases discussed in the text. Note the logarithmic scale.

measurements of arbitrarily high precision. However, actual precision attained may well be limited by the processes we ignored (e.g. the atmosphere) or the terms we eliminated in deriving Equation (9.27) – flat-field uncertainties, for example, scale as the first power of the exposure time, and can eventually dominate.

In the **background-limited** case, the background term

$$n_{\text{pix}} \left(1 + \frac{n_{\text{pix}}}{p_{\text{b}}} \right) (\dot{b}_{\text{e}} + \dot{d}_{\text{e}})$$

is not insignificant compared to the electron rate from the source. This is usually the case that is most interesting to the observer, since it describes the limits of detection in a given situation. In the background-limited case, the SNR ratio still increases as the square root of the observing time, but now there is a penalty:

$$\text{SNR} = \left\{ 1 + n_{\text{pix}} \left(1 + \frac{n_{\text{pix}}}{p_{\text{b}}} \right) \frac{(\dot{b}_{\text{e}} + \dot{d}_{\text{e}})}{\dot{N}_*} \right\}^{\frac{1}{2}} \sqrt{\dot{N}_*} \sqrt{t} = \{1 + B_*\}^{\frac{1}{2}} \sqrt{\dot{N}_*} \sqrt{t} \qquad (9.29)$$

The factor $\{1 + B_*\}$ becomes large under any occurrence of the following conditions: low source brightness, high sky brightness, high dark rate, large digital aperture, or small sky sample. A frequent situation is that of a faint source and bright sky, the **sky-limited** case. Remember also that Equation (9.29) ignores many sources of uncertainty due to both random and systematic effects. Detection limits derived from this expression should thus be taken to be optimistic.

The final case is the one in which the read noise is large. Here the SNR initially increases linearly with time, but eventually reaches the $\sqrt{t}$ dependence of either the bright-star or the sky-limited case.

Summary

- Digital images are ordered sets of numbers that can represent the output of an array of sensors or other data. Concepts:

pixel	*fill factor*	*undersampling*
pixel value	*gray-scale map*	*detector response*
ADU	*DN*	

- An important advantage of digital images is that they can be mathematically manipulated to remove defects and extract information. Concepts:

image arithmetic	*data cube*	*RGB color model*
CMYK	*false color*	*image functions*
digital filtration	*image convolution*	*kernel*
Gaussian kernel	*Laplacian kernel*	*boxcar*
unsharp mask	*median filter*	

- Digital images from a CCD can be processed to remove the effects of the detector and telescope. Concepts:

raw image	*bias frame*	*rejection algorithm*
overscan	*dark response*	*dark rate*
CCD gain	*linearity correction*	*chopping secondary*

- The flat-field correction very often limits photometric precision of a detector. Concepts:

flat-field image	*twilight flat*	*dark sky flat*
dome flat	*compound flats*	*shift and stare*
dither	*illumination correction*	

- Preprocessing images from an array requires subtraction of the dark signal and bias, then division by the normalized flat. Treatment of data from infrared arrays is slightly different because of the strong and variable sky background.

- Fringing is a variation in the sky background intensity due to interference effects in thin layers of a detector. Fringes can be removed if a flat without fringes is available.

- Combining images requires alignment, which requires both identification of feature coordinates and transformation of images. Concepts

centroid	*point-spread function*	*PSF fitting*
image alignment	*translation*	*rotation*
trim	*canvas*	*image mosaic*
magnification	*distortion*	*shift and add*
unsharp mask	*median filter*	

- Special methods for combining images can compensate for the loss of resolution due to interpolation, and can compensate for bad pixels. Concepts:

nearest pixel	*resampling*	*bilinear interpolation*
interlace	*pixel flag*	*drizzle*
image mask	*clean image*	

- Digital aperture photometry is a technique for measuring apparent brightness from a digital image. Concepts:

digital aperture	*sky annulus*	*PSF*

- The CCD equation gives the theoretical relation between the exposure time and expected SNR in digital aperture photometry, given source and sky brightness and detector and telescope characteristics. Concepts:

read noise	*digitization noise*	*background-limited*
photon-noise limited		

Exercises

1. Derive expressions for – and compute values of – the coefficients a, b, and c in Equation (9.7) for the detector whose calibration data fits the solid line labeled "second order" in Figure 9.10b.

2. The table at left below gives the coordinates and pixel values near a faint star on an array image. The small array at right is a sample of the nearby background. Find the x, y coordinates of the centroid of the star image using the criteria outlined in Section 9.4.1 of the text. Use a spreadsheet.

$y\backslash x$	1	2	3	4	5	6	7		
8	23	20	17	19	18	17	23		
7	18	25	20	18	26	18	19	16	19
6	20	27	33	30	27	23	18	14	16
5	19	31	40	34	28	22	25	13	11
4	26	29	53	51	28	28	21	21	18
3	22	26	40	32	33	18	24	16	17
2	23	30	26	24	26	23	14	20	18
1	16	19	20	18	18	17	16		

3. Suggest a strategy, similar to that in the latter part of Section 9.4.3, for combining N unaligned images to create a single mosaic image, $\mathbf{C}_{OR}$, that contains the combined data for every observed location in the collection

4. Show that in Equation (9.21), the formulae for the uncertainties in digital aperture photometry due to an uncertainty in the linearity correction will be given by

$$s_{\rm L}^2 = \sum_{x,y} \left(\sigma_{\rm L}(r)A[x,y]r[x,y]\right)^2 + \frac{n_{\rm pix}^2}{p_{\rm b}}\bar{\sigma}_{\rm L}^2 \bar{r}_{\rm b}^2 > P\sigma_{\rm L}^2(\bar{r}_{\rm a})\bar{r}_{\rm b}^2 + \frac{n_{\rm pix}^2}{p_{\rm b}}\bar{\sigma}_{\rm L}^2 \bar{r}_{\rm b}^2$$

where the quantities are those defined in Section 9.5.3. Explain why uncertainties in linearity are less troublesome if one is comparing stars of nearly equal brightness.

5. Show that the variance due to flat-field uncertainty in Equation (9.24) is

$$s_{\rm f}^2 = \sum_{x,y} \left(A[x,y](n[x,y] + b_{\rm e})\right)^2 \sigma_{\rm f}^2 + \frac{n_{\rm pix}^2}{p_{\rm b}}\bar{\sigma}_{\rm f}^2 b_{\rm e}^2$$

6. On a 20-second exposure, a star with magnitude $B = 15$ produces an SNR = 100 signal with a small telescope/CCD combination. Assuming this is a photon-noise limited case, how long an exposure should be required to produce the same SNR for star with $B = 13.6$?

7. A star with $V = 21.0$ is known to produce a count rate of 10 electrons per second for a certain telescope/detector combination. The detector read noise is 4 electrons per pixel, and the dark rate is zero. Compute the exposure time needed to reach a SNR = 10 under the following conditions:

   (a) dark sky and good seeing: aperture radius = 3.5 pixels, sky brightness = 1.4 electrons per pixel per second;

   (b) moonlit sky and poor seeing: aperture radius = 5.0 pixels, sky brightness = 4 electrons per pixel per second.

8. A certain CCD has a gain of 2.4 electrons per ADU, a read noise of 7 electrons per pixel, and a dark current of 2.5 ADU per pixel persecond. In the V filter, the sky brightness averages 8 ADU per second. An astronomer wishes to observe a nebula whose average brightness is expected to be 7 ADU per pixel persec per second over a digital aperture area of 100 pixels. Compute the expected SNR for measurements of the nebula's brightness on exposures of (a) 1 second, (b) 10 seconds and (c) 100 seconds.

Chapter 10
Photometry

> The classification of the stars of the celestial sphere, according to different orders
> of magnitude, was made by ancient astronomers in an arbitrary manner, without
> any pretension to accuracy. From the nature of things, this vagueness has been
> continued in the modern catalogs.
>
> — François Arago, *Popular Astronomy*, Vol I, 1851

Astronomers have measured apparent brightness since ancient times, and, as is
usual in science, technology has acutely influenced their success. Prior to the
1860s, observers estimated brightness using only their eyes, expressing the
results in the uncannily persistent magnitude system that Ptolemy[1] introduced
in the second century. As Arago notes, the results were not satisfactory.

In this chapter, after a brief summary of the history of photometry, we will
examine in detail the surprisingly complex process for answering the question:
how bright is that object? To do so, we will first introduce the notion of a defined
bandpass and its quantitative description, as well as the use of such bandpasses in
the creation of standard photometric systems. Photometry is most useful if it represents the unadulterated light from the object of interest, so we will take some pain
to describe how various effects might alter that light: spectrum shifts, absorption by
interstellar material, and the characteristics of the observing system. We will pay
particular attention, however, to the heavy burden of the ground-based photometrist:
the influence of the terrestrial atmosphere and the techniques that might remove it.

10.1 Introduction: a short history

The history of photometry is brief compared to that of astrometry, due to the
symbiotic absences of scientific interest and appropriate instrumentation. John

---

[1] The magnitude system may very well predate Ptolemy. Ptolemy's catalog in the Almagest (*c.*137
CE) may be based substantially on the earlier catalog of Hipparchus (*c.*130 BC), which has not
been preserved. It is unclear which astronomer — Ptolemy, Hipparchus, or another — actually
introduced the scale. Moreover, Ptolemy is largely silent on the method actually used to establish
the visual brightness estimates he recorded. Although Ptolemy tends to assign stars integral
magnitudes, 156 stars (out of 1028) are noted as slightly (one third of a magnitude?) brighter
or fainter than an integral value.

B. Hearnshaw (1996) provides a book-length history of astronomical photometry up to 1970. Harold Weaver (1946) gives a shorter and more technical account of developments up through World War II. A definitive history of the charge-coupled device (CCD) era remains unwritten.

To what degree will two stars assigned the same magnitude by a naked-eye observer actually have the same brightness? Modern measurements show pre-telescopic catalogs (e.g. Ptolemy and Tycho, both of whom were more interested in positions than in brightness) have an internal precision of about 0.5 magnitudes. Even the most skilled naked-eye observer can do little better: al Sufi in the ninth century devoted great attention to the problem and achieved a precision near 0.4 magnitudes. At the eyepiece of a telescope, several observers (e.g. the Herschels and, less successfully, the Bonner Durchmusterung observers Argelander and Schonfeld) produced better results (0.1 to 0.3 magnitudes) with a method of careful comparison to linked *sequences* of brightness standards.

After a suggestion by the French physicist François Arago (1786–1853), Karl Friedrich Zöllner (1834–1882) built the first optical/mechanical system for astronomical photometry in 1861. Many similar instruments soon followed. An observer using one of these *visual photometers* either adjusts the brightness of a comparison until it matches that of the unknown star, or dims the telescopic brightness of the unknown star until it disappears. Zöllner's instrument, for example, used crossed polarizers to adjust the image of an artificial star produced by a kerosene lamp.

Because the unknown need not be near a standard sequence in the sky, the visual photometer was efficient. Moreover, these devices were more *precise*, because brains are much better at judging equality (or complete extinction) than at making interpolations, especially interpolations based on memory of a sequence. Finally, the visual photometer was more *accurate* since making a mechanical adjustment gives a quantifiable measure fairly independent of a particular astronomer's eye and brain.

Astronomers got busy. Edward Pickering, at Harvard, for example, built a two-telescope "meridian photometer," which used crossed polarizers to equalize the images of two real stars. Between 1879 and 1902, Harvard visual photometrists measured the magnitudes of about 47,000 stars with a precision of about 0.08 magnitudes, and with an accuracy (based on modern measurements) of better than 0.25 magnitudes. Astronomers could now confidently examine the mathematical relationship between brightness and the ancient magnitude scale. Although several fits were proposed, by 1900 everyone had settled on the now familiar "Pogson normal scale":

$$\Delta m = -2.5 \log(b_1/b_2)$$

where b_1 and b_2 are the brightness of objects 1 and 2. The ancient scale turned out to be quite non-uniform in the logarithm: for example, the average brightness ratio between Ptolemy's magnitude 1.0 and 2.0 stars is 3.6, but between his

5.0 and 6.0 stars it is 1.3. The telescopic scales (e.g. Argelander) are closer to Pogson normal.

While the Harvard visual work progressed, photography matured. In 1850, William Cranch Bond and John Whipple, also at Harvard, photographed a few of the brightest stars. The invention of dry photographic plates (1871) increased convenience and sensitivity; eventually (around 1881) stars were recorded that were too faint to be seen by eye in any telescope. Many influential astronomers appreciated the vast potential of this new panoramic detector, and with virtually unprecedented international cooperation launched the *Carte du Ciel* project to photograph the entire sky and measure the brightness of every star below magnitude 11.0 (see Chapter 4). Astronomers soon learned to appreciate the difficulties in using photographs for quantitative photometric work, and it was not until the period 1900–1910 that several workers (notably Schwarzschild, Wirtz, Wilkins, and Kapteyn) established the first reliable ***photographic magnitude scales***. After the introduction (1910–1920) of physical photometers for objectively measuring images on plates, photography could yield magnitudes with uncertainties in the range 0.015–0.03 magnitudes. Such precision required very great care in the preparation, processing, and reduction of plate material, and could usually only be achieved in differential measurements among stars on the same plate.

In the first sustained photoelectric work, Joel Stebbins and his students at Illinois and Wisconsin performed extensive and precise photometry, first with selenium cells (1907), but soon with the vacuum photocell. Poor sensitivity at first limited the observations to very bright stars, but in 1932, when Albert Whitford and Stebbins added a vacuum-tube amplifier to the detector circuit, detection limits on their 0.5-meter telescope improved from 11th to 13th magnitude. The real revolution occurred in the 1940s, when the ***photomultiplier tube*** (***PMT***), developed for the military during World War II, became the astronomical instrument of choice for most precision work. It had very good sensitivity and produced uncertainties on the order 0.005 magnitudes in relative brightness.

The years from 1950 to 1980 were intensely productive for ground-based photoelectric work. Harold Johnson was an important pioneer in this era, first using the RCA 1P21 photomultiplier to define the UBV system, and later using red-sensitive photomultipliers to define an extended broadband system through the visual-near-infrared atmospheric windows.

Although astronomers still use photomultipliers for specialized work today, the CCD and other modern solid-state detectors have superceded them. In the optical, CCDs have superior efficiency, better stability, and a huge multiplex advantage (i.e. they can record many objects simultaneously, including standards). For ground-based differential work, CCD photometric precision on bright sources is generally set by photon-counting statistics (e.g. Equation (9.25)) or by uncertainties in calibration. For all-sky photometry and infrared work, the atmosphere imposes more serious limitations –0.01 magnitude uncertainty is often regarded as routine. Photometry from spacecraft with solid-state devices, on the other hand, offers the potential of superb precision in both differential and all-sky work. For example, the Kepler space mission for detecting occultations by extrasolar planets, presently (2010) nearing launch, hopes to achieve uncertainties below 10 μmag over time scales of several weeks.

Observations from space are very, very costly, however, so ground-based photometry continues to be a central astronomical activity.

10.2 The response function

A photometric device is sensitive over a restricted range of wavelengths called its **bandpass**. We distinguish three general cases of bandpass photometry to fit three different scientific questions.

10.2.1 Types of photometry

Single-band photometry. Suppose, for example, you suspect an extra-solar planet will move in front of a certain star, and you are interested in the occultation's duration and the fraction of the star's light blocked. You need only use a single band, since a geometric effect like the occultation of a uniform source will be identical at every wavelength. You would probably make a sequence of monitoring observations called a **time series**, a tabulation of brightness as a function of time, and you would tend to choose a wide band to maximize signal and minimize the required exposure time and telescope size.

Broadband multi-color photometry. On the other hand you might want to know not just the brightness of a source, but also the general shape of its spectrum. Broadband multi-color photometry measures an ultra-low-resolution spectrum by sampling the brightness in several different bands. Although there is no strict definition, a "broad" band is generally taken to mean that the width of the band, $\Delta\lambda$, divided by its central wavelength, λ_c, is greater than 7%–10%, or, equivalently, the spectroscopic resolving power $R = \lambda_c/\Delta\lambda < 10$–15. Broadband systems choose bands that admit the maximum amount of light while still providing valuable astrophysical information. For example, the UBVRI system, the most common broadband system in the optical, uses bandwidths in the range 65–160 nm ($R = 4$–7). It provides information on surface temperature for a wide variety of stars, and more limited information on luminosity, metal content, and interstellar reddening.

The terminology recognizes each band as a "color", so "two-color photometry" measures magnitudes in two separate bands: B and V, for example. For both historical and practical reasons, one traditionally reports the results of n-color photometric measurements by giving one magnitude and $(n-1)$ color indices. The magnitude tells the apparent brightness, and the indices tell about other astrophysical variables like surface temperature. The term "color", as shorthand for "color index" has thus come to have a second meaning – *color is the difference between two magnitudes*. So for example, the results of "two-color photometry" in B and V will be reported as a V magnitude and *one* $(B-V)$ color.

Narrow- and intermediate-band photometry. Although multi-color narrow-band photometry (roughly $R > 50$) can provide information about the shape of the spectrum, its intent is usually to isolate a specific line, molecular band, or other feature. The strategy here exchanges the large signal of the broadband system for a weaker signal with more detailed spectroscopic information. Common applications include the measurement of the strength of absorption features like Balmer-alpha or sodium D, or of the ratio of the intensities of emission lines in gaseous nebulae. Intermediate-band photometry ($15 < R < 50$) measures spectroscopic features that cannot be resolved with broader bands, but avoids the severe light loss of the very narrow bands. Examples of such features include discontinuities in spectra (for example, the "Balmer discontinuity" due to the onset of continuous absorption by hydrogen in stellar atmospheres at a wavelength of 364.6 nm), or very broad absorption features due to blended lines or molecular bands (for example, the band due to TiO in the spectra of M stars that extends from 705 to 730 nm).

10.2.2 Magnitudes

Recall that for some band (call it P), the **apparent magnitude** of the source as defined in Chapter 1 is just

$$m_P = -2.5 \log(F_P) + C_P = -2.5 \log \int_0^\infty R_P(\lambda) f_\lambda d\lambda + C_P \qquad (10.1)$$

Here m_P is the bandpass magnitude; F_P is the energy flux (the irradiance) within the band; f_λ is the monochromatic flux (also called the flux density or the monochromatic irradiance – it has units of watts per square meter of area per unit wavelength, or W m^{-3}). We choose the constant C_P to conform to some standard scale (e.g. the magnitude of Vega is zero in the visual system). The function $R_P(\lambda)$ describes the **response** of the entire observing system to the incident flux: it is the fraction of the energy of wavelength λ that will register on the photometer. We usually assume that f_λ is measured *outside* the Earth's atmosphere.

Photon detectors count photons, rather than measure energy directly. Recall that the **monochromatic photon flux** $\phi(\lambda)$ (number of photons per second per square meter of area per unit wavelength) is related to f_λ:

$$\phi(\lambda) = \frac{\lambda}{hc} f_\lambda$$

Photon detectors do *not* directly measure the quantity F_P in Equation (10.1) but report a signal proportional to the **photon flux within the band**:

$$\Phi_P = \int_0^\infty R_{PP}(\lambda)\phi(\lambda)\mathrm{d}\lambda = \frac{1}{hc}\int_0^\infty R_P(\lambda)f_\lambda\lambda\mathrm{d}\lambda$$

Here $R_{PP}(\lambda)$ is the **photon response**: the fraction of photons of wavelength λ detected by the system. This suggests that photon-counting detectors and energy-measuring detectors will measure on the same magnitude scale if

$$m_P = -2.5\log(\Phi_P) + C_{PP} = -2.5\log(F_P) + C_P$$

which requires

$$R_{PP}(\lambda)\propto\frac{R_P(\lambda)}{\lambda}$$

Although directly *measured* magnitudes are bandpass magnitudes, it makes perfect sense to talk about and compute a **monochromatic magnitude**. This is defined from the monochromatic flux:

$$m_\lambda = -2.5\log(f_\lambda) + C'(\lambda) = -2.5\log\frac{hc\phi(\lambda)}{\lambda} + C'(\lambda) \tag{10.2}$$

Here again, the value of the function $C'(\lambda)$ is arbitrary, but is often chosen so that the monochromatic magnitude of Vega or some other (perhaps fictitious) standard is a constant at every wavelength. In this case, $C'(\lambda)$ is a strong function of wavelength. Sometimes, however, the function $C'(\lambda)$ is taken to be a constant, and the monochromatic magnitude reflects the spectrum in energy units. You can think of the monochromatic magnitude as the magnitude measured with an infinitesimally narrow band. Conversely, you can think of intermediate or broadband photometry as yielding a value for m_λ at the effective wavelengths of the bands, so long as you recognize the energy distribution referenced is one of very low spectroscopic resolution.

10.2.3 Response function implementation

How is a band response implemented in practice? Both practical limits and intentional controls can determine the functional form of the responses $R_P(\lambda)$ or $R_{PP}(\lambda)$.

The **sensitivity of the detector** clearly limits the range of wavelengths accessible. In some cases, detector response alone sets the bandpass. Ptolemy, for example, based his magnitude system simply on the response of dark-adapted human vision, sensitive in the band 460–550 nm. In other cases, the detector response defines only one edge of the band. Early photographic magnitudes, for example, had a bandpass whose long-wavelength cutoff was set by the insensitivity of the photographic emulsion longward of 450 nm.

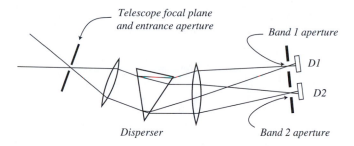

Fig. 10.1
A spectrophotometer.
Each aperture defines the
range of wavelengths
that pass to its detector. It
is possible to alter the
wavelengths sampled by
rotating the dispersing
element or translating the
apertures. In this case,
the instrument is known
as a spectrum scanner.

A ***filter*** – an element placed in the optical path to restrict transmission – is the usual method for intentionally delimiting a band. A ***bandpass filter*** defines both ends of the band by blocking all wavelengths except for those in a specific range. A filter can serve as a ***high-pass*** or ***low-pass*** element by defining only the lower or upper cutoff of a band. Filters that limit the transmission of all wavelengths equally are termed ***neutral-density filters***.

Another strategy for photometry is to use a dispersing element to create a spectrum. Sampling discrete segments of a spectrum with one or more photo-detectors is equivalent to multi-band photometry. Such instruments are termed ***spectrophotometers***. A spectrophotometer (see Figure 10.1) generally defines bandpasses by using apertures, slots, or detectors of the proper size to select the desired segment of the spectrum. Multi-pixel solid-state detectors like CCDs blur the distinction between a spectrophotometer and a spectrograph: taking a CCD image of a spectrum is equivalent to letting each pixel act as an aperture that defines a band.

For ground-based observations, ***atmospheric transmission***, $S_{atm}(\lambda)$, limits the wavelengths that are accessible, and may completely or partially define a response function. Absorption in the Earth's atmosphere set the short wave-length cutoff of early photographic photometry at 320 nm, for example. In the infrared, absorption by water vapor is significant and variable. Figure 10.2 shows the approximate atmospheric transmission in the near infrared from 0.8 to 2.6 μm expected at a high elevation site. Also marked on the plot are the half-widths of the Johnson J and K bands as defined by filter transmission only. In these bands the atmosphere will set the long cutoff of J and the short cutoff of the K band, and variations in the atmosphere may change the shape of the overall photometric response function.

Normally, however, magnitudes are defined outside the Earth's atmosphere, and an astronomer must usually remove atmospheric effects during data reduction.

As an example of response definition, Figure 10.3 shows how four different factors interact to produce the response of the Johnson U band:

1. The transmissions of the filter – Corning glass number 9863 in Johnson's original definition.

Fig. 10.2 Atmospheric transmission in the near infrared. Transmission curve is based on a model of the atmosphere at an elevation of 2.0 km, and will change with changes in water-vapor content. Light-gray lines locate the Johnson J and K photometric band-filter sensitivity (FWHM). Dark-gray lines show the sensitivity of the MKO filters for J, H, and K. The Johnson band definitions are much more susceptible to water-vapor variation than are the MKO definitions.

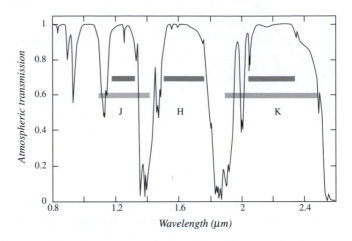

Fig. 10.3 Response function (shaded) for the Johnson U band. The function $R(\lambda)$ is the product of (1) the filter transmission, (2) the detector quantum efficiency with either a quartz or a glass window, and (3) the transmission of the atmosphere (two extremes, 4 mm and 2 mm of O_3, are indicated). The telescope and optics transmission usually do not affect the shape of $R(\lambda)$.

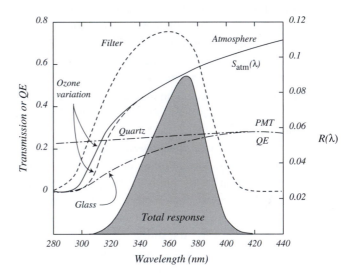

2. The quantum efficiency (QE) of the detector as a function of wavelength. In this case, the detector was a particular photomultiplier, the RCA 1P21 (now obsolete), which had an S-4 photocathode. The glass window of early tubes was later replaced with fused quartz, changing the short wavelength transmission.

3. The transmission of the atmosphere, $S_{atm}(\lambda)$. Photometry in this band assumes that the object is at the zenith, and that the ozone partial pressure is 3 mm. Changes in ozone concentration or zenith angle change the shape of $R_U(\lambda)$. For a PMT with a quartz window, the atmosphere sets the short wavelength cutoff. This feature of the U-band definition can be troublesome.

4. Transmission of the telescope optics. This is not plotted in the figure, since the reflectivity of freshly deposited aluminum is nearly constant in this region, with a value of around 0.92. Use of glass lenses, windows, or silver surfaces would change the shape of the response function.

10.2.4 Response function description

You will encounter various terms used to describe the response function. For example, for most response functions, there will be a single maximum value, R_{max}, which occurs at the **peak wavelength** λ_{peak}. Likewise, there are usually (only) two half-maximum points. These can be taken as specifications of the wavelengths at which transmission begins and ends, λ_{low} and λ_{high}:

$$R(\lambda_{peak}) = R_{max} \qquad .$$

$$R(\lambda_{low}) = R(\lambda_{high}) = R_{max}/2 \qquad .$$

Given the half maxima, we can then define one measure for the width of the response by computing the **full width at half-maximum**:

$$\text{FWHM} = \lambda_{high} - \lambda_{low} \qquad .$$

The half-maximum points also determine the **central wavelength** of the band, which may be more representative of its mid point than λ_{peak}:

$$\lambda_{cen} = (\lambda_l + \lambda_{high})/2$$

A somewhat more sophisticated and possibly more useful measure of the width of a particular response function is the **bandwidth**:

$$W_0 = \frac{1}{R_{max}} \int R(\lambda) d\lambda$$

Likewise, a somewhat more sophisticated measure of the center of a band is its **mean wavelength**, which is just

$$\lambda_0 = \frac{\int \lambda \cdot R(\lambda) d\lambda}{\int R(\lambda) d\lambda}$$

Figure 10.4 illustrates these relations. For a symmetric function,

$$\lambda_{peak} = \lambda_{cen} = \lambda_0 \qquad .$$

Perhaps even more informative is the **effective wavelength** of the response to a particular source. The effective wavelength is a weighted mean wavelength (weighted by the source flux) and indicates which photons most influence a particular measurement:

$$\lambda_{eff} = \frac{\int \lambda \cdot f_\lambda \cdot R(\lambda) d\lambda}{\int f_\lambda \cdot R(\lambda) d\lambda}$$

Fig. 10.4 Definitions of the middle and width of a band. The curve shows the function $R(\lambda)$. The mean wavelength divides the area under the curve into two equal parts (shaded and unshaded). The dark-gray rectangle has a width equal to the bandwidth and an area equal to the area under the curve.

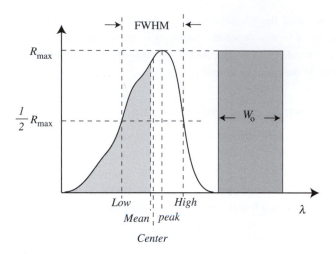

Fig. 10.5 (a) Effective wavelengths for two different sources in the same band. The solid curves apply to a hot source, and the dotted curves apply to a cool source with the same magnitude in the band. (b) Definition of the isophotal wavelength: the area of the hatched rectangle is the same as the shaded area under the curve. The dashed curve is the response function.

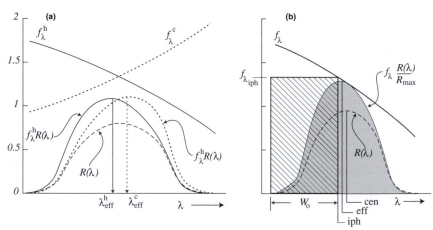

Figure 10.5 illustrates that different sources will in general have different effective wavelengths.

It is tempting to think of any bandpass measurement as equivalent to a measurement of the monochromatic flux at wavelength λ_{eff} multiplied by the bandwidth, W_0. This is nearly correct in practice, and for broadband photometry of stars (provided spectra are sufficiently smoothed) using this equivalence produces an error of a percent or less. To be strictly accurate with such an equivalence, we need to introduce yet another definition for the "middle" of the band. This one is called the *isophotal wavelength*, λ_{iph}. The isophotal wavelength is the one for which we have

$$W_0 \cdot f_{\lambda_{\mathrm{iph}}} = \frac{1}{R_{\max}} \int f_\lambda \cdot R(\lambda) \mathrm{d}\lambda$$

As with the effective wavelength, the exact value of the isophotal wavelength will depend on the spectrum of the source.

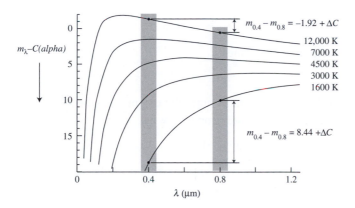

Fig. 10.6 Color indices for blackbodies. Curves are generated by taking the logarithm of the Planck function. Note that monochromatic magnitudes increase downwards. Spectra have been shifted vertically by arbitrary amounts for clarity. In this figure, $\Delta C = 0$.

10.2.5 Color indices

Multi-band photometry can measure the shape of an object's spectrum. It is convenient to think of the bands as sampling the monochromatic flux of a smoothed spectrum at their isophotal wavelengths. For example, Figure 10.6 shows the spectra of several blackbodies whose temperatures range from 1600 K to 12,000 K. The vertical scale of the figure shows the monochromatic magnitude in a system in which the constant in Equation (10.2) is independent of wavelength. Remember, this is *not* the usual case in astronomical photometry, where the spectrum of some standard object (e.g. Vega, which is similar to a blackbody with temperature of 9500 K), would be a horizontal line in a plot of m_λ as a function of λ. In the figure, we assume two bands, one with a mean wavelength at 0.4 μm, the other at 0.8 μm. It is clear that the arithmetical difference between these two magnitudes for a particular spectrum depends on the average slope of the spectrum, which in turn depends on the source's temperature. The convention is to speak of the difference between any two bandpass magnitudes used to sample the slope of the spectrum as a ***color index***.

For blackbodies, at least, the color index is not just useful, but definitive – its value uniquely measures the body's temperature. By convention, you compute the index in the sense:

$$\text{index} = m(\text{shorter}\lambda) - m(\text{longer}\lambda)$$

As mentioned earlier, astronomers usually symbolize the color index as the magnitude difference, sometimes enclosed in parenthesis. In the case of Figure 10.6, we might write the index as $(m_{0.4} - m_{0.8})$. In the case of the Johnson–Cousins red and infrared bands, the index would be written $(m_R - m_I)$, or more commonly $R - I$.

The behavior of the color index at the long and short wavelength extremes of the Planck function is interesting. In the Rayleigh–Jeans region (i.e. where $\lambda kT \gg hc$) you can show that

$$m_\lambda = \log T + C(\lambda) \tag{10.3}$$

so that the color index becomes

$$(m_{\lambda_1} - m_{\lambda_2}) = C(\lambda_1) - C(\lambda_2) = \Delta C$$

a constant independent of temperature. For example, in the Johnson broadband system, a blackbody of infinite temperature has color indices

$$(U - B) = -1.33, (B - V) = -0.46$$

At short wavelengths, the **Wien approximation** for the surface brightness of a blackbody holds:

$$B(\lambda, T) \approx \frac{2hc^2}{\lambda^5} \exp\left(-\frac{hc}{\lambda kT}\right)$$

So the color index is

$$(m_{\lambda_1} - m_{\lambda_2}) = \frac{a}{T}\left(\frac{1}{\lambda_1} - \frac{1}{\lambda_2}\right) + C(\lambda_1) - C(\lambda_2) \tag{10.4}$$

Thus, at very small temperatures or wavelengths, the index is a linear function of $1/T$.

10.2.6 Line and feature indices

Real objects almost always have more complex spectra than do blackbodies, with features of astrophysical significance that may include absorption and emission lines, bands, and various discontinuities. Multi-band photometric indices can measure the strength of such features.

Two bands often suffice to measure the size of a discontinuity or the strength of a line, for example. In Figure 10.7a, bands C and D sample the continuum on the short and long wavelength sides of a sharp break in a spectrum. The index $(C-D)$ will be sensitive to the size of the break – but note two features of the index:

First, the actual relation between the size of the break and the numerical value of the $(C-D)$ index depends on the constants employed in the definition

Fig. 10.7 Definition of indices to measure the strength of (a) a spectrum discontinuity, and (b) an absorption line. Monochromatic magnitudes are defined so that the constant in Equation (10.2) is independent of wavelength.

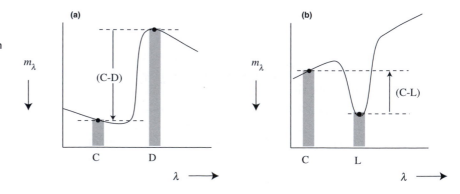

of the bandpass magnitudes in Equation (10.1). It might be convenient to have $(C - D) = 0$ when the break vanishes, but this may violate the convention that all indices should be zero for the spectrum of some standard object. (Examine Figure 1.5 – Vega has several non-zero spectrum discontinuities, yet all its indices are zero in some systems.)

Second, positioning the bands is important. The sensitivity of the index to the size of the break will diminish if either bandpass response includes light from the opposite side of the break. Likewise, if a band is located too far away from the break, unrelated features in the spectrum can affect the index. Obviously, it will be easier to position narrow bands than wide bands, but narrow bands give weaker signals.

A similar index can measure the intensity of an absorption or emission line (Figure 10.7b). Here one narrow band is centered on the feature, and the other on the nearby continuum. The magnitude difference measures the line strength. This strategy is common in detecting and mapping objects with strong emission lines in their spectra: for example, the astronomer takes two CCD exposures – one through a filter centered on the emission line in question, the second through one centered on the nearby continuum. Digital subtraction of the two registered and properly scaled images produces zero signal except in pixels where a source is emitting radiation in the line.

Figure 10.8 illustrates an alternative strategy for measuring a line index. Two bands – one broad, the other narrow – are both centered on the line. The narrow band is quite sensitive to the strength of the line, while the broad band is relatively insensitive, since most of the light it measures comes from the continuum. The index

$$\text{line index} = m_{\text{narrow}} - m_{\text{wide}}$$

tracks the strength of the absorption, in the sense that it becomes more positive with stronger absorption. One widely used line index of this sort is the β index, which measures the strength of the Balmer beta line of hydrogen, usually useful for luminosity or temperature classification of stars.

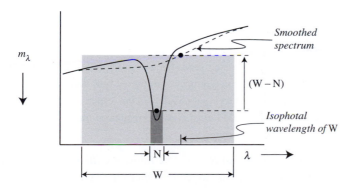

Fig. 10.8 A line index computed from wide and narrow bands centered on the same absorption line.

Fig. 10.9 Three bands can measure the curvature of the spectrum. In both (a) and (b), the index $2(X - C)$ tracks the monochromatic magnitude's departure from linearity.

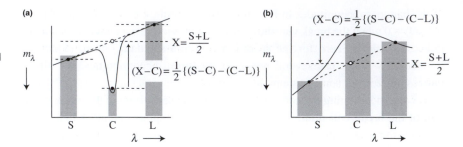

Finally, consider a third kind of index. Three bands can measure the *curvature* (i.e. the second derivative, rather than the first) of a spectrum. Curvature can arise on a relatively small scale because of a sharp absorption or emission line, or on a large scale because of broad or diffuse features (molecular bands in gases, or absorption features in the reflection spectra of crystalline solids, for example). Figure 10.9 illustrates two situations with three (a) equally and (b) unequally spaced bands at a short, central, and long wavelength (S, C, and L). If we consider just the monochromatic magnitudes, and if the bands are equally spaced as in Figure 10.9a, the index

$$\text{curvature} = (m_S - m_C) - (m_C - m_L) = S + L - 2C$$

will be zero if the logarithmic spectrum is linear, and positive if the central band contains an absorption feature. *The curvature index depends on the difference between two color indices.* In practical systems the index will still track curvature even if bands are not equally spaced, and even if $C'(\lambda)$ in Equation (10.2) is *not* a constant.

10.3 The idea of a photometric system

The term ***photometric system*** implies at least two specifications:

1. The wavelength response for each band – that is, the shape of the function $R_P(\lambda)$ in Equation (10.1)
2. Some method for standardizing measurements made in those bands. This is important for two reasons:
   - Each observer needs to know the value for the constant C in Equation (10.1) that will assure agreement of his magnitudes with those of all other observers.
   - The differing hardware produces some variety in the response functions in practice, so a method for standardization must allow correction of the inevitable systematic effects due to imperfect matching.

The first specification, that of $R_P(\lambda)$, determines the ***instrumental*** or ***natural system***. The first and second together determine the ***standard system***.

Observations in the natural system alone can be quite useful (e.g. determining the period of a variable star), but only by placing magnitudes on the standard system can two astronomers confidently combine independent measurements.

Standardization might involve observations of laboratory sources, e.g. a blackbody of known temperature and therefore known absolute flux in W m^{-2}. Almost always, though, a single astronomical object or set of objects is a much more practical standardizing source. Almost all standard systems today rely upon some network of constant-brightness standard objects distributed around the sky. If everyone agrees on a list of stars and their corresponding magnitudes, anyone can calibrate measurements made in their instrumental system by observing the standards and the unknowns with the same procedures. Because systematic differences will most likely arise if the spectrum of the star observed is different from the spectrum of the standard star, most systems strive to define a set of standards that includes a wide variety of spectral types.

As we have seen with the Johnson U, K, and J bands, the definition of the bandpass sometimes involves the atmosphere. Atmospheric absorption, however, is variable, and removing this variation should be part of the standardization procedure.

Because standardization is so essential to a photometric system, some astronomers have devised *closed photometric systems*, in which a relatively small group of observers carefully controls the instruments and data reduction, maximizing internal consistency. Many space-based observations (e.g. HIPPARCOS), and many ground-based surveys (e.g. the Sloane Digital Sky Survey) constitute closed systems. An *open photometric system*, in contrast, is one in which all astronomers are encouraged to duplicate the defined natural system as best they can, and, through reference to a published list of standard stars, add to the pool of observations in the system.

10.4 Common photometric systems

Astronomers have introduced several hundred photometric systems. Bessel (2005) gives an extensive review of the most common systems. Here we examine only a few of the most widely used as an introduction to the operation of most.

10.4.1 Visual and photographic systems

The dark-adapted human eye determines the band of the *visual photometric system*. In the earliest days of astronomy, the standardization procedure required that magnitudes measured in the system be consistent with the ancient catalogs (e.g. Ptolemy, al Sufi, and Bayer). The introduction of optical/mechanical visual photometers led to the establishment of *standard sequences* of stars, including (initially) the *north polar sequence* and (later) many secondary sequences (the

Table 10.1. *Bandpasses of historical importance*

Band	Symbol	Band definition	λ_{peak}, nm	FWHM
Visual	m_{vis}	Mesotopic[*] human eye	515–550	82–106
International photographic	m_{pg}, IPg	Untreated photographic emulsion + atmosphere	400	170
International photovisual	m_{pv}, IPv	Orthochromatic emulsion + yellow filter	550	100

[*] Visual photometry of stars uses a mixture of photopic (color, or cone) and scotopic (rod) vision, with the shift from cones to rods occurring with decreasing levels of illumination. The effective wavelength of the eye thus shifts to the blue as light levels decrease (the Purkinje effect); see Appendix B3.

48 Harvard standard regions and the 115 Kapteyn selected areas were perhaps the best studied).

In the early twentieth century, astronomers defined two bands based on the properties of the photographic emulsion (Table 10.1). The poor properties of the photographic emulsion as a photometric detector, and lack of very specific definitions, limited the success of this system. The ***international photographic band*** is sensitive in the near ultraviolet–blue region. The response of the ***international photovisual band***, somewhat fortuitously, roughly corresponds to that of the visual band (i.e. the human eye, sensitive to green–yellow). The IAU in 1922 set the zero point of both magnitudes so that 6th magnitude A0 V stars[2] in the north polar sequence would have (roughly) the same values as on the old Harvard visual system. This meant that the color index,

$$\text{color index} = m_{pg} - m_{pv}$$

should be zero for A0 stars, negative for hotter stars, and positive for cooler stars.

Many other photographic systems exist. The photovisual magnitude originally depended on "orthochromatic" plates, which were made by treating the emulsion with a dye to extend its sensitivity to about 610 nm. Other dyes eventually became available to extend photographic sensitivity to various cut-offs ranging through the visible and into the near infrared. Twentieth-century astronomers devised many filter–emulsion combinations and set up standard

---

[2] A0 V is the spectral type of Vega, which is *not* in the north polar sequence. Because of the early decision to keep visual magnitudes roughly consistent with the ancient catalogs, the photographic and photovisual magnitudes of Vega turn out to be close to zero. The importance of Vega stems in part from its brightness, which makes it a good candidate for absolute (i.e. watts per square meter per meter of wavelength) measurement of specific irradiance.

star sequences in a variety of photography-based systems. All these are mainly of historic interest.

10.4.2 The UBVRI system

By far the most widely used ground-based photometric system prior to the present has been the Johnson–Cousins UBVRI system (Table 10.2 and Figure 10.10). Johnson and Harris (1954) defined the UBV portion first, based on the response of the RCA 1P21 photomultiplier, a set of colored glass filters, and a list of magnitudes for a relatively small number of standard stars scattered around the celestial sphere. The V band closely corresponds to the international photovisual band and its zero point was set so that $V = m_{pv}$ for the standards in the north polar sequence. The U and B bands correspond to short- and long-wavelength segments of the photographic band, and to be consistent with the international system, their zero points are set so that the colors $U - B$ and $B - V$ are zero for A0 V stars.

After some pioneering work at longer wavelengths by Stebbins, Kron and Whitford, Harold Johnson and his collaborators in the period 1960–1965 extended the UBV system to include bands in the red (R_J) and near infrared (I_J), as well as the longer infrared bands (JHKLMNQ) discussed in the next section. Modern work with CCDs, however, has tended to replace the R_J and I_J with the R_C and I_C bands specified by Cousins and his collaborators (see Table 10.2 for the differences). In current practice, the lists of Arlo Landolt (1983, 1992) and Menzies *et al.* (1989, 1991) define the standard stars for the UBV(RI)$_C$ system.

Modern CCD observers sometimes have difficulty replicating the original photomultiplier-based instrumental system. A complicating factor is the great variation in CCD spectral response due to differing surface treatments, gate material, gate structure, backside illumination, etc. The U band causes the

Table 10.2. *The Johnson–Cousins UBVRI system.* The R_J and I_J data are from Colina *et al.* (1996). All other widths are from Bessel (1992). Effective wavelengths and monochromatic fluxes for a zero-magnitude, zero-color star are from the absolute calibration of Vega and Sirius by Bessell *et al.* (1998). Vega has $V = 0.03$ on this system

	U	B	V	R_C	R_J	I_C	I_J
λ_{eff}, nm	366	436	545	641	685	798	864
FWHM	66	94	88	138	174	149	197
f_λ at λ_{eff} in units of 10^{-12} W m^{-2} nm^{-1} for $V = 0$	41.7	63.2	37.4	22.6	19.2	11.4	9.39

Fig. 10.10 Normalized response functions for the UBVRI system. Also shown are the monochromatic magnitudes for a representative A0 and G2 dwarf. Note the importance of the Balmer discontinuity near 370 nm in the A0 spectrum, and the break due to metal absorption near 400 nm in the G2 spectrum.

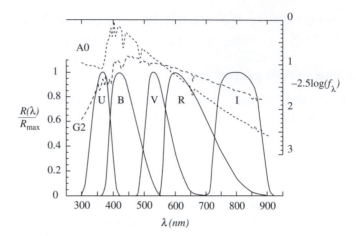

most trouble, partly because it is defined by the atmosphere at a particular altitude (see Section 10.2.3), and partly because of generally poor CCD response in the ultraviolet. Close matches are possible with a good knowledge of the individual CCD response and a careful choice of filters. For details, see Bessel (1990).

This multi-band system was designed with the rough spectral classification of stars in mind. Figure 10.10 shows the responses of the normalized $UBV(RI)_C$ bandpasses superimposed on spectra of an A0 and a G2 dwarf (i.e. matching, respectively, Vega and the Sun). The $U - B$ index is clearly sensitive to the Balmer discontinuity (present very obviously in the A star at 370 nm, and much reduced in the G star). The discontinuity – and hence the $U - B$ index – depends upon luminosity, at least for hot stars. The other indices are primarily sensitive to temperature (and therefore spectral type). The $B - V$ color is more sensitive to metal abundance than are $V - R$ or $R - I$, and fails as a useful index for M stars because of molecular band absorption. (In astrophysics, a "metal" is any element other than hydrogen or helium.) Because of its long baseline and relative insensitivity to chemical abundances, the $V - I$ index is the most purely temperature-sensitive index in this system ($V - K$ is even better, for the same reason). Appendix A10.1 tabulates the colors of various spectral types. The system is useful for measuring the photometric properties of objects besides normal stars: Solar System bodies, supernovae, galaxies, and quasars have all been extensively observed.

10.4.3 The broadband infrared system: JHKLMNQ

The broadband infrared system (Table 10.3) might be regarded as an extension of the UBVRI system, and shares a common zero point (so the colors of an un-reddened A0 V star are zero). Detectors in this region cannot be silicon CCDs, but must be infrared arrays or single-channel infrared-sensitive devices.

Table 10.3. *The broadband infrared system.* JHKL from Bessell *et al.* (1998), M band from Rieke and Lebofsky (1985), and N and Q from Rieke *et al.* (1985)

	J	H	K	L	M	N	Q
λ_{eff}, μm for A0 stars	1.22	1.63	2.19	3.45	4.8	10.6	21
FWHM	0.213	0.307	0.39	0.472	0.46	3–6	6–10
f_λ at λ_{eff} in units of 10^{-11} W m^{-2} μm^{-1} for $V = 0$	315	114	39.6	7.1	2.2	0.96	0.0064

Table 10.4. *Mauna Kea (MKO) filter characteristics.* Central wavelengths of L' and M' are significantly different from L and M, hence the renaming. Note that these are filter characteristics: actual bandpass responses will depend on detector, atmosphere, telescope optics, etc

	J	H	K	L'	M'
λ_{cen}, μm	1.24	1.65	2.20	3.77	4.67
FWHM	0.16	0.29	0.34	0.70	0.22

Subtraction of background can be a very serious problem in the infrared, as discussed in the previous chapter.

A more important complication is the fact that, for the ground-based infrared, bandpass definitions can depend very critically on atmospheric conditions (mainly the amount of water vapor encountered along the line of sight). Different observatories with identical hardware can experience different infrared window sizes and shapes if they are at different altitudes (extending to space observatories). The same observatory can experience similar bandpass variations due to changing humidity.

Different observatories have thus defined infrared bands differently, and the values in Table 10.3 merely represent the typical choices prior to the twenty-first century. The IAU in 2000 recommended a preferred natural system for JHK – the Mauna Kea Observatory near-infrared system (see Table 10.4). The MKO system attempts to minimize sensitivity to water vapor while optimizing the signal-to-noise ratio (SNR), usually by narrowing the FWHM.

A second important characteristic of infrared photometric systems stems partly from their relative immaturity: the standard star magnitudes for these bands are not as well defined as in the CCD region. The situation is best in the JHK bands, where at least three different and largely non-overlapping (and

Table 10.5. *The four-color and β system*

Name	u Ultraviolet	v Violet	b Blue	y Yellow	H_β wide	H_β Narrow
λ_{eff}, nm for A0 stars	349	411	467	547	489	486
FWHM, nm	30	19	18	23	15	3.0

still evolving) lists of standard stars have been in common use. These are beginning to converge on a common system consistent with measurements derived from the MKO-near-infrared bandpasses.

10.4.4 The intermediate-band Strömgren system: uvbyβ

Bengt Strömgren designed this intermediate-band system in the late 1950s, and David Crawford and many others developed it observationally in the 1960s and 1970s. They published several lists of standard stars during these years. The system avoids many of the shortcomings of the UBV system, and aims to classify stars according to three characteristics: temperature, luminosity, and metal abundance. Classification works well for stars of spectral types B, A, F, and G, provided the photometry is sufficiently accurate. Photometrists frequently supplement the four intermediate-band colors, uvby, with a narrow band index, β, which tracks the strength of absorption in the Balmer beta line. The β index greatly improves the luminosity classification for hotter stars, and is a good temperature indicator for cooler stars.

Emission in all of the four intermediate bands depends on temperature, but in addition, emission in the u and v bands is depressed by the presence of metals in a star's atmosphere. Also, the u band is depressed by the Balmer discontinuity, a temperature-dependent feature which is strongest for A0 stars, but which also depends on luminosity. To indicate astrophysical information, then, Strömgren photometry is generally presented as a y magnitude, a $(b-y)$ color, and two curvature indices. The $(b-y)$ color closely tracks temperature in the same way as the Johnson $B-V$ (in fact, $b-y \approx 0.68(B-V)$ over a large range of stellar types), but $(b-y)$ is somewhat less sensitive to abundance effects and is more useful at lower effective temperatures than is $B-V$. The two curvature indices are

$$c_1 = (u-v) - (v-b)$$

$$m_1 = (v-b) - (b-y)$$

The c_1 index measures the strength of the Balmer discontinuity, and in combination with temperature from $(b-y)$ yields information about luminosity. It is an improvement over the Johnson $(U - B)$, partly because the U filter straddles the Balmer discontinuity. The m_1 index measures metal abundance. The precise relationships between the indices and the astrophysical parameters are more complex than suggested here, but they have been well calibrated for spectral types hotter than K0.

10.4.5 Other systems

Many other photometric systems find less widespread use than those just described, and it is helpful to describe a few examples.

Photometry from space need not contend with any of the atmospheric and many of the background issues that complicate photometry from the ground. Within the parameters of a given detector, space observatories permit much greater freedom to base bandpass design on purely astrophysical considerations. The NICMOS2 camera on the Hubble Space Telescope (HST), for example, carried about 30 filters, many centered at bands completely inaccessible from the ground.

It is nevertheless very important to be able to tie space observations to ground-based measurements. The HIPPARCOS space mission, for example, used a two-filter broadband system closely related to B and V, while some of the NICMOS filters correspond to the JKLMN bands. The primary CCD camera for the HST (the WFPC/WFPC2), had slots for 48 filters, but those most commonly used closely matched the UBVRI system. The HST standard magnitudes, incidentally, are defined so that a source with constant f_v has zero colors.

We can expect the introduction of novel ground-based systems to continue. New CCD-based systems might even replace well-established photomultiplier-based systems. For example, the Sloan Digital Sky Survey (SDSS), the auto-mated ground-based program that is expected to produce photometry for over 10^8 stellar and non-stellar objects, uses a five-color system (see Table 10.6) designed to make optimal use of silicon CCD sensitivity. The SDSS database will be larger than all the PMT-based UBVRI observations accumulated since the 1950s. Since the SDSS colors give as good or better astrophysical informa-tion, the SDSS may (or may not) eventually displace UBVRI as the dominant broadband system in the visual.

Table 10.6. *The bands for the SDSS five-color system*

	u'	g'	r'	i'	z'
λ_{cen}, nm	354	477	623	762	915
FWHM	57	139	137	153	95

10.5 From source to telescope

> ...Slowly the Emperor returned –
> Behind him Moscow! Its onion domes still burned...
> Yesterday the Grand Army, today its dregs!
> ...They went to sleep ten thousand, woke up four.
>
> — *Victor Hugo*, Russia, 1812, Trans. Robert Lowell

A grand army of photons leaves a source, but many are lost on their march to our telescope. This section follows one regiment of that army to consider its fortunes in detail. The goal of photometric reduction will be to reconstruct the original regiment from its dregs – to account for all losses and transformations during its long journey in the cold.

At least four different effects can alter the photons on their way to the telescope:

- wavelength shifts
- extragalactic absorption
- Galactic and Solar System absorption
- atmospheric absorption.

10.5.1 Wavelength changes

The regiment that leaves the source is $\phi_E(\lambda_E)d\lambda_E$, that is, all those photons with wavelength between λ_E and $\lambda_E + d\lambda_E$ emitted in one second in the direction that would place them in a unit area of our telescope aperture. The subscript E just means "emitted". The dregs of the regiment are the members of that original group that actually survive at the top of our atmosphere, which we will call $\phi(\lambda)d\lambda$. In the general case, we allow for the possibility that both numbers and their wavelengths can change.

We first consider the consequences of wavelength change by itself. Because of the Doppler effect, or because of the expansion of the Universe, or because of various relativistic effects, the wavelength of each photon on arrival may differ from its original value, λ_E. The new value is given by

$$\lambda_o = (1 + z)\lambda_E$$

where z is the **redshift parameter** ($z = (\lambda_o - \lambda_E)/\lambda_E$) of the source. Because of this wavelength change, the photons emitted into wavelength interval $d\lambda_E$ will arrive spread out over wavelength interval $d\lambda_o = (1 + z)d\lambda_E$. Since we consider only the effect of wavelength change, and ignore the absorption effects listed above, we can say that the *number* of photons is conserved, that is:

$$\phi(\lambda)d\lambda = \phi_E(\lambda_E)d\lambda_E$$

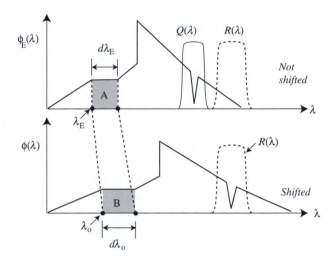

Fig. 10.11 Wavelength shifts and photometry. The upper panel shows an unshifted spectrum, and the lower panel shows the same spectrum shifted in wavelength by a redshift parameter $z = 0.2$. Photons originating in area A arrive in area B. Photons measured in band R originate in bandpass Q.

We have dropped the subscript for the observed wavelength. Thus, the observed and emitted monochromatic photon flux are related by (see Figure 10.11)

$$\phi(\lambda) = \frac{1}{1+z}\phi_E\left(\frac{\lambda}{1+z}\right) = f_\lambda \frac{\lambda}{hc} \tag{10.5}$$

The function f_λ is the observed monochromatic flux density of Chapter 1. Although photon number is conserved, the monochromatic flux density is not:

$$f_E(\lambda_E) = f_E\left(\frac{\lambda}{1+z}\right) = \frac{hc\phi_E(\lambda_E)}{\lambda_E} = (1+z)^2 f_\lambda \tag{10.6}$$

In practice, we must deal with an observer who makes magnitude measurements in a band. An observer who uses a bandpass with photon response $R(\lambda)$ will measure the magnitude

$$m_R = -2.5 \log \int R(\lambda) \frac{\phi(\lambda)}{\lambda} \, d\lambda + C_R$$

$$C_R = -2.5 \log \int R(\lambda) \frac{g_\lambda}{hc} \, d\lambda$$

where g_λ is the spectrum of a photometric standard of magnitude zero. We would like to understand how m_R relates to a magnitude measured for these same photons before their wavelength shift. These photons began their journey in a wavelength band different from R; call this band Q (again, see Figure 10.11). Measuring a magnitude in band Q on the unshifted spectrum gives

$$m_Q = -2.5 \log \int Q(\lambda) \frac{\phi_E(\lambda)}{\lambda} d\lambda + C_Q$$

$$= -2.5 \log \left[(1 + z) \int Q(\lambda) \frac{\phi(\lambda(1 + z))}{\lambda} d\lambda \right] + C_Q$$

$$C_Q = -2.5 \log \left[\int Q(\lambda) \frac{g_\lambda}{hc} d\lambda \right]$$

If $m_R = m_Q$ then the observation in band R directly gives the magnitude the source would have in band Q if there were no shift. This equality is generally possible if R has the shifted shape of Q (i.e. $R((1 + z)\lambda) = Q(\lambda)$). However, since R and Q must often be standard bands in some system, it is hard to meet this condition, so we must consider the difference:

$$m_R - m_Q = 2.5 \log(1 + z) + C_R - C_Q + 2.5 \log \left[\frac{\int Q(\lambda) \frac{\phi(\lambda(1 + z)) d\lambda}{\lambda}}{\int R(\lambda) \frac{\phi(\lambda)}{\lambda} d\lambda} \right] \qquad (10.7)$$

For objects in our own galaxy, z is quite small, and one almost always uses $R(\lambda) = Q(\lambda)$, $C_R = C_Q$. In that case, the first three terms in Equation (10.7) add to zero. The last term describes the effect of photons shifting into, out of, and within the band. In the case of narrow bands near sharp spectral features, even small Doppler shifts can produce large differences between $\phi((1 + z)\lambda)$ and $\phi(\lambda)$. In such a case, one could compute m_Q from an observed m_R using Equation (10.7), provided enough is known about $\phi(\lambda)$.

For distant objects, z becomes large because of the expansion of the Universe. Equation (10.7) again suggests that it should be possible, given knowledge of z and $\phi(\lambda)$, to use an observed bandpass magnitude to compute the magnitude *that would be observed* (in the same or in a different band) *if the source had redshift* $z = 0$. Hubble called this kind of magnitude correction the **K correction**. Although different authors define the K correction in slightly different ways, Hogg *et al.* (2002) give a good general introduction.

Wavelength shifts will affect the colors of galaxies with large z. Application of Equation (10.7) for two different bands gives an expression for the color change as a function of z. Having observed a color, you can solve that expression for z, so long as you can approximate the spectrum of a galaxy. These **photometric redshifts** from observed colors are valuable estimators of galaxy distance because they do not require observationally difficult spectroscopy of very faint objects.

10.5.2 Absorption outside the atmosphere

Space is not empty. Interstellar gas and dust in our own galaxy absorb and scatter light. Absorption (in which the photon ceases to exist) and scattering (in which the photon changes direction) are physically distinct processes, but they have the same effect on the regiment of photons headed towards our telescope – they remove photons from the beam. It is common to refer to both processes simply as "absorption." Absorption not only reduces the overall number of photons that arrive at the telescope – an effect sometimes called *extinction* – but it also alters the shape of the spectrum.

Diffuse gas absorbs photons to produce *interstellar absorption lines and bands*. In the optical, the sodium D doublet is usually the strongest interstellar line, and in the ultraviolet, the Lyman-alpha line at 121.6 nm is usually strongest. Especially at short wavelengths, gas will also produce continuous absorption and absorption edges due to ionization. A strong feature at 91.2 nm due to ionization of hydrogen is very prominent, for example. Absorption by dust will generally alter the overall shape of the spectrum, and depending on its composition, add a few very broad features. In the region 0.22–5.0 μm, dust scatters short-wavelength photons more strongly than long-wavelength photons, so the resulting change in the shape of the spectrum is termed *interstellar reddening*. In our notation, assume that

$S_{ism}(\lambda)$ = the fraction of photons of wavelength λ that are transmitted by the interstellar medium within our own galaxy.

$S_{exg}(\lambda)$ = the fraction of photons arriving at observed wavelength λ that are transmitted by the interstellar medium outside our own galaxy

Note that because of the cosmological redshift, absorptions described by $S_{exg}(\lambda)$ involve photons that had wavelength $\lambda/(1 + z')$ when they were absorbed by material with redshift parameter z'. (This produces the phenomenon of the *Lyman-alpha forest* in the spectra of distant objects: multiple absorption lines due to Ly α at multiple redshifts.) The photon flux that reaches the top of the Earth's atmosphere, then, is just

$$\phi(\lambda) = S_{ism}(\lambda)S_{exg}(\lambda)\phi_0(\lambda) = S_{ism}(\lambda)S_{exg}(\lambda)\frac{1}{1+z}\phi_E((1+z)\lambda) = f_\lambda \frac{\lambda}{hc}$$

We will call $\phi(\lambda)$ the *photon flux outside the atmosphere* and $\phi_0(\lambda)$ the *photon flux outside the atmosphere corrected for interstellar absorption*. In extragalactic astronomy, it is frequently difficult to estimate $S_{exg}(\lambda)$, so it may be important to distinguish between $\phi_0(\lambda)$ and the flux

$$\phi_G(\lambda) = S_{exg}(\lambda)\phi_0(\lambda) \qquad ,$$

where $\phi_G(\lambda)$ is the flux corrected for Galactic absorption but not for extragalactic absorption.

10.5.3 Absorption by the atmosphere

The Earth's atmosphere removes photons from the stream directed at our telescope, both through scattering and through true absorption. As before, we refer to both processes as "absorption", and note that atmospheric absorption will both reduce the apparent brightness of the source spectrum as well as alter its shape. We therefore refer to *atmospheric extinction* and *atmospheric reddening*. The atmosphere also introduces some sharper features in the spectrum, the *telluric lines and bands*.

Extinction is a strong function of wavelength. At sea level, three opaque regions define two transmitting windows. Rayleigh scattering and absorption by atoms and molecules cause a complete loss of transparency at all wavelengths shorter than about 300 nm. This sets the short end of the *optical–infrared window*. The second opaque region, from absorption in molecular bands (primarily due to H_2O and CO_2), begins at around 0.94 μm, has a few breaks in the near infrared and mid infrared, then extends from 30 mm to the start of the *microwave–radio window* at around 0.6 cm. The radio window ends at around 20 m because of ionospheric absorption and reflection.

Atmospheric extinction has a profound influence on life. The atmospheric infrared opacity prevents the Earth's surface from radiating directly into space and cooling efficiently. This so-called greenhouse effect is responsible for maintaining the average surface temperature at about 30 K higher than it would be without the atmosphere. Short-wavelength electromagnetic radiation is quite detrimental to biological systems, and none of the forms of life presently on Earth could survive if exposed to the solar gamma-ray, X-ray, and shortwave-ultraviolet radiation that is presently blocked by the atmosphere. Had life here originated and evolved to cope with an environment of either low temperatures or hard radiation, we would all be very different creatures indeed.

The wavelength dependence of extinction has an equally profound effect on astronomical life. Astronomy began by peering out at the Universe through the narrow visual window and evolved over many centuries to do a better and better job in that restricted region of the spectrum. Astronomy only discovered the radio window in the middle of the twentieth century. Yet later in that century, spacecraft (and aircraft) finally provided access to the entire spectrum. Only with the introduction of decent infrared arrays in the 1980s could astronomers take advantage of the gaps in the near-infrared atmospheric absorption available at dry high-altitude sites. Atmospheric absorption has made optical astronomy old, radio astronomy middle-aged, and gamma-ray, X-ray, and infrared astronomy young.

Quantitatively, we can postulate an atmospheric transmission function

$S_{atm}(\lambda, t, e, a)=$ the fraction of photons of wavelength λ that are transmitted by
the Earth's atmosphere at time t, elevation angle e, and azimuth a

The photon flux that actually reaches the telescope is then

$$\phi_A(\lambda) = S_{atm}(\lambda, t, e, a)\, S_{ism}(\lambda)\, S_{exg}(\lambda)\, \frac{1}{1+z}\, \phi_E((1+z)\lambda) = f_\lambda^A \frac{\lambda}{hc}$$

The rate at which energy ultimately gets detected in an infinitesimal band will be

$$dE_{sig} = aT'_P(\lambda)f^A_\lambda d\lambda = aT'_P(\lambda)\frac{\phi_A(\lambda)}{\lambda}d\lambda$$

Here a is the effective collecting area of the telescope, and $T'_P(\lambda)$ is a function that quantifies the overall wavelength-dependent efficiency of the instrument. It includes such things as wavelength-sensitive reflectivity and transmission of optical elements, filter transmission, and detector quantum efficiency. Integrating the previous equation, we can express the raw **instrumental magnitude** measured inside the atmosphere:

$$m^A_P = -2.5\log\int T'_P S_{atm}(\lambda)\frac{\phi(\lambda)}{\lambda}d\lambda + C'_P$$
$$= m^O_P + A_{atm} + A_{ism} + A_{exg} + C^z_P$$
$$= m_P + A_{atm}$$

Here the A parameters represent the atmospheric, Galactic, and extragalactic absorption, in magnitudes; C^z_P is the correction for wavelength shift; and C'_P is the constant that sets the zero point of the instrumental magnitude scale. The quantity m_P, the **instrumental magnitude outside the atmosphere**, depends on the telescope and photometer but is independent of the atmosphere. The quantity m^O_P is the instrumental magnitude in the emitted frame corrected for all absorption effects.

We can write m_P as

$$m_P = -2.5\log\int T_P(\lambda)\frac{\phi(\lambda)}{\lambda}d\lambda + C_P$$

Here, $T_P(\lambda)$ and C_P characterize the instrumental system located outside the atmosphere.

10.5.4 Photometric data: reduction strategy

The idea now is to remove all the effects outlined above and reconstruct the original stream of photons. Assume we have array data from a ground-based observatory. The data reduction steps will be:

1. Preprocess images to remove instrumental effects: non-linearities, sideband effects, dark, bias, and flat field. Correct for charge-transfer efficiency (CTE), fringing, and geometric effects like scale variation as needed. If appropriate, process (combine, shift and add, drizzle) to improve the SNR, and to remove flaws like bad pixels, mosaic gaps, and cosmic rays.

2. Perform digital aperture or area photometry on the processed images. This will measure the instrumental magnitude inside the atmosphere, m^A_P.

Fig. 10.12 Absorption geometries. (a) A plane-parallel slab. We assume the top of the atmosphere is at $h = 0$. Note that $ds = \sec(z)dh$, where z is the local zenith angle Figure (b) illustrates the fact that lower layers are more important in a spherical atmosphere. Figure (c) shows that the angle z increases with depth in a spherical shell. Refraction effects have been ignored.

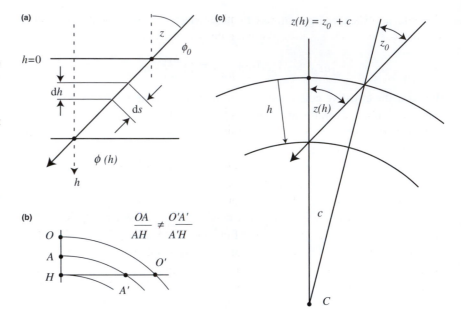

3. Remove the effects of atmospheric absorption: compute the instrumental magnitudes outside the atmosphere, m_P.
4. Transform instrumental magnitudes and indices to a standard system, m_P^{STD}, if needed.
5. Derive astrophysical and astronomical corrections and parameters, as needed:
   (a) corrections for absorption: A_{ism}, A_{exg};
   (b) the correction for wavelength shift, C_P^z;
   (c) astrophysical conditions of the source: temperature, metallicity, stellar population age, distance, diameter, etc.

We described the first two steps in Chapter 9. Step 5 is beyond the scope of an introductory text, so we will concern ourselves now with steps 3 and 4.

10.6 The atmosphere

After the basic preprocessing and aperture photometry, the ground-based photometrist may wish to remove, or at least minimize, the effects of absorption by the terrestrial atmosphere. Under some circumstances, this is an impossible task. In other cases, astronomers can be confident that their inside-the-atmosphere measurements will yield excellent estimates of the outside-the-atmosphere values.

10.6.1 Absorption by a plane-parallel slab

Figure 10.12a shows a stream of photons traversing a horizontal slab of absorbing material (air, for example). The photons travel at angle z with respect to the vertical. We assume that the density and absorbing properties of the material change with h, the depth in the material, but are independent of the other coordinates. We assume that if a flux of $\phi'(\lambda, h)$ travels over a path of length ds, the material will absorb a certain fraction of the photons. We write this absorbed fraction as

$$\frac{d\phi'(\lambda, h)}{\phi'(\lambda, h)} = -\alpha(\lambda, h)ds = -\sec(z)\alpha(\lambda, h)dh \qquad (10.8)$$

where we introduce the function $\alpha(\lambda, h)$ to describe the absorption per unit distance. We can apply this result to the Earth's atmosphere by identifying z as the zenith angle of the source. However, the geometric and optical properties of the real, spherical atmosphere mean that the value of the ratio $d\phi'/\phi'$ is more complicated than the extreme right-hand side of Equation (10.8) suggests. There are two effects involved. First, as you can see from Figures 10.12b and c, because the atmosphere has spherical rather than plane symmetry, the angle z is not a constant, but is an increasing function of h. Second, the actual angle at any height will be even greater than that given by the spherical model because of atmospheric refraction. Taking both effects into account and assuming we have an observatory at depth H in the atmosphere, the solution to Equation (10.7) is

$$\phi_A(\lambda) = \phi(\lambda)e^{-\int_o^H \sec(z(h))\alpha(h)dh} = \phi(\lambda)e^{-\tau(\lambda, H)X} \qquad (10.9)$$

Here $\phi_A(\lambda)$ and $\phi(\lambda)$ are the monochromatic photon fluxes inside and outside the atmosphere, respectively. We introduce two new functions on the right-hand side of Equation (10.9). First, the *optical depth at the zenith*:

$$\tau(\lambda, H) = \int_0^H \alpha(h)dh$$

Physically, this definition implies that the monochromatic brightness at the zenith changes by the factor $S_{atm} = \exp(-\tau)$ due to absorption. Second, we introduce the concept of the *air mass*:

$$X(\lambda, z) = \frac{1}{\tau(\lambda)} \int_0^H \sec(z(h))\alpha(h)dh \approx X(z) \approx \sec(z)$$

The air mass along a particular line of sight is a dimensionless quantity. It tells how much more absorbing material lies along that line than lies towards the zenith. The approximation $X = \sec(z(H))$ is good for small zenith angles. (The error is less than 1% for $z < 70°$, corresponding to an air mass of less than 3.) For larger zenith distances, the formula

$$X(z') = \sec(z')[1 - 0.0012(\sec^2 z' - 1)]$$

is a much better approximation. Here z' is the "true" zenith angle – the angle, $z(h = 0)$, between the observer's vertical and the optical path outside the atmosphere – which can be computed from the object coordinates and the sidereal time.

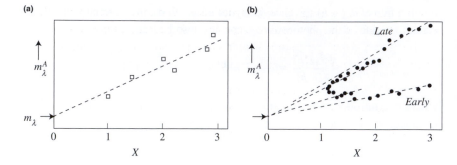

10.6.2 Bouguer's law

From Equation (10.9), we can represent the monochromatic magnitude on the instrumental scale as

$$m_\lambda^A = -2.5 \log\left[\frac{hc}{\lambda}\phi_A(\lambda)\right] = -2.5 \log\left[\frac{hc}{\lambda}\phi(\lambda)\right] + 2.5\tau(\lambda)X(z)\log(e)$$

$$m_\lambda^A = m_\lambda + 1.086\tau(\lambda)X$$

We have omitted the constant for magnitude zero-point definition; m_λ^A is the magnitude as observed inside the atmosphere, and m_λ is the magnitude in the same system outside the atmosphere. Finally, we define the ***monochromatic extinction coefficient***, $k(\lambda) = 1.086\tau(\lambda)$, and rewrite the previous equation as

$$m_\lambda^A(X) = m_\lambda + k(\lambda)X \qquad (10.10)$$

This expression, which states that the apparent magnitude is a linear function of air mass, is known as **Bouguer's[3] law** (or sometimes, Lambert's law). Bouguer's law suggests the method for determining the value of the extinction coefficient, and thus a method for converting apparent magnitudes inside the atmosphere to magnitudes outside the atmosphere. The astronomer simply measures the brightness of some steady source (the ***extinction source***) at at least two different air masses – then, in a plot of magnitude as a function of air mass, Equation (10.10) tells us that the slope of the straight-line fit is $k(\lambda)$ and the y-intercept is m_λ; see Figure 10.13. Once she knows $k(\lambda)$, the astronomer can compute outside-the-atmosphere magnitudes for any other stars by making a

---

[3] Pierre Bouguer (1698–1758), a French Academician, was celebrated in his day for leading an expedition to Peru in 1735 to measure the length of a degree of latitude. The expedition conclusively demonstrated Newton's hypothesis that the Earth was oblate. Bouguer derived his law for atmospheric absorption by investigation of the general problem of light transmission through a medium. He also holds the distinction of being the first quantitative photometrist in astronomy – in 1725 he measured the relative brightnesses of the Sun and Moon by comparison to a candle flame.

single observation and applying Bouguer's law. With one powerful and elegant stroke, the astronomer has removed the absorbing effects of the atmosphere.

The power and elegance of Bouguer's law depends on the persistence of two conditions during the time over which observations are made:

(1) that $k(\lambda)$ is stationary – does not change over time,
(2) that $k(\lambda)$ is isotropic – does not change with location in the sky.

If both these conditions hold, observers will say that the atmosphere is **photometric** and feel powerful and capable of elegance. If the conditions are violated (visible clouds are one good indication), observers will recognize that certain kinds of photometry are simply impossible. There are intermediate situations – Figure 10.13b shows observations in which condition (1) is violated – extinction here changes gradually over several hours. As long as the changes are carefully monitored, the astronomer can still hope to recover the outside-the-atmosphere data via Bouguer's law.

Condition (2) is always violated because of the spherical nature of the atmosphere: absorption by lower layers becomes relatively more important at large zenith angles (Figure 10.12b), and total extinction as well the extinction versus wavelength function will change. This effect is not significant at smaller (< 3) air masses, so usually can be (and is) ignored. In general, it is a good idea to avoid *any* observations at very large air masses – the likelihood of encountering non-uniformities is greatly increased, as are all other atmospheric effects like seeing distortions, differential refraction, and background brightness levels.

10.6.3 Sources of extinction

Figure 10.14 plots $k(\lambda)$ for a typical clear (cloud-free) sky in the 0.3–1.4 μm region. As illustrated, the value is the sum of contributions from four different

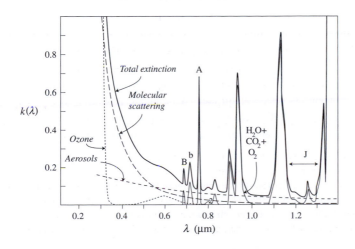

Fig. 10.14 A model for the contributions to the extinction coefficient. Aerosol and water-vapor absorption are highly variable. This is a low-resolution plot, so band structure is smoothed. Letters mark strong telluric Fraunhoffer features and the photometric J-band window.

processes, each of which has a characteristic spectral dependence. The processes are:

Rayleigh scattering by molecules. In this process, a photon encounters a molecule of air and is redirected. The probability of scattering is much greater for short-wavelength photons (for pure Rayleigh scattering, extinction is proportional to λ^{-4}). Molecular scattering explains why the sky is blue, since multiply scattered photons from the Sun will tend to be those of shortest wavelength. Molecular scattering is a component of the extinction that is stable over time, and its magnitude scales directly with the atmospheric pressure – higher altitudes will have more transparent skies.

Absorption by ozone. Continuous absorption by the O_3 molecule in the ultraviolet essentially cuts off transmission shortward of 320 nm. Ozone also absorbs less strongly in the Chappuis bands in the visible near 600 nm. Atmospheric ozone is concentrated near the stratopause at around 48 km above sea level, so the benefit of high-altitude observatories is limited with respect to ozone extinction. Ozone abundance is subject to seasonal and global variations, but does not appear to vary on short time scales.

Scattering by aerosols. Aerosols are suspensions of small solid or liquid particles (particulates) in air. Particulates range in diameter from perhaps 50 μm down to molecular size. Aerosol particulates differ from water cloud drops by their much longer natural residence time in the atmosphere. In fact, the way most aerosols are removed is by reaction with water droplets in clouds[4] and subsequent precipitation. Several different processes inject particulates into the atmosphere. Sea spray and bursting bubbles introduce salt. Winds over deserts introduce dust. Volcanoes inject ash and sulfur dioxide (a gas that interacts with water vapor to form drops of sulfuric acid). Burning fossil fuel and biomass introduce ash, soot, smoke, and more sulfur dioxide. The wavelength dependence of aerosol scattering depends largely on the size of the particle, and the typical wide range of sizes present (salt particles tend to be large; smoke particles, small) usually produces a relatively "gray" extinction (a λ^{-1} dependence is typical). A pale-blue (rather than deep-blue) sky indicates high aerosol extinction. Sometimes aerosols can produce striking color effects, including the lurid twilight colors from stratospheric volcanic ash and the "green sky" phenomenon due to Gobi Desert dust. Aerosol scattering can be quite variable, even on a short time scale, and different components reside at different atmospheric levels. Although salt, dust, and industrial pollution mainly stay in the lower layers (a scale height of 1.5 kilometers is representative), some volcanic eruptions and intense forest fires can

---

[4] Aerosol particles are crucial to the *formation* of water clouds – water vapor condenses into droplets much more readily if aerosols provide the "seed" surfaces on which condensation can proceed. Without such seeds, very clean air can supersaturate and reach a relative humidity well over 100%.

inject aerosols into the stratosphere, where they may persist for weeks or even
years, since there is no rain in the stratosphere.

Molecular-band absorption. The main molecular *absorbers* are water vapor
and carbon dioxide, although oxygen has a few relatively narrow features,
and we have already discussed the ozone bands. Water vapor and CO_2 bands
demarcate the relatively transparent windows in the near and middle infrared.
Carbon dioxide is well mixed with altitude in the atmosphere, but water vapor
is concentrated near the surface and varies with temperature, time, and location. At sea level, the amount of vapor in one air mass corresponds to about
10 mm of liquid, on average. On Mauna Kea, one of the best conventional sites,
the average is about 1 mm of precipitable water. At the south pole, which
benefits from both high elevation and low temperature, values approach
0.15 mm. Stratospheric observatories carried by balloons or aircraft enjoy even
lower values.

10.6.4 Measuring monochromatic extinction

Ground-based astronomers use a variety of methods for removing the effects of
atmospheric extinction. We look at a few cases here, and then examine the
complications introduced by heterochromatic (broadband) systems. To keep
things simple at the start, we assume that we are observing a monochromatic
magnitude and that our instrument is sensitive to exactly the same wavelength as
some standard system.

Case 1: assume a mean extinction
At remote high-altitude sites, the extinction at shorter wavelengths is usually
due almost entirely to Rayleigh scattering, and is therefore stable. Under these
conditions, it may be safe simply to use the average (or better still, the median)
extinction coefficient determined by other observers for the same site over the
past years. This is a particularly reasonable approach if one is doing differential
photometry, or if standard and program stars are observed at nearly the same air
mass at nearly the same time.

Case 2: use known outside-the-atmosphere magnitudes
If you wish to determine the extinction yourself from the Bouguer law, this is
extremely simple method applies – *if* you happen to know m_λ, the magnitude
outside the atmosphere of some constant star. In this case, just point your telescope at the star, note the air mass, X, and take an image. You measure $m_\lambda^A(X)$,
you know m_λ, so just solve Equation (10.10) for $k(\lambda)$. The method becomes
more robust if you know values of m_λ for more than one star.

Case 3: draw the Bouguer line from observations

If you don't know m_λ, then you need to do a little more work and generate the Bouguer line. You take two or more exposures of the same field of stars over a wide range of air masses. This, of course, requires waiting for the zenith distance of your field to change. Many observers record the extinction field every 90 minutes or so. You plot the resulting instrumental magnitudes as a function of air mass, and if the night is **photometric,** you obtain a plot like Figure 10.13a.

Choose your extinction field carefully. With a good choice and a wide-field CCD, you may well have several stars (let's say 30) that yield good SNRs on a short exposure. Your data then will have greater statistical significance than a plot for just a single star in determining $k(\lambda)$. To do the computation of the Bouguer line, you might combine all 30 fluxes at each air mass and fit *one* straight line. You will often find it economical to make the extinction field identical to the program field or to a field containing standard stars.

Case 4: variable extinction and multi-night data

What if the extinction changes? If you observe over many consecutive nights, change is likely. If you are fortunate, the extinction will change slowly over time and uniformly over the entire sky. In this case, some modification of the previous method will yield $k(\lambda, t)$. For example, if you are sure of the outside-the-atmosphere magnitudes of some of your constant stars, then simply monitoring the constant stars will give instantaneous values of $k(\lambda, t)$.

If you do not have the instrumental outside-the-atmosphere magnitudes, there is still hope: if you have the *standard* magnitudes of two constant stars, observing them at different air masses measures $k(\lambda, t)$. The usual practice is to take one frame containing a standard star near the meridian (the "D" frame), and then immediately take a frame containing the second standard at large air mass in the east (the "M" frame). Provided that the extinction coefficient is the same in all directions, Bouguer's law gives the difference in the instrumental magnitudes of M and D, measured inside the atmosphere, as

$$\Delta m_{\mathrm{MD}}^{\mathrm{A}} = m_{M\lambda}^{\mathrm{A}}(X_M) - m_{D\lambda}^{\mathrm{A}}(X_{\mathrm{D}}) = m_{M\lambda} - m_{D\lambda} + k(\lambda, t)(X_{\mathrm{M}} - X_{\mathrm{D}})$$
$$\Delta m_{\mathrm{MD}}^{\mathrm{A}} = \Delta m_{\mathrm{MD}} + k(\lambda, t)\Delta X_{\mathrm{MD}}$$

For monochromatic magnitudes, the magnitude difference in the instrumental system should be equal to the difference in the standard system, i.e. $\Delta m_{\mathrm{MD}}^{\mathrm{STD}} = m_{M\lambda} - m_{D\lambda}$, so that we can write

$$k(\lambda, t) = \frac{\Delta m_{\mathrm{MD}}^{\mathrm{A}} - \Delta m_{\mathrm{MD}}^{\mathrm{STD}}}{\Delta X_{\mathrm{MD}}}$$

All the quantities on the right-hand side are either known or measured. Because of the possibility of extinction change, many photometrists adopt the strategy of observing "MD" pairs every few hours through the night.

Case 5: use all the data

The most general methods make use of all available information, and include data for all nights in which sources are observed in common. Every frame taken during the run of several nights is affected by extinction and therefore contains information *about* extinction. One approach might work as follows: derive values for $k(\lambda)$ from the best nights – those for which it is possible to make good linear fits to the extinction data – and compute the outside-the-atmosphere magnitudes for *every* constant star (not just those used for the fits). You then should have a large set of extra-atmospheric magnitudes that you can use to find the extinction as a function of time for the more marginal nights. Cloudy nights in which the extinction changes rapidly will be suitable only for differential work. The extinction problem is well suited to a least-squares solution with constraints imposed by standard stars. See the discussion in chapter 10 of Sterken and Manfroid (1992) for a good introduction.

10.6.5 Heterochromatic extinction

The previous discussion strictly applies only for monochromatic magnitudes. For bandpass magnitudes, we must rewrite Equation (10.10) as

$$k_{\mathrm{P}}X = m_{\mathrm{P}}^{\mathrm{A}} - m_{\mathrm{P}} = -2.5 \log\left\{ \frac{\int T_{\mathrm{P}} f_\lambda S_{\mathrm{atm}} \mathrm{d}\lambda}{\int T_{\mathrm{P}} f_\lambda \mathrm{d}\lambda} \right\}$$

or

$$k_{\mathrm{P}}X = -2.5 \log\left\{ \frac{\int T_{\mathrm{P}} f_\lambda \exp\left[\frac{-1}{1.086} k(\lambda) X\right] \mathrm{d}\lambda}{\int T_{\mathrm{P}} f_\lambda \mathrm{d}\lambda} \right\} \qquad (10.11)$$

where k_{P} is the extinction coefficient for band P. In general, the function on the right-hand side of Equation (10.11) is *not* linear in X, and will depend strongly on the shape of the function f_λ. We should expect, therefore, that it must be the case that $k_{\mathrm{P}} = k_{\mathrm{P}}(X, SpT)$, where the variable SpT indicates the spectral shape of the object observed. For most narrow-band and some intermediate photometry, the variation of k_{P} with X and SpT is so small that it can safely be ignored. For wider bands, however, this is not the case.

Think of the variation in k_{P} as due to two different but related effects. First, as we had seen earlier, the effective and isophotal wavelengths of a bandpass depend on the spectrum of the source. This effect is present even in outside-the-atmosphere photometry. We should expect that the extinction measured, say, for a red star will differ from the extinction measured for a blue star, since the center of the bandpass is different for the two. Because of this effect, the Bouguer plot of apparent magnitude versus air mass will give straight lines of different slopes for stars of different spectral shapes. The second, more

Fig. 10.15 The Forbes effect. (a) A model source spectrum, bandpass transmission T_{BP}), and extinction coefficient. The extinction is due entirely to a strong feature near the blue edge of the band. (b) The flux actually detected at the telescope is shown as a function of wavelength for four different air masses. Note the relatively small change between $X = 1$ and $X = 2$ (shaded regions) compared to the change between $X = 1$ and $X = 0$. (c) The Bouguer diagram for the data in (b), illustrating the non-linear relationship and the difference between the actual extra-atmospheric magnitude (filled circle) and the intercept of a linear fit to observable data (open circle).

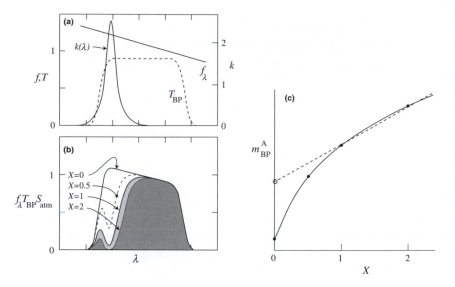

invidious problem arises because atmospheric extinction itself changes the shape of the spectrum that reaches the telescope. This effect – called the **_Forbes effect_** – means you actually observe different spectra for the same star at different air masses. (Alternatively, you can think of the effect as changing the shape of the bandpass response as air mass changes.) A Bouguer plot of apparent magnitude versus air mass will therefore give a _curved_ line. The Forbes effect, as illustrated in Figure 10.15, is particularly problematic if strong atmospheric absorption affects some parts of the photometric band more than others. This is the case for Johnson U and many of the wider infrared bands (review Figures 10.1 and 10.3). In such cases, the magnitude change in going from $X = 2$ to $X = 1$ can be considerably less than in going from $X = 1$ to $X = 0$. In some cases, both the width of the band and its effective wavelength can change dramatically at the smaller air masses. Use of outside-the-atmosphere magnitudes when the Forbes effect is present is a little tricky, and sometimes depends on having a good model of the response function, the atmosphere, and the unknown source. For precise work, therefore, it is best to use bands that exclude strong atmospheric absorptions (e.g. the MKO near-infrared system).

10.6.6 Second-order extinction coefficients

A solution for Equation (10.11) would be possible if we could make a good approximation of both the monochromatic extinction function $k(\lambda)$ and the shape of the spectrum f_λ, either numerically, or perhaps with a Taylor series expansion of each function. The required functions or their derivatives are rarely known, so what is usually done is to assume that the photometric color, which gives some sense of the shape of the source spectrum, will account for most variations in the broadband extinction, and write

$$k_P = k'_P + k''_P \cdot (ci) \tag{10.12}$$

where (ci) represents some "appropriate" photometric index like $(B-V)$, and k''_P is called the **second-order extinction coefficient**. For example, the coefficient for the Johnson V band might be written as

$$k_V = k'_V + k''_{V,\,BV} \cdot (B-V)$$

The color in this case is $(B-V)$, but might also have been chosen to be $(V-R)$ or $(V-I)$. It is also quite common to use the instrumental colors instead of the standard colors, since instrumental colors will be available for many more extinction stars. Instrumental magnitudes are usually written in lower case for the Johnson–Cousins system, so the instrumental color in the above equation, for example, would be $(b-v)$. For broadband extinction, therefore, one has for the ith observation of star j

$$m^A_{P,i,j} = m_{P,j} + \left(k'_P + k''_P \cdot (ci)\right) X_{i,j} \tag{10.13}$$

Various approaches can then lead from the data (the $(X_{i,j}, m^A_{P,i,j})$ pairs) to values for the outside-the-atmosphere instrumental magnitudes, $m_{P,j}$, and for the two extinction coefficients, k'_P and k''_P. The second-order coefficient is difficult to determine from a few nights' observations. Fortunately, this coefficient should not vary much if the primary sources of extinction do not include absorption bands due to water vapor. It is common practice, therefore, to use mean values – established over time for a particular instrument and site – for the second-order coefficients.

All the cases discussed in the previous section on monochromatic data can be applied to correcting heterochromatic data for extinction, with the understanding that a second-order extinction coefficient may need to be determined. Also, understand that if you use an outside-the-atmosphere color index in Equation (10.13), you have not accounted for the Forbes effect at all. If the Forbes effect is severe, the form of Equation (10.13) is probably inadequate. You may then need to fit the extinction data with a function of X whose form has been derived from a model (similar to Figure 10.15) of how the flux in the bandpass changes with air mass. This requires detailed knowledge of both the bandpass shape and absorption behavior, as well as the spectrum of the source.

10.6.7 Indices or magnitudes?

The traditional method for reporting n-color photometric data is to give one magnitude and $n-1$ indices. It has also been traditional to make the extinction computations not for n magnitudes, but for one magnitude and $n-1$ indices.

Suppose, for example, we observe in the Johnson B and V bands. If we have a B frame and a V frame taken at about the same air mass, we can write Equation (10.13) once for each band, and then subtract the equations, yielding

$$(b - v)^A = (b - v) + \left\{ (k'_B - k')_V + \left(k''_{B,BV} - k''_{V,BV} \right) \cdot (B - V) \right\} X$$

Combining the coefficients, we have

$$(b - v)^A = (b - v) + \left\{ k'_{B-V} + k''_{B-V} \cdot (B - V) \right\} X \tag{10.14}$$

Here the new first- and second-order coefficients describe the effects of extinction *on the index*. There is an objective reason for analyzing extinction data via Equation (10.14) rather than via Equation (10.13): if either instrument sensitivity or atmospheric aerosol extinction drifts during an observing run, the effect on the observed color indices will be minor compared to the effect on the individual magnitudes.

This reasoning is less compelling in modern observing. For one thing, CCDs and infrared arrays are much less prone to sensitivity drift than photomultipliers. Moreover, the requirement that all instrumental bands be observed at the same air mass is somewhat restrictive, and may prevent the use of all the available extinction data.

10.7 Transformation to a standard system

Assume we have observed an outside-the-atmosphere instrumental magnitude for some source, either by correcting a ground-based observation for extinction or by direct observation with a space telescope. If we wish to compare our result with those of other observers, we must all use the same photometric system. Almost always, this means executing the next step in the photometric reduction procedure, the transformation from the instrumental to the standard system. This **transformation** will depend on (1) differences between the response of our instrumental system and that of the standard system and (2) the shape of the spectrum of the source.

10.7.1 The monochromatic case

Suppose we observe with indefinitely narrow-band filters whose central wavelengths, λ_1 and λ_2, may be slightly different from the standard wavelengths, λ_{S1} and λ_{S2}. The difference between the first standard magnitude and our instrumental result is

$$m^{STD}_{\lambda_{S1}} - m_{\lambda_1} = -2.5[\log f_{\lambda_{S1}} - \log f_{\lambda_1}] + C^{STD}_1 - C_1 \tag{10.15}$$

The last two terms in Equation (10.15) quantify the difference between the efficiency of our telescope/photometer and the standard instrument. These will be the same for every source observed. The first term on the right, in brackets, you will recognize as something like a color index. This prompts us to rewrite (10.15) as:

$$m_{\lambda_{S1}}^{STD} \approx m_{\lambda_1} + \alpha_{12}(m_{\lambda_1} - m_{\lambda_2}) + \alpha_1 = m_{\lambda_1} + \alpha_{12}(ci)_{12} + \alpha_1 \qquad (10.16)$$

In this equation, α_{12} is called the **color coefficient** for the transformation and α_1 is called the **zero-point constant**. Since indices are more closely related to physical variables than are individual magnitudes, in an n-band system, astronomers traditionally work with one magnitude transformation and $n - 1$ index transforms, of the form

$$(m_1 - m_2)^{STD} \approx m_{\lambda_1} - m_{\lambda_2} + (\alpha_{12} - \alpha_{21})(m_{\lambda_1} - m_{\lambda_2}) + \alpha_1 - \alpha_2$$

Redefining constants and simplifying the notation slightly:

$$(ci)_{12}^{STD} \approx \gamma_{12}(ci)_{12} + \beta_{12} \qquad (10.17)$$

We can justify Equations (10.16) and (10.17) if the first term of Equation (10.15) is a linear function of some color index of the instrumental system (or of the standard system). From Figure 10.16, you can see that Equation (10.16) is valid only if $(ci)_{12}$ measures the slope of the logarithmic spectrum *between λ_1 and the standard wavelength*, λ_{S1}. Unfortunately, at the resolution of narrow-band photometry, many spectra have almost discontinuous features like lines. Many narrow-band systems may have indices that give a good value for the local derivative at each of the standard wavelengths. Thus, the relation in Equation (10.16) is rarely exact.

  In practice, astronomers almost always assume a transformation equation like (10.16) or (10.17) but will use empirical methods to establish the color coefficient and zero point, or even the functional form for the transformation itself. The function might include a higher-order color term, or use some other photometric criterion to improve the accuracy of the transformation.

  The general approach is simple: you measure m_{λ_1} and $(ci)_{12}$ for a range of standards whose spectra you expect to resemble those of the unknown sources under investigation. Your plot of $(m_{\lambda_1}^{STD} - m_{\lambda_1})$ as a function of $(ci)_{12}$, as in Figure 10.17, then establishes a relation like Equation (10.16). In the example in the figure, the data invite a quadratic fit, so you would add a term with a **second-order color coefficient** to the transformation equation. Another reasonable empirical fit might use two different linear relations, one for lower and another for higher values of the color index. If a grid of standards establishes the transformation, then the transformation can introduce errors if the spectrum of an unknown source is sufficiently different from any element of the grid.

Fig. 10.16 Color transformation. Narrow-band measurements at wavelengths 1 and 2 can predict the monochromatic magnitude at the standard wavelength (filled circle at wavelength S) if the color index measures the slope between wavelengths S and 1. If the (ci) does not measure the slope – as in wavelengths 1 and 3, an error results.

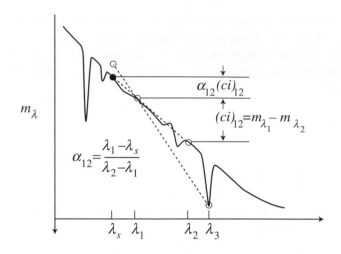

Fig. 10.17 Empirical determination of the transform coefficients from a plot of standard minus instrumental magnitudes as a function of color index. This data could be fitted with a quadratic function (A) or with two different linear functions: (B) for $ci < w$ and (C) for $ci > w$.

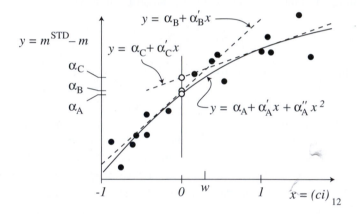

10.7.2 The heterochromatic case

The problem of transformation to a broadband standard system is similar to the monochromatic case: we just replace the fluxes in Equation (10.15) with the appropriate integrals:

$$m_P^{STD} - m_P = -2.5 \left[\log \int T_{P,STD} f_\lambda d\lambda - \log \int T_P f_\lambda d\lambda \right] + C_P^{STD} - C_P$$

Except for the zero-point terms, the transformation to the standard system will reflect differences between the instrument response of your system, T_P, and the instrument response of the standard system, $T_{P,STD}$. In the broadband case, differences can be due to any differences in the detailed shape of the response function, especially those resulting in different central wavelengths and bandwidths. Because a broadband system examines a smoothed version of the spectrum, and because the bands tend to be close together relative to the scale of the smoothing, the approximation of Equation (10.16) will tend to be better in broadband than in most narrow-band systems.

Procedures are similar to those discussed for the monochromatic case: you plot, as a function of color, the difference between instrumental and standard magnitudes for a set of standards that spans the range of spectra you expect in the unknowns. If the functions T_P and $T_{P, STD}$ are identical, you will find a horizontal line at the level of the zero-point correction. If the functions differ (the usual case!), you fit the data to find the color term or terms in the transformation. The fit may be a straight line, or may require higher-order terms, and will apply safely only to those kinds of spectra sampled by the standard stars you have observed.

Summary

- The history of photometry has imposed the magnitude scale and the definition of several important broadband photometric systems.

- Photometric bandpass response functions are generally categorized as broad-, intermediate-, or narrow-band. A response can be implemented by filters, detector sensitivity, atmospheric transmission, or some combination of these. Concepts:

resolving power	*response function*	*photon response function*
high-pass filter	*peak wavelength*	*central wavelength*
mean wavelength	*effective wavelength*	*isophotal wavelength*
FWHM	*bandwidth*	*photon flux*
zero point		*bandpass magnitude*

- Photometric indices, which are linear combinations of bandpass magnitudes, quantify characteristics of an object's spectrum. Concepts:

color index	*blackbody spectrum*	*monochromatic magnitude*
line index	*curvature index*	*feature index*

- A standard photometric system specifies both the response functions of its bands as well as some method for standardizing measurements. Concepts:

open system	*closed system*	*instrumental system*
visual magnitude	*standard sequence*	*north polar sequence*
ci	*international system*	*photovisual magnitude*
UBVRI	*Cousins system*	*JHKLMNQ*
MKO filters	*Strömgren system*	*uvbyβ*
c_1 and m_1		*SDSS system*

- A shift in an object's spectrum caused by the Doppler effect or cosmological expansion will produce a photometric change that can be corrected if an astronomer has sufficient information (the K correction).

(continued)

Summary (*cont.*)

- Conversely, the color change observed in a very distant object can lead to an estimate of its redshift.

- Absorption by material outside the atmosphere can produce both reddening and absorption lines and bands in a spectrum.

- Absorption by material inside the atmosphere can produce both reddening and telluric absorption lines and bands. Concepts:

  optical–infrared window   *microwave–radio window*
  instrumental magnitude   *magnitude outside the atmosphere*

- Photometric data reduction proceeds in steps: (1) preprocessing, (2) digital photometry, (3) atmospheric extinction correction, (4) transformation to a standard system, and (5) further corrections and analysis.

- Bouguer's law is the basis for the correction for atmospheric extinction:

$$m_\lambda^A(X) = m_\lambda + k(\lambda)X$$

Concepts:

optical depth	*air mass*	*extinction coefficient*
ozone bands	*molecular bands*	*Rayleigh scattering*
aerosols	*Bouguer line*	*mean extinction*
second-order	*monochromatic*	*heterochromatic*
extinction	*extinction*	*extinction*
Forbes effect		

- Transformation to the standard system requires observation of standard objects using instruments identical to those used for the unknowns. Concepts:

  zero-point constant   *standard star/extinction star*
  color coefficient   *second-order color coefficient*

Exercises

1. Show that for a response function with a boxcar or triangular profile, the bandwidth = FWHM, but that for a Gaussian, the bandwidth < FWHM.

2. The table below gives the response function for a photometric bandpass, as well as the flux distributions for two sources. Characterize this system by computing (use a spreadsheet) all of the following:

   wavelength at peak transmission
   the FWHM
   bandwidth

mean wavelength

effective wavelength for each source

isophotal wavelength for each source

λ (nm)	$R_{BP}(\lambda)$	f_λ^A	f_λ^B
500	0	1.70	0.37
505	0.04	1.56	0.47
510	0.24	1.43	0.57
515	0.41	1.31	0.67
520	0.5	1.20	0.78
525	0.55	1.10	0.89
530	0.64	1.00	1.00
535	0.77	0.92	1.12
540	0.88	0.84	1.24
545	0.96	0.77	1.37
550	0.99	0.70	1.50
555	1	0.64	1.64
560	0.81	0.57	1.78
565	0.5	0.52	1.92
570	0	0.46	2.07

3. A photometer on a spacecraft employs a grating-and-slot arrangement (see Figure 10.1) such that all radiation with wavelength between 2.0 and 4.0 µm is detected.

   (a) Assume the detector is a perfect bolometer, so that 50% of the energy between 1.0 and 3.0 µm is detected, independent of wavelength. In other words, $R_{BP}(\lambda)$ is a "boxcar" with mean wavelength 3.0 µm and bandwidth 2.0 µm. Compute the effective wavelength of this band for a hot star with $f_\lambda = A\lambda^{-4}$.

   (b) Now assume the detector is replaced with an infrared photon detector with uniform quantum efficiency such that 50% of the incident photons at each wavelength in the band are detected. Again, compute the effective wavelength of this band for a hot star with $f_\lambda = A\lambda^{-4}$. Note that you will need to devise an expression for the energy response function of the system.

4. Show that Equations (10.3) and (10.4) follow from the Rayleigh–Jeans and Wien approximations to the Planck law.

5. Gabriel very carefully constructs a filter for his CCD photometer so that the response function matches the standard bandpass of the Johnson V color very precisely. He observes two very well-established standard stars whose catalog data are given below. Gabriel discovers that with his CCD, no matter how carefully he observes, he always finds one star is brighter than the other: its image always contains more total analog-to-digital units (ADUs) on the CCD. Liz suggests to him that this is

because the CCD is a photon-counting device. (a) Explain her reasoning. (b) If Liz is correct, which star should be the brighter on the CCD and why?

	V	B – V
Star 1	9.874	0.058
Star 2	9.874	0.861

6. An astronomy student obtains two images of a galaxy, one in the B band, the other in the V band. Outline the image arithmetic operations the student would execute in order to produce a map of the $(B - V)$ color index for the galaxy. Failure to subtract the constant background sky for each image would cause problems in the map. For which parts of the map would these problems be most serious? On the other hand, would subtracting the background sky introduce any problems in the map? If so, which parts, and why?

7. Investigate the website for the Sloan Digital Sky Survey. In what ways is the SDSS five-color system superior to the UBVRI system?

8. An MOS capacitor observes two sources in the band 400–600 nm. Source A has a spectrum such that the distribution of photons in the 400–600 nm band is given by $n_A(\lambda) = A\lambda^3$. Source B has a distribution of photons given by $n_B(\lambda) = B\lambda^{-2}$ in the same band. If the two sources generate photoelectrons at exactly the same rate, compute their brightness ratio. You may assume the detector's quantum efficiency is not a function of wavelength.

9. Speculate, in terms of the Forbes effect, why it might be useful to define the standard magnitude as one measured at 1 air mass, rather than at zero air mass. What difficulties might be inherent in this choice?

10. An observer uses the B and V filters to obtain four exposures of the same field at different air masses: two B exposures at air masses 1.05 and 2.13, and two V exposures at air masses 1.10 and 2.48. Four stars in this field are photometric standards. Their standard magnitudes are given in the table below, as are the instrumental magnitudes in each frame.

	(B – V)	V	b(1)	b(2)	v(1)	v(2)
Air mass			1.05	2.13	1.10	2.48
Star A	–0.07	12.01	9.853	10.687	8.778	9.427
Star B	0.36	12.44	10.693	11.479	9.160	9.739
Star C	0.69	12.19	10.759	11.462	8.873	9.425
Star D	1.15	12.89	11.898	12.547	9.522	10.002

(a) Compute the extinction coefficients for the instrumental system: k'_b, k''_b, k'_v, and k''_v. Hint: at each air mass, Equation (10.10) holds. Write an equation for the difference between the magnitudes at the two air masses (e.g. an equation for $b(2) - b(1)$). Examination of this equation will suggest a method for computing the coefficients. You may find it helpful to enter the data from the table into a spreadsheet in performing the computations.

(b) Compute the instrumental magnitudes of each star at zero air mass

(c) Compute the transformation coefficients, α_V, α_{B-V}, β_{B-V}, and γ_{B-V}, using the method outlined in Section 10.7.

11. A photometric bandpass whose response function is shown in curve A below measures the strength of the emission feature shown in curve B. In the figure, the source has zero velocity. Compute the change in the brightness measurement, in magnitudes, that would result if the source were given a radial velocity of 300 km s^{-1}.

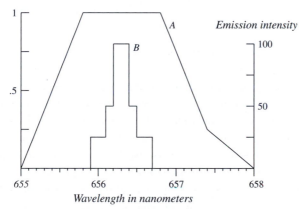

Chapter 11
Spectrometers

The dark D lines in the solar spectrum allow one therefore to conclude, that sodium is present in the solar atmosphere.

— Gustav Kirchhoff, 1862

This news [Kirchhoff's explication of the Fraunhofer solar spectrum] was to me like the coming upon a spring of water in a dry and thirsty land. Here at last presented itself the very order of work for which in an indefinite way I was looking — namely to extend his novel methods of research upon the sun to the other heavenly bodies.

— William Huggins, 1897

Beginning in 1862, Huggins used a spectroscope to probe the chemical nature of stars and nebulae. Since then, spectrometry has been the tool for the observational investigation of almost every important astrophysical question, through direct or indirect measurement of temperature, chemical abundance, gas pressure, wavelength shift, and magnetic field strength. The book by Hearnshaw (1986), from which the above quotes were taken, provides a history of astronomical spectroscopy prior to 1965. Since 1965, the importance of spectroscopy has only increased. This chapter introduces some basic ideas about spectrometer design and use. Kitchin (1995, 2009) and Schroeder (1987) give a more complete and advanced treatment, and Hearnshaw (2009) provides a history of the actual instruments.

Literally, a ***spectroscope*** is an instrument to look through visually, a ***spectrometer*** measures a spectrum in some fashion, and a ***spectrograph*** records the spectrum. Astronomers are sometimes particular about such distinctions, but very often use the terms interchangeably. This chapter introduces the basics of the design and operation of spectrometers in astronomy. We confine our discussion to the class of instruments that use dispersive elements, and examine those elements in detail: prisms, surface relief gratings, and volumetric phase holographic gratings. Some useful spectrometers operate without a slit or small aperture, but the most commonly used designs employ a slit or optical fiber aperture. Accordingly, we will give most of our attention to the important design parameters for the slit/fiber spectrometer, and to the special requirements astronomical applications impose on the design. Finally, we will discuss the particular problem of turning raw spectroscopic data, usually in the form of a digital image, into useful astrophysical information.

11.1 Dispersive spectrometry

We note at the outset that there are two methods for generating spectra: dispersing light of different wavelengths into different directions, and analyzing the wavelength distribution of light without such dispersion. We will not treat non-dispersive spectroscopy, but refer to the discussion of Fabry–Perot and Michelson interferometers in chapter 4 of Kitchin (2008) for a quantitative discussion.

Figure 11.1a shows a rudimentary dispersive spectrometer. An abstract telescope–spectrograph combination – represented by a featureless "box" – accepts a heterochromatic ray, which we assume to be on the optical axis, and disperses it so that photons or rays of wavelength λ are sent in direction θ, while those of wavelength $\lambda + \mathrm{d}\lambda$ are sent in direction $\theta + \mathrm{d}\theta$. The ***angular dispersion*** (a concept introduced for prisms in Chapter 5) is simply $\mathrm{d}\theta/\mathrm{d}\lambda$. Figure 11.1b indicates that a useful spectrometer must bring all the rays of wavelength λ to the same point, P, at object distance s_c, where a detector can measure their intensity. Waves of wavelength $\lambda + \mathrm{d}\lambda$ will focus at a different spot, a distance $\mathrm{d}x$ away on the detector, and the ***linear dispersion*** is defined as

$$\frac{\mathrm{d}x}{\mathrm{d}\lambda} = s_c \frac{\mathrm{d}\theta}{\mathrm{d}\lambda}$$

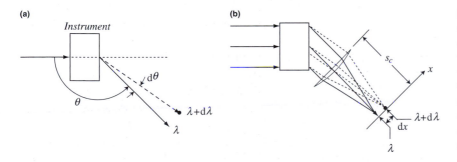

Fig. 11.1 (a) Angular dispersion: light from a distant source enters a telescope–spectrograph which disperses different wavelengths to different directions. (b) Linear dispersion: after dispersion all rays of the same wavelength are brought to a focus at image distance, s_c. Images of wavelength λ and $\lambda + \mathrm{d}\lambda$ are separated by distance $\mathrm{d}x$ in the focal plane.

Astronomers often find the reciprocal of the above quantity to be more intuitive, and may say "dispersion" when the number they quote is actually the ***reciprocal linear dispersion***, or ***plate factor***, p:

$$p = \frac{d\lambda}{dx} = \left[s_c \frac{d\theta}{d\lambda} \right]^{-1}$$

A "high dispersion spectrometer" is one in which the linear dispersion is large and p is small. The units of p are usually nanometers or Ångstroms (of wavelength) per millimeter of distance in the focal plane.

The size of the image produced by perfectly monochromatic rays that focus at point P cannot be indefinitely small. The image will be smeared out in the x direction to a linear width, w_0, a distance that depends on the angular size of the source, on the geometric details of the optics, as well as on other processes that limit optical resolution: diffraction, atmospheric seeing, optical aberrations, and errors in the dispersion process. Dispersion means that the width w_0 corresponds a range of wavelengths, $\delta\lambda_0$. This quantity, called the ***spectral purity of the optics***, measures the spectrometer's ability to resolve details in the spectrum. However, the ***effective spectral purity*** of the complete instrument also depends on the detector's ability to resolve linear detail. If w_d = width of 2.3 detector elements, then

$$\delta\lambda = \text{spectral purity} = \text{the larger of:} \begin{cases} \delta\lambda_0 = w_0 \dfrac{d\lambda}{dx} \\ \delta\lambda_d = w_d \dfrac{d\lambda}{dx} \end{cases}$$

If two emission lines, for example, are closer together than $\delta\lambda$, they will overlap so much that their separate identities cannot be discerned. If we take $\delta\lambda$ to be the full width at half-maximum (FWHM) of the monochromatic instrumental profile, then the definition of resolution rests on the Rayleigh criterion introduced in Chapter 5. Figure 11.2 schematically illustrates the ***instrumental profiles*** (a plot

Fig. 11.2 Instrumental profiles of two emission lines observed at different spectral purities, slit image widths, and plate scales. The tic marks on the horizontal axis indicate the digital resolution (pixel width) in the dispersion direction for critical sampling of the profiles.

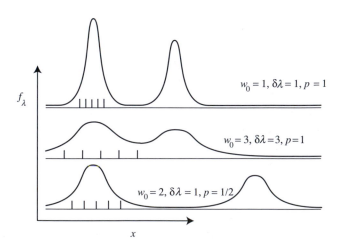

of intensity vs. x or θ) of two monochromatic spectral features and illustrates the distinction between spectral purity and dispersion. Astronomers commonly refer to the spectral purity as the **resolution**. The related dimensionless parameter, the **resolving power** (a concept you may recall from Chapter 3), also measures the spectrometer's ability to record detail in the spectrum:

$$R = \frac{\lambda}{\delta\lambda}$$

Astronomical spectrometers range in resolving power from very low (R = a few hundred) to very high (R = 100,000 or more).

11.2 Dispersing optical elements

11.2.1 Prisms

We have already discussed the angular dispersion of prisms in Section 5.2. We saw that for a prism

$$\frac{\partial\theta}{\partial\lambda} \cong -\frac{4K_2 \sin(A/2)}{\lambda^3 \cos\alpha} \tag{11.1}$$

Here A is the apex angle of the prism, α is the angle of incidence in the minimum deviation configuration, and K_2 is a constant ranging from about 0.003 to about 0.017 μm^{-2}, for various types of glass.

Prisms find use in astronomical spectrographs, but seldom (except in the near infrared) as the primary disperser. Their weight and relative expense are minor disadvantages. More serious are their low transmission in ultraviolet, their low angular dispersion at long wavelengths, and their highly non-linear variation of angular dispersion with wavelength.

11.2.2 The diffraction grating

Despite the name, this disperser depends on the interference of diffracted light waves to produce dispersion. Its simplest form, the **amplitude grating**, is a set of closely spaced parallel lines uniformly arranged across a flat surface (see Figure 11.3). This pattern can be either a series of slit-like openings – a **transmission grating** – or a series of separate, tall, but very narrow, mirrors or facets – a **reflection grating**. To produce appreciable diffraction effects, the widths of the slits or mirrors should be on the order of a few wavelengths, so in the optical, astronomical gratings typically have between 100 and 3000 lines per millimeter.

Figure 11.3 illustrates the principles of the grating. Adjacent grating facets are a distance σ apart – the **groove spacing** or **grating constant**. Gratings are conventionally described by giving the reciprocal of σ, the **groove frequency**, usually quoted in lines per millimeter. In panel (a) of the figure, a reflection grating, a plane wave strikes the grating at angle α and reflects at angle θ. Although there will be a preference for reflected rays to travel in the direction

Fig. 11.3 (a) A reflection grating. Grating facets are tall, narrow mirrors extending perpendicular to the plane of the paper. Light striking between the facets is not reflected. The figure traces three parallel rays that strike the centers of adjacent facets. (b) A transmission grating. See the text for further explanation.

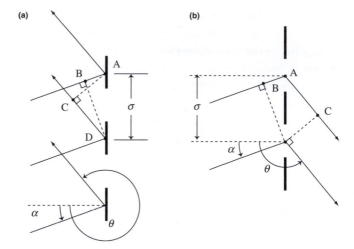

$\theta = -\alpha$ specified by geometrical optics, diffraction effects allow rays to spread out from the narrow facets in all directions. All angles are measured counterclockwise from the grating normal.

Wave interference means that rays of a particular wavelength, λ, will only reflect (diffract) from the grating at particular angles. To see this, consider the two rays in Figure 11.3a that strike the centers of adjacent facets. Both rays strike at angle α and diffract at angle θ. Upon leaving the grating, the optical path lengths of the two rays differ by the amount $\Delta\tau = \overline{AB} - \overline{CD}$. The two waves will constructively interfere only if $\Delta\tau$ is some integral multiple of the wavelength:

$$\Delta\tau = m\lambda, \quad m = 0, \pm 1, \pm 2, \ldots$$

From the figure, we have $\overline{AB} = \sigma \sin \alpha$ and $\overline{CD} = \sigma \sin (2\pi - \theta) = -\sigma \sin \theta$, so for constructive interference we require

$$\sin \theta + \sin \alpha = \frac{m\lambda}{\sigma} \tag{11.2}$$

This rather general result also applies to the transmission grating, and is known as the **grating equation**. The integer m is called the **order**. For a large number of facets, interference effects suppress all diffracted rays that fail to satisfy the grating equation. Unlike the dispersing prism, gratings cause red light to deviate more than blue. Differentiating the grating equation tells us that the angular dispersion of a grating is

$$\frac{\mathrm{d}\theta}{\mathrm{d}\lambda} = \frac{m}{\sigma \cos \theta} \tag{11.3}$$

Since $\cos \theta$ changes only slowly with λ, the angular dispersion of a grating is roughly constant with wavelength, again in contrast to the behavior of prisms. From Figure 11.3, it is clear that high angular dispersions can be achieved by

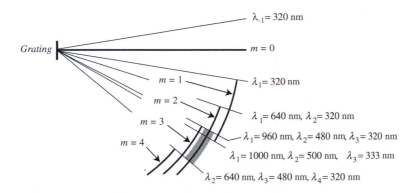

Fig. 11.4 The angular overlap of grating orders. Positions of the blue edges (taken to be at 320 nm) of orders −1 through +4 are shown. The thick gray arc shows the free spectral range of the second order, assuming $\lambda_{\text{max}} = 640$ nm.

either selecting higher orders or by increasing the number of lines per millimeter on the grating.

An important characteristic of diffraction gratings (and a disadvantage relative to prisms) is dispersion into multiple orders. An analysis of grating efficiency (see the discussion in Kitchen, 2008, pp. 367–372, or Schroeder, 1987, pp. 243–247) shows that most of the light from a simple amplitude grating gets diffracted into the $m = 0$ order, where $\theta = -\alpha$, and where there is no dispersion. Figure 11.4 illustrates an additional problem: the non-zero orders overlap. A particular value of θ corresponds to a different wavelength for each order, i.e. if λ_m is the wavelength in direction θ produced by order m, then $\lambda_1 = 2\lambda_2 = 3\lambda_3 = n\lambda_n$, or

$$m\lambda_m = (m + 1)\lambda_{m + 1}$$

The overlapping of multiple orders means that some method – a blocking filter or a detector of limited spectral sensitivity – must be used to eliminate unwanted orders. For example, suppose we limit the response of a detector to wavelengths shorter than some value, λ_{max} (for example, a silicon detector might have $\lambda_{\text{max}} = 1100$ nm, or perhaps grating efficiency is low longward of 850 nm). If we attempt to observe the spectrum in order m with this detector, the spectrum from order $m + 1$ overlaps λ_{max}, and at the same value of θ, deposits photons of wavelength

$$\lambda_{m+1} = \frac{m}{(m + 1)} \lambda_{\text{max}}$$

For example, first-order light at 1100 nm mixes with second-order light at 550 nm. We therefore insert a filter to block all light with wavelengths shorter than λ_{m+1} to eliminate the overlap. A quantity called the *free spectral range* quantifies the resulting restriction. The free spectral range is just the range of wavelengths not blocked, that is

$$\Delta\lambda_{\text{FSR}} = \lambda_{\text{max}} - \frac{m}{(m + 1)} \lambda_{\text{max}} = \frac{\lambda_{\text{max}}}{(m + 1)} \tag{11.4}$$

11.2.3 Blazed gratings

Amplitude gratings, pictured in Figure 11.3, operate by blocking (reducing the amplitude of) waves whose phases would destroy the constructive interference produced by diffraction in the periodic structure. For example, if the two rays pictured constructively interfere in first order, then the (blocked) ray that strikes between the facets would destructively interfere with them. In any particular order, then, amplitude gratings are inefficient, due to (1) this necessary blocking by the grating as well as (2) diffraction of most light into zeroth as well as other orders. **Phase gratings** produce dispersion effects similar to amplitude gratings, but operate by periodically adjusting the phase of the diffracted waves. They can minimize both disadvantages of the amplitude grating.

The **blazed reflection grating** is the most commonly employed example of a phase grating. As illustrated in Figure 11.5a, the surface of the blazed grating has a sawtooth-shaped profile, with each diffracting facet tilted at an angle ε (the **blaze angle**), measured counterclockwise with respect to the plane of the grating. This is sometimes called an **echellette grating**, from the French *échelle* (stair or ladder). The goal is to arrange the tilt so that all rays diffracted from a single facet are in phase. From the figure, it is clear that this will occur if there is specular reflection from the facet, i.e. if the angles of incidence and reflection with respect to the facet normal are equal: $\beta_1 = -\beta_2 = \beta$. This condition means that

$$\alpha = \beta + \varepsilon$$
$$\theta = 2\pi + \varepsilon - \beta$$
$$\alpha + \theta = 2\varepsilon$$

From the figure, it is clear that the conditions for constructive interference of rays diffracted from adjacent facets are identical to those we had for the amplitude grating, so we can apply the grating equation, (11.2), but substitute $\beta + \varepsilon = \alpha$ and $(\varepsilon - \beta) = \theta$:

$$\sin(\beta + \varepsilon) + \sin(\varepsilon - \alpha) = \frac{m\lambda}{\sigma}$$

which reduces to

$$\sin(\varepsilon) = \frac{m\lambda}{2\sigma\cos(\beta)}$$

or

$$\lambda_b = \frac{2\sigma}{m}\sin(\varepsilon)\cos(\alpha - \varepsilon) \tag{11.5}$$

Once a grating is manufactured with a particular blaze angle, Equation (11.5) suggests an associated **blaze wavelength** in order m, which depends slightly

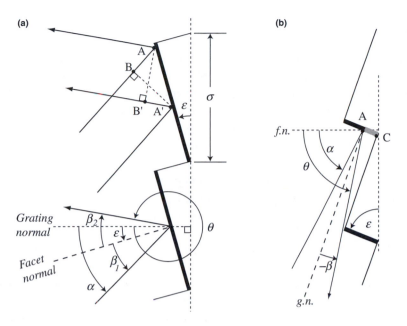

Fig. 11.5 (a) An echellette, or blazed reflection grating, and conditions for constructive interference. Heavy lines are the facets, at blaze angle ε, and spacing σ. All angles measured counterclockwise. (b) An echelle grating with large blaze angle ε. The gray region AC is shadowed at the blaze wavelength if the grating is illuminated as shown. This self-shadowing reduces grating efficiency.

on the angle of incidence. A more complete analysis (e.g. Möller, 1988, chapter 3) shows that the result of blazing is to shift the maximum efficiency of the grating from order 0 to order m. At the blaze wavelength, the grating will be completely illuminated (and most efficient) when mounted in the **Littrow configuration**, with $\beta = 0$ and $\alpha = \varepsilon$, so that incoming light is dispersed back on itself (not always the most convenient arrangement). Except for echelles (see below) blazed gratings are usually designed to work in order $m = \pm 1$, where their efficiency function (fraction of incoming light dispersed into the order as a function of wavelength) has a somewhat asymmetrical shape; see Figure 11.6.

Blazed transmission gratings also have sawtooth-shaped surface profiles, and achieve a phase shift along a facet by virtue of the changing optical path length in the high-index material. At the blaze wavelength, the transmission is in the direction given by Snell's law for refraction, and light is shifted from zeroth order to the design order.

11.2.4 Echelles

To produce a large angular dispersion, Equation (11.3) suggests operating at high order (m large) and with dispersed rays nearly parallel to the grating surface (θ near 90°). For a blazed grating, where $\theta = \varepsilon - \beta$, this suggests a design like the one in Figure 11.5b, where the blaze angle is very large and the incident and diffracted rays are nearly perpendicular to the reflecting facet (i.e. β small). Very coarse gratings facilitate operation at large m by limiting the right-hand side of the grating equation to physically meaningful values. Echelle gratings

Fig. 11.6 Representative
grating efficiencies of
three different phase
gratings with a blaze
wavelength of λ_b =
500 nm.

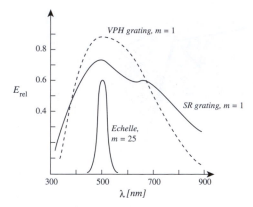

Fig. 11.6 Representative grating efficiencies of three different phase gratings with a blaze wavelength of λ_b = 500 nm.

used in astronomy typically have groove frequencies between 10 and 100 lines per millimeter, and operate in orders in the range 25–150.

Each value of θ in the echelle output contains photons of multiple wavelengths, one for each order. Equation (11.4) means that the free spectral range of any echelle order is very small. Rather than use filters to isolate this tiny range for a single order, the strategy is to separate all orders with a ***cross-disperser***. The cross-disperser can be a grating of low dispersing power or a prism; in either case, this second element disperses the echelle output (or, sometimes, input) in a direction perpendicular to the echelle dispersion. Figure 11.7 shows a representative set-up. With a camera (omitted for clarity in the figure) in place, the echelle-plus-cross-disperser combination produces multiple short spectra, one for each order, and stacked perpendicular to the dispersion direction on the output. Figure 11.8 is an example of the detector output. Most high-resolution astronomical spectrometers are based on echelles.

11.2.5 Volumetric phase gratings

So far we have been discussing phase gratings that rely on surface relief (SR) to produce periodic phase shifts. The ***volume phase holographic (VPH) grating*** produces phase shifts through spatial variations in the index of refraction, rather than surface relief. Researchers have produced VPH gratings in a number of configurations, including both transmitting and reflecting devices, but the form most useful in astronomical applications is the transmission grating illustrated in Figure 11.9. The grating is a thin slab or film of transparent material in which the "lines" are actually parallel planes of higher and lower index of refraction – holographic techniques are used to imprint the index modulation throughout the volume of the slab. In most current devices, the plane-to-plane index

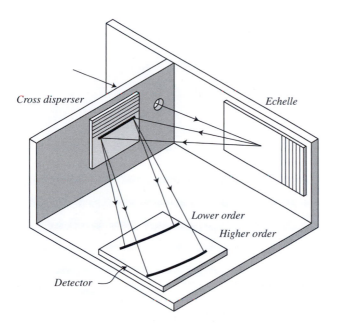

Fig. 11.7 An echelle and cross-disperser, showing the central ray and separation of two orders for two different diffracted rays from the echelle. For clarity, the figure omits all other rays, orders, and the spectrograph camera.

variation is roughly sinusoidal with an amplitude of 0.1 or less in the index, and with a peak-to-peak frequency of 200–6000 planes per millimeter.

The VPH gratings used in astronomy are usually in the "normal" or "Littrow" configuration illustrated in the figure, where the planes of enhanced index run perpendicular to the surface of incidence. The film thickness, d, typically ranges from a few microns up to 0.2 mm. If the spacing between fringe maxima is σ, then diffraction by the periodic index structure and the resulting interference effects disperse light exactly as described by the basic grating Equation (11.2):

$$\sin\theta + \sin\alpha = \frac{m\lambda}{\sigma}$$

However, as in the blazed SR gratings, there is now an additional condition – this one imposed by the fact that the grating extends through the volume of the film. The result is that efficiency will be enhanced by **Bragg diffraction** (refer to any introductory physics text) by the planes of constant index. Bragg diffraction occurs in a medium of index n_g if a ray of wavelength λ is incident on the grating fringe at internal angle α_g, defined as

$$n_g \sin\alpha_g = \frac{m\lambda}{2\sigma}$$
$$\sin\alpha_B = \frac{m\lambda_B}{2\sigma}$$

where we have applied Snell's law at the grating surface to get the external angle of incidence. The efficiency of the VPH grating is usually a maximum at the

Fig. 11.8 A cross-dispersed echelle spectrogram. Higher orders (shorter wavelengths) are at the top of the figure. The top panel shows the entire image, approximately 90 orders, with the lowest orders (shortest wavelengths) towards the top. The lower panel is an enlargement of the central region of the top panel (CCD image courtesy of Allison Sheffield).

Bragg angle α_B for the wavelength λ_B, which, from the grating equation, diffracts into direction $\theta = \alpha_B$ for first order. Incident light at other wavelengths does not satisfy the Bragg condition, and is diffracted with lower efficiency. The actual efficiency curve depends in a complex fashion on σ, on d, and on the shape of the index modulation and its amplitude, Δn. For many astronomically relevant gratings, efficiency will be maximum if

$$\mathrm{d} \approx \frac{\lambda_B}{2\Delta n} \qquad (11.6)$$

Since present technology limits Δn to less than 0.1, this requires d to be relatively large for high efficiencies. However, the width of the efficiency profile depends strongly on the ratio σ/d, so grating design is a compromise between high efficiency (large d) and broad spectral coverage (small d).

Volumetric phase holographic gratings have a number of advantages over their SR counterparts:

- A VPH grating can have a very high efficiency at the Bragg wavelength.
- The *"superblaze"* property: the blaze wavelength (i.e. the Bragg wavelength) of a VPH grating can be selected by tilting the grating to the appropriate Bragg angle.
- Because of their holographic production method, VPH gratings are less liable to imprecision in line spacing and parallelism, can have smaller values of σ, and can be physically quite large compared to ruled gratings.
- Current VPH films are encapsulated in a rugged glass assembly, so unlike SR gratings, can be cleaned and treated with anti-reflection coatings.

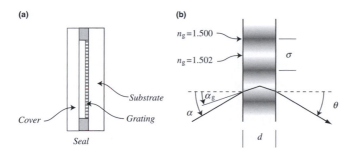

(a)

Cover — Seal

Substrate

Grating

(b)

$n_g = 1.500$

$n_g = 1.502$

σ

α_g

α

θ

d

Fig. 11.9 (a) A VPH grating assembly showing the film-like grating deposited on a glass substrate and sealed against humidity. (b) The path of a ray at the Bragg wavelength and angle of incidence ($\alpha = \theta$) in first order for a normal, or Littrow transmission VPH grating. In typical VPH gratings the refractive index varies sinusoidally, as represented by lighter and darker grays in the figure. In non-Littrow gratings, the planes of constant index are at an angle to the grating normal. We ignore refraction effects of the substrate and cover glass.

The important negative aspects of VPH gratings are:

- The wavelength bandpass can get narrow for high line-density VPH gratings.
- It is currently impossible to produce the VPH equivalent of an echelle – operation at high orders is not well understood.
- A VPH grating costs more than an equivalent SR grating.
- A spectrograph must have special design to take advantage of the superblaze capability of VPH gratings.

11.2.6 Grating manufacture

Palmer (2002) gives a complete treatment of the operation and construction of surface-relief gratings.

Manufacturers produce ***ruled gratings*** by drawing a diamond-tipped cutting tool across the optically flat surface of a soft metal blank. The shape of the diamond controls the shape of the grooves, and the engine that moves the tool (or blank, or both) controls groove depth, spacing, and parallelism. Current engines maintain precision through interferometric methods. The capacity of the engine, and, more critically, the wear of the diamond cutter, limit grating size, so very large ruled gratings are not possible. The ruled metal blank is usually used as a master to mold replica gratings – these are resin imprints of the master, subsequently mounted on a rigid substrate and aluminized. Transmission gratings are not aluminized, but do demand resins that cure to high optical quality.

Production of ***holographic gratings*** is perhaps more elegant. The manufacturer uses a laser to create two or more monochromatic plane waves whose interference creates a pattern of light and dark lines on a flat surface. The surface is coated with photoresist and the development of the photoresist, etching, and reflection coating produces an SR grating. Holographic gratings generally have greater precision, lower surface scattering, and potentially larger size than ruled gratings. The simple holographic technique produces a grating surface with a sinusoidal cross-section, so these devices are not strongly blazed and usually have a lower efficiency than ruled gratings. However, ***ion etching*** or more sophisticated holography can produce an echellette surface pattern on a holographic grating. Very coarse holographic gratings are difficult to produce.

Mosaics of ruled or holographic gratings are also difficult to produce, but feasible, often by bonding multiple replicas on a monolithic surface. Several examples are in operation at large telescopes.

It is quite easy to deposit a holographic grating on a curved surface, and *concave holographic gratings* find some use in astronomy. It is even possible to vary groove spacing and parallelism to remove optical aberrations. A *flat-field concave grating* will image a spectrum on a plane surface without additional optics.

As mentioned above, VPH gratings also rely on laser interference effects for their production. In this case, the manufacturer illuminates parallel bright and dark planes in the interior volume of a transparent material. Several materials are under investigation, but current devices use a suspension of ammonia or metal dichromate in gelatin (dichromated gelatin, or DCG). When chemically processed, DCG exhibits a change in index due to differential shrinkage that depends on its exposure to light. Processed DCG must be of limited thickness, and its density enhancements are degraded by humidity, so VPH gratings are usually mechanically stabilized and sealed between glass plates.

11.3 Spectrometers without slits

11.3.1 The objective prism

The objective prism spectrograph mode (probably first employed by Fraunhofer) is conceptually simple. As illustrated in Figure 11.10, a prism is placed directly in front of a telescope objective. Parallel rays from distant objects are dispersed by the prism and focused by the objective. A completely monochromatic source will produce a simple image in the focal plane, but a polychromatic source will produce a spectrum – a different image at each wavelength. A two-dimensional detector can thus record, in a single exposure, a separate spectrum for every star in the field – a large *multiplex advantage*. For this reason, astronomers frequently have mounted objective prisms on wide-field telescopes like Schmidts. Because of the confusion and light loss produced by multiple orders, objective *gratings* are seldom employed in astronomy. Figure 11.11 is an example of an objective prism view of the Pleiades star cluster.

Resolving power and the instrumental profile will depend on both the prism and telescope parameters. A telescope that normally produces monochromatic stellar images of angular diameter $\delta\theta$ should, when equipped with an objective prism, have spectral purity

$$\delta\lambda_0 = \delta\theta \frac{d\lambda}{d\theta} \cong \delta\theta \frac{\lambda^3 \cos\alpha}{4K_2 \sin(A/2)} \simeq \frac{\delta\theta}{0.02A} \lambda^3 [\mu m] \qquad (11.7)$$

The numerical values on the right-hand side assume a thin prism of flint glass.

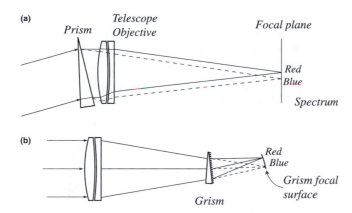

Fig. 11.10 (a) An objective prism on a refracting telescope. On Schmidt telescopes, the prism would be mounted just in front of the corrector plate. (b) A non-objective grism. In practice, the relative distance between the grism and the focal plane is usually much smaller than illustrated here.

For example, assume a 1-meter $f/3$ Schmidt is equipped with a prism with apex angle 1°. According to Equation (11.7), the spectral purity expected in 1 arcsec seeing at 500 nm is $\delta\lambda_0$ = (1 arcsec)(0.125 μm^3)/[(0.02 μm^2)(3600 arcsec)] = 1.7×10^{-3} = 1.7 nm, which improves to 0.85 nm at 400 nm. The optical resolving power is R = 500/1.7 = 300 at 500 nm and 470 at 400 nm. The plate factor at 500 nm will be $p = (1/f)(d\lambda/d\theta)$ = 113 nm mm^{-1}. Seeing sets the resolving power in this case. If this same spectrometer were placed in space, stellar image sizes should be around 0.12 arcsec, and we would find R = 2400 at 500 nm.

Remember that for any spectrometer the effective resolving power may not be optics- or seeing-limited, but could be set by the detector resolution. The pixel size of most astronomical charge-coupled devices (CCDs) is in the range 13–26 μm, and sampling theory requires a minimum of 2 pixels per resolution element. In the above example, for instance, if we had 13 μm pixels then $\delta\lambda_d$ = $2.3 \times (13 \times 10^{-3}) \times [p]$, or 3.4 nm at 500 nm, nearly twice the value for $\delta\lambda_0$ in the seeing-limited case. Since we realize the detector now limits the resolution, we should expect R = 150 at 500 nm.

Spectra from objective prisms are generally very low resolution, but are well suited for survey work that requires rough spectral classification, color determination, or identification of peculiar spectra in a field that contains many objects.

11.3.2 The non-objective prism and grism

Objective prisms must be the same size as the telescope aperture, an expensive requirement, and an impossible one for apertures larger than around 1 meter. A prism placed near the focal plane (the "non-objective" configuration) in the converging beam of a telescope will be smaller, but generally produces spectra severely compromised by coma and other optical aberrations.

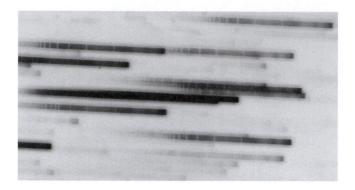

Fig. 11.11 An objective-prism spectrogram of the Pleiades star cluster. This illustration is a small section (about 5%) of an original photographic plate taken with a Schmidt telescope and is of relatively high dispersion (p = 10 nm mm⁻¹ at 400 nm). Note over- and under-exposed spectra, and overlapping spectra; and also the change in linear dispersion and projected spectrum brightness with wavelength (red is towards the right).

The **grism** substantially reduces the aberrations of non-objective spectra. (The grism is a grating–prism combination. A similar device, incorporating a grating on a lens surface, is called a **grens**.) The grism usually consists of an SR transmission grating mounted on the hypotenuse of a right-angle prism. The apex angle of the prism is generally chosen so that rays at the grating blaze wavelength, λ_B, converge on the optic axis (Figure 11.10b):

$$\sin A = \frac{\lambda_B}{\sigma(n-1)}$$

As illustrated, the focal plane of the grism is slightly tilted from the Gaussian plane. A telescopic camera can quickly convert into a low-resolving-power ($R < 2000$) spectroscope by the insertion of a grism in the beam, and many large telescopes, including space telescopes, routinely provide grisms mounted on filter wheels. Volumetric phase holographic grisms, which are capable of relatively high angular dispersions, have begun to appear.

11.4 Basic slit and fiber spectrometers

Slitless spectrometers exhibit some serious disadvantages: overlapping spectra in crowded fields, contamination by background light, resolution limits imposed by seeing, limited use for extended objects, and lack of convenient wavelength calibration. The slit or fiber-fed spectrometer addresses all these issues. The basic notion is to restrict the light that reaches the dispersing element to only that light from the small angular area of interest on the sky.

Figure 11.12 sketches the layout of a simple slit spectrometer with all-transmitting optics. The rectangular slit has width w_s in the plane of the diagram and height h, and is located at the focus of a telescope of aperture D_{TEL} and effective focal length f_{TEL}. The slit limits light entering the spectrometer to a narrow rectangle on the sky, and (as long as the angular slit width is smaller than

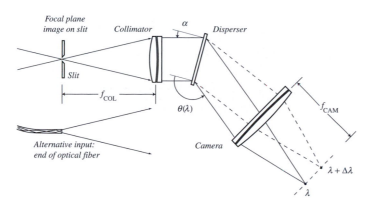

Fig. 11.12 A simple, all-transmission slit spectrometer. An alternative input is the end of an optical fiber, which would replace the slit at the focus of the collimator. The opening of the slit or fiber is w_s, measured in the vertical direction in this diagram.

the seeing disk), the slit width limits the spectral purity of the spectrometer. The angular size of the slit on the sky is

$$\phi_s = \frac{w_s}{f_{TEL}}$$

Alternatively, an optical fiber with one end in the focal plane of the telescope could bring light to the spectrometer. In this case, w_s becomes the diameter of the fiber core. Following the light though the spectrometer on the figure, note that rays at the slit (or fiber end) find themselves at the focus of the **collimator** lens. They emerge from that lens as a set of parallel rays. To avoid either losing light off the edge of the collimator or making the collimator larger than needed, we require it to have a focal ratio

$$\Re_{COL} = \frac{f_{COL}}{D_{COL}} = g_f \Re_{TEL} = g_f \frac{f_{TEL}}{D_{TEL}} \tag{11.8}$$

The factor g_f here accounts for the fact that an optical fiber can degrade the focal ratio of the beam it transmits. In the case of a slit, $g_f = 1$, but in the case of a fiber-fed system g_f may be somewhat smaller, depending on fiber length and quality. The general import of Equation (11.8) is that *the focal ratio of the collimator should match that of the telescope/fiber.*

The **throughput** of a spectrometer is an important measure of its quality. At a particular wavelength, throughput is just the fraction of light incident in the focal plane of the telescope that actually reaches the spectrometer's detector. Light lost at the edge of a slit, fiber, or collimator decreases throughput.

Continuing through Figure 11.12, the **collimated beam** (i.e. parallel rays) then strikes the dispersing element at angle α. The beam disperses to angle $\theta(\lambda)$, and a camera lens (diameter D_{CAM} and focal length f_{CAM}) finally focuses the rays as a spectrum on the detector. If the disperser is a grating (the usual case in astronomy) the reciprocal linear dispersion is

$$p = \frac{d\lambda}{dx} = \frac{1}{f_{CAM}} \frac{d\lambda}{d\theta} = \frac{\sigma \cos \theta}{m f_{CAM}} \tag{11.9}$$

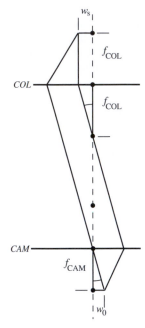

Fig. 11.13 Ray trace in the absence of a disperser, showing the width of the image of the spectrograph slit or fiber in the focal plane of the camera.

For a prismatic spectrometer, one would substitute an expression like Equation (11.1) for $d\theta/d\lambda$ above. The spectral purity depends on w_0, the projected size of the slit or fiber on the detector. Without a disperser present (see Figure 11.13), $w_0 = w_s \, f_{CAM}/f_{COL}$. Inserting the disperser will modify the slit image in the dispersion direction by a projection effect that changes the image size by a factor known as the ***anamorphic magnification***, $r_{an} = d\theta/d\alpha = \cos\alpha/\cos\theta$. Thus, the width of the slit image on the detector will be

$$w_0 = r_{an} w_s \frac{f_{CAM}}{f_{COL}} = r_{an}\phi_s \frac{f_{TEL}}{f_{COL}} f_{CAM} = r_{an}\phi_s \frac{D_{TEL}}{D_{COL}} f_{CAM} \qquad (11.10)$$

Here we have substituted $\phi_s = w_s/f_{TEL}$, the input slit size in angular units, and have made use of Equation (11.8). The optical limit on spectral purity for the grating spectrograph is therefore

$$\delta\lambda_0 = w_0 p = r_{an}\phi_s \frac{D_{TEL}}{D_{COL}} \frac{\sigma\cos\theta}{m} \qquad (11.11)$$

and the resolving power of the spectrometer is

$$R = \frac{\lambda}{\delta\lambda_0} = \frac{\lambda}{r_{an}\phi_s} \frac{D_{COL}}{D_{TEL}} \frac{m}{\sigma\cos\theta} \qquad (11.12)$$

We have used only first-order geometric optics to derive Equations (11.11) and (11.12), so this result ignores diffraction effects and aberrations. Nevertheless, this expression is central for the practical design of spectrometers. There are three important comments:

First, for ground-based stellar work, you frequently achieve large R by setting the angular width of the slit on the sky, ϕ_s, to less than the full width at half-maximum (FWHM) of the seeing disk, ϕ_{see}. In the cases where the seeing disk does not over-fill the slit, however, we must replace ϕ_s with ϕ_{see} in the above expressions. In this case, the ***seeing-limited resolving power*** is greater than the slit-limited value. Fiber inputs enjoy no such gain in R due to improved seeing, since the entire output end of the fiber is illuminated no matter what the size of the input image. In the usual case where $\phi_s < \phi_{see}$, light will be lost off the edges of the slit or fiber, and throughput reduced. In the case of the slit, at least, there are devices – ***image slicers*** – that redirect this light back though the slit. The direct connection between telescopic image size and spectrograph resolution means that ***adaptive optics*** or any other steps that improve seeing will benefit spectrometer resolution. For space telescopes and adaptive optics systems, the absolute limit on the resolving power is set by the FWHM of the Airy disk: $\phi_0 = 1.22\lambda/D_{TEL}$.

Substituting this into Equation (11.12), and making use of the definition of r_{an} we have

$$R_0 = \frac{mD_{COL}}{1.22\sigma \cos \alpha} \approx \frac{mW}{\sigma} = mN \qquad (11.13)$$

where W is the length of the grating and N is the number of lines ruled on it.

Second, note that the projected width of the slit, fiber, or seeing disk, must satisfy

$$w_0 \geq 2d_{px},$$

where d_{px} is the pixel spacing in the direction of the dispersion. Failure to satisfy this condition results in under-sampling and consequent reduction in R. Making w_0 very much larger than 2 pixels wastes detector length and usually increases noise.

Finally, suppose you have designed a very successful spectrometer for a 1.0-meter telescope. Equations (11.11) and (11.12) say that to apply the same design (ϕ, m, σ, and θ all held constant) to a larger telescope, *the entire instrument must be scaled up in proportion to the size of the primary.*

The diameter of the collimator must increase in direct proportion to D_{TEL}, and its focal ratio must match that of the new telescope. The length of the grating increases with the diameter of the collimator, as does the diameter of the camera. On the larger telescope, one would try to reduce the focal length (reduce the focal ratio) of the camera to preserve plate scale and avoid gigantic detectors.

Thus, on a 30-meter telescope, the same spectrometer design needs to either be something like 30 times larger (and something like $30^3 = 27,000$ times more massive), or suffer from lower resolving power. A number of novel strategies (see, for example, Dekker *et al.*, 2000) can reduce the effective slit width and "fold" the optics to reduce mass and escape this scaling rule. Nevertheless the scaling rule explains the attractiveness of large gratings, as well as part of the popularity of echelles (large m/σ) for even moderate-sized telescopes.

11.5 Spectrometer design for astronomy

11.5.1 An example configuration

Special constraints that may be unimportant in a laboratory environment are crucial in the design of astronomical spectrometers. For example, telescope optics determine the ideal focal ratio of the spectrometer collimator; the need to study faint objects sets a high premium on throughput; and the desire to mount the instrument on the telescope favors compact and rigid designs. Off-the-shelf instruments can be a poor choice for astronomy, and, especially for larger systems, astronomical spectrometers tend to be one-of-a kind devices built for a specific telescope.

Designing any astronomical spectrometer properly begins with some notion of the scientific questions the device will address. This sets the range of wavelengths to be investigated as well as the minimum value of R required, with an inevitable trade-off between the value of R and the faintness of the objects to be investigated. A reformulation of Equation (11.12) can guide further decisions:

$$R = \frac{1}{D_{TEL}\phi_s} \frac{\lambda m}{\sigma} W = \frac{1}{D_{TEL}\phi_s} (\sin\alpha + \sin\theta) W \tag{11.14}$$

Here we have used the length of the disperser, $W = D_{COL}/\cos\alpha$, instead of the diameter of the collimator, and in the rightmost expression, have made use of the grating equation.

The astronomer has no control over the seeing at his site or the diameter of his telescope. The remaining design parameters in the middle expression in Equation (11.14) all have to do with the choice of the dispersing element. Until the advent of VPH gratings, modern astronomical spectrometers tended to employ blazed reflection gratings (either echelles or single gratings) because of the difficulty in producing efficient transmission gratings and the drawbacks of prisms. The grating will exert the strongest constraints on the design, because there is a limited choice of sizes and manufacturers. The choice of echelle or single grating is a fundamental one. This choice appears in Equation (11.14) as the value of the dimensionless expression $\lambda m/\sigma$ (which at 500 nm equals 0.1 to 1.5 for a conventional first-order grating, slightly higher than this for a VPH grating, and even larger (1 to 3) for an echelle).

For a given resolution, the echelle requires a smaller grating and therefore produces a more compact spectrometer. The echelle also has the advantage that its coverage of the spectrum, in the form of a one-for-each-order stack of strips, makes efficient use of the square format of most CCDs. The echelle has disadvantages as well – besides requiring a more complex alignment process, the additional reflection/dispersion by the cross-disperser creates more scattered light and lowers overall efficiency. Data reduction, which requires extraction of the orders and reassembly to form a complete spectrum, can become complicated.

For a given resolution, a single grating will be more efficient and minimize scattered light. Its simplicity makes it relatively versatile compared with the echelle, because it is easy to swap out one grating for another of differing blaze wavelength and/or grating constant. It will, however, require a larger collimator for the same resolving power and result in a less compact spectrometer. The overall length of the detector limits the range of wavelengths recorded, and since detectors (and certainly detector mosaics) are likely to be the most expensive single component in the spectrometer, this should be an important consideration.

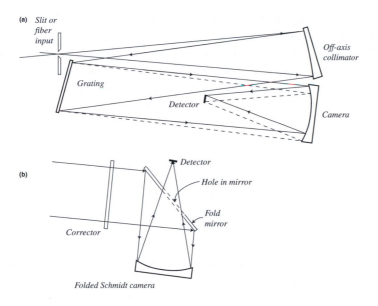

(a) Slit or fiber input

Off-axis collimator

Grating

Detector

Camera

(b) Detector

Hole in mirror

Fold mirror

Corrector

Folded Schmidt camera

Fig. 11.14 (a) A simple slit spectrometer with reflective optics in a near-Littrow configuration. (b) A folded Schmidt camera that might replace the prime focus camera in (a).

In the case of both the single grating and the echelle, the projected width of the grating determines the aperture of the camera: D_{CAM} must be at least as large as $W \cos \theta$, plus an allowance for the spread in θ due to dispersion. The camera focal length determines the plate scale, and generally one wants the shortest possible camera focal length consistent with adequate sampling. For large systems, this usually means a very fast camera, and because many rays will be off-axis, a Schmidt camera is a popular choice. Figure 11.14 is a schematic of a simple but representative all-reflecting spectrometer. The collimator is an off-axis paraboloid with its focus at the slit. The blaze angle and tilt of the grating will determine the central wavelength of the spectrum.

The camera is located just short of impinging on the collimated beam, in an arrangement termed the ***near-Littrow*** configuration, which has the grating in as close to the efficient $\alpha = \theta$ position as possible. Since grating efficiency deteriorates rapidly as $|\alpha - \theta|$ increases, a useful alternative is the ***quasi-Littrow*** configuration. In this configuration, $\alpha = \theta$ in the plane of the central ray to and from the collimator in Figure 11.14, but the grating is tilted upwards, so the camera would lie above the collimated beam in the figure. The quasi-Littrow has the disadvantage that the slit image is not perpendicular to the dispersion direction in the final spectrum. For the optics of the camera itself, Figure 11.14a shows a simple prime focus camera, and (b) shows a ***folded Schmidt*** camera, which has the advantage of easy access to the detector, as do cameras based on versions of the ***Schmidt–Cassegrain*** idea. For very large detectors like CCD mosaics, an ***all-refractive camera*** is generally more efficient than these Schmidt-based approaches.

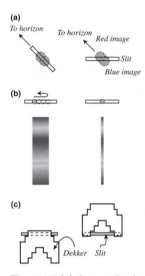

Fig. 11.15 (a) Atmospheric refraction and slit input. In the right-hand panel, the slit is not in the direction of atmospheric dispersion, so light lost at the slit is systematically from the extreme red and blue ends of the spectrum. In the left-hand panel, light lost is less overall and not systematic with wavelength. (b) Trailed (left) and untrailed (right) spectra – the trailed version requires a longer exposure, but is easier to examine visually. (c) The use of a dekker in front of a long slit to mask and reveal different parts of the slit.

The basic layout of Figure 11.14 is subject to vast variation. Cost, as always, is a major consideration: the lower cost of widely available optical components in standard sizes is attractive, but restricts the design options. Spherical mirrors (if focal ratios are slow) can replace paraboloids. A commercial photographic lens is often a useful option for the camera in a small spectrometer if only visual wavelengths are of interest.

For large telescopes, however, the benefits of an optimized design outweigh its cost. That cost, in turn, often argues for building a versatile spectrograph with several options for resolving power and wavelength sensitivity. It is not uncommon, for example, to split the collimated beam with a mirror or dichroic filter and construct separate grating/camera arms optimized for short and long wavelengths. You can consult the ESO, Keck, Subaru, AAO, and Gemini websites to get a sense of both the variety and complexity of spectrograph designs for large telescopes.

Spectroscopy in the thermal infrared has significant additional requirements. The telescope, sky, and spectrograph emit significant radiation longward of 2.5 μm in wavelength, so all mid-infrared spectrographs are cooled, beginning with the slit (a cold slit greatly reduces sky background), and completely encapsulated in a chamber that is either evacuated or filled with an inert gas. Special observing techniques are required to remove the bright background due to the telescope mirrors and sky.

11.5.2 Slit orientation and spectrum widening

Dispersion by the atmosphere means that star images are actually tiny spectra with the red end oriented towards the horizon (see Section 5.4.3). For a ground-based spectrometer with a single slit that is narrower than the seeing disk, it is best to orient the slit in the direction of atmospheric dispersion to avoid systematic loss of some parts of the spectrum, and some loss of resolution; see Figure 11.15a.

In earliest days of spectroscopy, astronomers examined spectra visually, and quickly noticed it was much easier to recognize emission and absorption features if the spectrum was widened in the direction perpendicular to the dispersion (typical widening might be ten times w_0). In a widened spectrum the monochromatic image of the slit really does look like a "line," and in the photographic recording of spectra, sufficient widening is important for the visual recognition of features. With digital detectors, spectrum widening is less crucial, but sometimes widening can improve signal-to-noise ratio (SNR) if the noise sources are primarily independent of exposure time, like read noise, or flat-field uncertainties. You can widen spectra through telescope motion or optical scanning that trails the image along the height of the slit during the exposure, or through insertion of a cylindrical lens that widens the image in one dimension only. See Figure 11.15b.

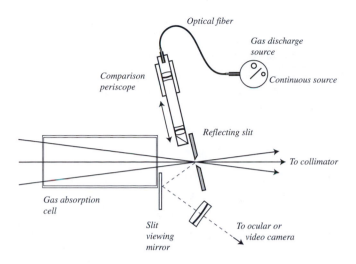

Optical fiber

Gas discharge source

Comparison periscope

Continuous source

Reflecting slit

To collimator

Gas absorption cell

Slit viewing mirror

To ocular or video camera

Fig. 11.16 Some strategies for viewing the slit and for acquiring comparison spectra. The optics in the lower part of the figure view reflections from the front of the slit. In the upper part of the figure, a periscope can slide down to cover the slit, and light from either a gas discharge (emission line spectrum) or solid filament (continuous spectrum) is fed in via an optical fiber and/or relay optics. The left side of the figure shows a cell with transparent windows. Gas in this cell imposes an absorption line spectrum on any light travelling from the telescope to the slit.

11.5.3 Getting light in

The astronomer needs to verify that the object of interest is indeed sending light into the slit or fiber aperture of the spectrometer for the duration of the exposure. For a single object, he might use an arrangement like the one sketched in Figure 11.16. In the figure, the jaws of the slit that face the telescope are reflective and tilt away from the normal to the optical axis, so that pre-slit optics let the astronomer view (possibly with a small video camera) the focal plane and slit. He can position the telescope to center the object on the slit and *guide* telescope tracking during the exposure: he monitors the light reflected from the edges of the slit and actively corrects for any telescope drift. Image acquisition and guiding are more complicated with multi-object and integral-field spectrometers.

In addition to light from astronomical objects, spectrometers need to accept light from calibration sources. *Wavelength calibration* (see below) often requires that light from a gas-discharge or spark lamp with a well-understood emission line spectrum can enter the slit on the same optical path as objects in the sky. *Flat-field* calibrations require the same for a continuous source, and the retracting periscope sketched in Figure 11.16 is one way to deliver the light from calibration sources. There are many others, some rather easily implemented with fiber optics. A *dekker* (Figure 11.15c) or pre-slit mask is useful for restricting the parts of the slit that receive light. The dekker illustrated, for example would permit comparison spectra to be recorded above and below an object spectrum.

Multi-object spectrometers allow simultaneous recording of the slit spectra of many objects in the field. For example, an astronomer is interested in obtaining spectra of every one of 50 galaxies in a distant cluster of galaxies. She does not allow light to enter the slit of the spectrograph directly but instead makes the following special arrangement illustrated in Figure 11.17a. She positions 50

Fig. 11.17 (a) A multi-object spectrometer input. Fibers are positioned in the focal plane – in this case using a drilled plate, and their output is stacked at the spectrograph slit. Spectra appear in a similar stack at the spectrometer output. Some fibers should be devoted to sampling sky background. (b) Multiple-slit input. Multiple slits cut in an occulting plate are placed in the focal plane of the telescope, and form the entrance apertures for an imaging spectrograph. The spectrograph output consists of spectra distributed according to position on the sky. A long slit (D) can sample both object and background. Objects B and C, whose spectra would overlap with A, are not provided with slits.

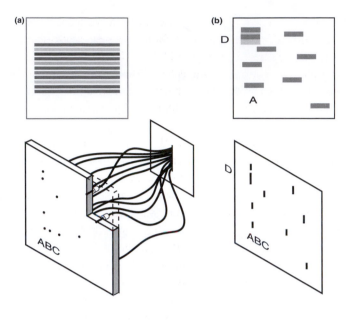

different optical fibers so that their output ends are arranged in a vertical stack along the slit, and so that the input end of each is positioned in the focal plane of the telescope. Each fiber captures the light from a different galaxy, and a single exposure produces 50 spectra stacked vertically. Saving a factor of 50 in telescope time justifies the tedious job of positioning the fibers. In the early days, multi-object inputs were fabricated by drilling holes in a metal plate at positions determined by an image of the field of interest, and then gluing a fiber into each hole (this is still done, although the procedure is often automated). To measure background, it is useful to have some fraction of the fibers at blank sky locations. Conveniently, several large spectrographs now provide automated fiber positioners, whose robotic arms can place the fiber ends at whatever locations are needed for a particular project.

Imaging spectrometers also allow simultaneous recording of multiple spectra. Suppose we modify the instrument in Figure 11.12 by replacing the grating with a plane mirror and removing the slit entirely. The image formed on the detector is then identical to the focal-plane image, magnified by the factor f_{COL}/f_{CAM}. If we enlarge the diameters of the collimator and camera to accommodate a beam considerably larger than the beam passed by the slit, and correct the optics to minimize aberrations in the plane of the detector, then this modified instrument will record a wide-field image of the sky. Furthermore, with the grating in place, the imaging spectrometer, like an objective prism or non-objective grism, will produce spectra of every object in the field. The *multiple-slit mask*, however, more powerfully exploits the advantages of an imaging spectrometer. The astronomer places an opaque plate in the telescope focal plane; the plate is pieced with slits or apertures at the locations of objects of interest. This is similar to the plate used for the fiber-fed multi-object spectrometer, but without the fibers – see Figure 11.17b. The multiple-slit mask produces spectra at multiple locations on the detector, but with no contamination from non-science objects, and minimal sky background. Data for sky subtraction can come from a long slit that coincides with both object and adjacent sky, or from slits cut at blank sky locations.

Integral-field spectrometers address the desire to examine the spectrum at *every* location in the image of an extended object like a galaxy, producing an intensity map in the form of a data cube with dimensions of x, y, and wavelength. An advanced panoramic detector with wavelength sensitivity – like the arrays of superconducting tunneling junctions presently under development – could produce such three-dimensional data. Superconducting tunnel junction (STJ) diodes are expected to provide R of around 500.

Another approach is to creating an (x, y, λ) data cube is to take repeated spectra of the object with a **long-slit spectrometer** – if the slit is oriented in the y direction, changes in the spectrum in that direction appear in the final spectroscopic image; see Figure 11.18a. The complete data cube can be built up by stepping the slit across the object in the x direction in subsequent exposures. This method of long-slit image scanning requires no special equipment other than spectrometer optics capable of forming an excellent image over the entire y dimension of the slit. It is, of course, time consuming.

Unlike this long-slit scanning, a true integral-field spectrometer constructs the (x, y, λ) data cube with a single exposure.

One example is the **image slicer**, which places an array of n long, narrow mirrors in the image plane, stacked side by side in the x direction (see Figure 11.18b). Each mirror acts like a slit, and each is tilted at a different angle. Subsequent reflections then reform each mirror's slice of the image on the slit of the spectrograph, but displaced in the y direction so that none of the slices overlap. A single exposure can then capture n spectra stacked on the output detector, each of which is equivalent to a single exposure in the long-slit scanning method – a (y, λ) data plane for each x value. Small versions of the image slicer have been in use in stellar work for many years as a method for slicing up the light from the seeing disk and fitting it into a narrow slit.

A second form of the integral-field spectrometer tiles the focal plane with n individual apertures, each connected via an optical fiber to a particular location on the long slit of the spectrograph; see Figure 11.18c. The result is a stack of n spectra, each corresponding to the (x, y) coordinate of the input aperture. Many systems use an array of lenslets to transfer the focal-plane image to the fiber cores – otherwise, much light is lost due to fiber cladding and packing gaps. Lenslet optics are also helpful to couple the fiber outputs to the spectrograph slit. Blank sky samples and feeds from wavelength-calibration sources are easily implemented.

Fig. 11.18 (a) A long-slit spectrometer input. The object is imaged directly on the slit, and the single spectrum produced records any variation in the y direction. (b) An image slicer. Multiple mirrors intercept different parts of the image, each acting like a long slit. Reflections from each mirror are re-assembled along a single very long slit at the spectrometer entrance. (c) An integral-field spectrometer in which the telescope image plane is tiled with lenslets that feed individual fibers. The fiber output is distributed along the spectrometer entrance slit.

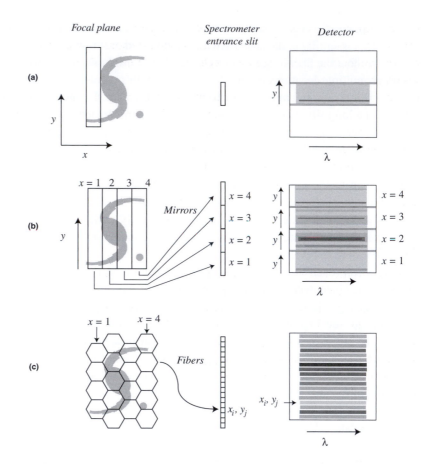

11.5.4 Spectrometer stability and mounting

We have seen that although much might be done to reduce the overall dimensions and weight of a spectrometer, its resolving power is directly proportional to the length of the grating employed, so some spectroscopic projects will always require large, heavy instruments.

In general, then, you will find spectrometers suitable for the study of the faintest objects at relatively low resolving power mounted at the Cassegrain or even prime focus, where the fast f number and limited number of pre-slit reflections encourages a compact design of limited weight but high efficiency. These spectrometers move with the telescope, so their parts experience variable gravitational stresses, and differential motion can produce systematic errors. These errors will appear as variations in the spectrum of the same object and differential shifts between the wavelength calibration and the object spectra.

Spectrometers mounted at the coudé focus are motionless with respect to the Earth, and an altazimuth mount moves those mounted at the Nasmyth focus only in the horizontal plane. The dimensions of these instruments are thus less constrained, and they can be quite massive, employing large gratings and optics. Moreover, coudé (and to a lesser extent, Nasmyth) mounts make it easier to control the thermal stability of the spectrograph and isolate it from vibrational disturbance.

Fiber optics permit the mechanical and thermal stability without the large f number and multiple reflections of the coudé focus. A fiber entrance mounted at the Cassegrain focus of the telescope can conduct light to a spectrograph housed in a motionless and environmentally stable enclosure. There is some penalty in using a fiber, since some light is lost in transmission and fibers will generally degrade the focal ratio (e.g. $f/8$ at fiber entrance might become $f/6$ at exit).

11.6 Spectrometric data

The astronomer collecting and reducing spectrometric data faces a task that shares some of the features described in earlier chapters for photometric and astrometric data, but he also faces some challenges that inspire unique practices. The detailed strategy for both observing and data reduction depends on the overall scientific goal: very precise radial-velocity studies have stringent stability and wavelength-calibration requirements, survey work is more concerned with efficiency, and spectrophotometry takes great care with flux calibration and atmospheric extinction correction.

11.6.1 Array preprocessing

The preferred detector is usually an array. For slit spectrographs, an instrument designer can save some expense by using a linear array, or one with reduced pixel count or resolution in the direction normal to the dispersion. However, given that an observatory is quite likely already to own one or more rectangular, square-pixel detectors, the design will often utilize a CCD or infrared array or mosaic similar to those used for direct imaging. A large square-ish array, of course, would be required for cross-dispersed echelle, multi-slit, and slitless spectrometry. The **bias**, **dark**, and **linearity** exposures for array detectors, and the corresponding corrections, are necessary steps in the preprocessing of spectrographic data, and proceed very much as in direct imaging

The **flat-field** correction needs special consideration. Here the calibration exposure requires a source whose image is as uniform as possible along the slit, and whose spectrum is continuous — a quartz halogen lamp and some sort of projection screen is the usual choice. The dark or twilight sky has a spectrum loaded with skyglow emission features and the scattered Fraunhofer solar spectrum, so it is an unsuitable calibration source. Slit length, grating tilt, and any other relevant parameter should be the same in the flat as in the object spectra, so that each pixel in the flat receives light of the appropriate wavelength. The flat-field image will depend on the spectrum of the lamp (usually unknown), its projected uniformity along the slit, the overall transmission efficiency of the spectrometer in both the wavelength and spatial dimensions, and the quantum efficiency of each pixel. Thus, the astronomer can expect to extract the pixel-to-pixel sensitivity from the flat, but not variations on a large wavelength or spatial scale.

Flat fields for long-slit, multiple-slit-mask, multi-object, and integral-field spectrometers can be treated as an extended case of single-slit flats. Usually a projection flat is not uniform over a wide field, so with these instruments one needs an additional calibration exposure, this one usually of the twilight sky. This exposure provides the large-scale "illumination correction" in the spatial dimension that accounts for effects like vignetting and variations in slit width or fiber efficiency. The flat-fielding problem for echelle spectra is similar to that for multi-object spectra, with the complication that an appreciable subtractive correction may be needed for scattered light.

A precise flat-field correction for slitless spectra is not possible, since the wavelength of light falling on a particular pixel in the program image cannot be predicted or repeated in the flat. Twilight flats, in which each pixel receives the thoroughly blended sky spectrum, are probably superior to projection flats in this case.

11.6.2 Spectrum extraction

Figure 11.19 is a sketch of a typical spectrum on the detector produced by a single star. We assume the slit is long enough to include a portion of the sky not significantly contaminated by the star's spectrum, and that the dark, bias, linearity, and flat-field corrections have been completed. The figure indicates a few night-sky emission features that show the orientation of the slit.

Note that the slit image is parallel to the y-axis of the detector. Usually you can, and should, adjust the detector orientation so that this is the case. If not, you can rotate the image digitally after the exposure, but remember that interpolation reduces resolution. With the slit image in the y direction, the dispersion direction should lie exactly along the x-axis – but it usually does not. Camera distortions

Fig. 11.19 Tracing the object and background spectra in the plane of the detector.

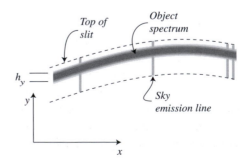

(not apparent over the short length of the slit) can tilt and curve the spectrum, as can atmospheric refraction and grating tilt on the axis normal to the grooves. Each order in an echelle spectrogram often exhibits considerable and distinct curvature and tilt.

You therefore trace, with a smooth curve, the y-position of the centroid of the spectrum at each x-value, and average the pixel values over the height, h_y, of the stellar spectrum. It would be appropriate to use a weighting scheme that accounts for the lower statistical values of less-well-exposed pixels at the edge of the image (or discards them), and rejection algorithms that remove cosmic rays and bad pixels. The result is a function, $A(x)$, the average intensity of the star plus sky spectrum as a function of x. If you want to subtract the background (the usual case), then you do a similar trace of the sky portion or portions of the image, and compute $B(x)$, the average background intensity as a function of x. Finally, the **extracted spectrum**, $I(x)$, is just the difference, $A(x) - B(x)$.

Different situations will demand some alterations in this basic procedure. For an extended object like a nebula that occupies the entire height of the slit, or for a single-fiber spectrograph, you must "nod" the telescope to a region of blank sky near the object and collect a sky spectrum on a separate exposure. In the infrared, background levels are so high and variable that a very rapid "chopping" to nearby sky and automated subtraction is often part of data acquisition. For multiple slit-mask, multi-object, and integral-field spectrographs, you should plan dedicated slits or fibers for the needed sky spectra.

The function $I(x)$ is the object spectrum, but modified by atmospheric and optics absorption, the grating efficiency, and detector sensitivity. For some purposes (e.g. a survey for particular spectral features) $I(x)$ is the final data product, but for others (e.g. spectral classification, measurement of relative line strengths) a **continuum-normalized** version of the spectrum is more useful. To produce the normalized version, you specify the continuum sections (regions with no absorption or emission features) of $I(x)$, and fit a smooth curve, $C(x)$, to only those parts of the spectrum. The normalized spectrum is just $I_N(x)= I(x)/C(x)$; see Figure 11.20.

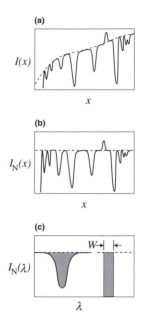

Fig. 11.20 (a) An extracted spectrum. The dashed-line curve is a smooth fit to the continuum. (b) A normalized version of the same spectrum. The dashed line has value unity. (c) The profile of a single absorption line in a wavelength-calibrated, continuum-normalized spectrum. The equivalent width, W, of the line measures the side of the rectangle shown, whose area equals the area between the continuum and the line profile.

11.6.3 Wavelength calibration

For most purposes, you will want to present the spectrum not as a function of CCD position, x, but as a function of wavelength. The usual – but not the only – method for doing so is to obtain the spectrum of an emission-line source (a gas-discharge tube or "arc") on a separate exposure. For the most secure calibration, the light from this **comparison source** must follow the same path though the spectrometer as the light from the object. If you suspect any shifts during the object exposure, you should take comparison exposures both before and after the object exposure. Some spectrographs place the comparison spectrum above and below the object spectrum on the same exposure by using a moveable dekker or by providing dedicated fibers.

You extract the comparison spectrum as you did the object spectrum. You can then pair the known wavelength of each emission line with the observed *x*-location of the center of that line in your spectrum. A functional fit to this data (usually a polynomial of some kind) is the ***dispersion solution***, $\lambda(x)$. If you have before-and-after comparisons, extract both and average them. You now can associate a wavelength with every *x*-value in your spectrum. The next step depends on your scientific goals. You might measure the *x*-values of features of interest and use the dispersion solution to compute their observed wavelengths. For other applications, you may wish to ***linearize*** the spectrum by interpolating pixel values so that there is a constant wavelength change from each pixel to the next. This produces the function $I(\lambda)$ in a convenient form.

One noteworthy departure from the above procedure is the use of a gas-absorption cell for wavelength calibration. The light from the object passes through a low-pressure cell containing a gas whose absorption spectrum exhibits many narrow absorption features. Currently, iodine vapor, I_2, which requires a cell only a few centimeters long and produces on the order of 10 lines per nanometer in the red, is the most popular choice. The cell is usually placed in front of the slit; see Figure 11.16. This arrangement has the advantage that the light paths through the spectrograph are identical for the object and comparison, and the very narrow iodine lines allow correction for changes in the slit profile due to guiding errors (the center of brightness of the illuminated slit varies if the star is not kept precisely centered) and for mechanical shifts during the exposure. Butler *et al.* (1996) describe the technique in detail as applied to the detection of extra-solar planets through the minute Doppler shifts a planet induces in its host star.

11.6.4 Flux calibration

The wavelength-calibrated spectrum $I(\lambda)$ has units of analog-to-digital unit (ADU) per wavelength interval, and reflects both the spectral energy distribution of the object and the detection efficiency of the atmosphere–telescope–spectrograph combination. If you wish to transform your data into either absolute or relative flux units, you must observe ***photometric standard stars***. There are several sets of such standards. The rationale is identical to the one described in the previous chapter for the monochromatic case in photometry, with the simplifying condition that each pixel is a detector in which the central wavelengths of the standard and the program object are identical. The procedure is to remove the effects of atmospheric extinction in both program and standard object, then transform the instrumental photon flux to the standard system, which in this case is a system that counts an absolute number of photons arriving outside the atmosphere. Recalling the notation of Chapter 10, the observed, inside-the-atmosphere photon count per unit time on a particular pixel will be

$$I(\lambda) = \phi(\lambda)\Delta\lambda A_{\text{TEL}} Q(\lambda) T(\lambda) S(\lambda) t$$
$$I_{\text{S}}(\lambda) = \phi_{\text{S}}(\lambda)\Delta\lambda A_{\text{TEL}} Q(\lambda) T_{\text{S}}(\lambda) S_{\text{S}}(\lambda) t_{\text{S}}$$

(11.15)

The second equation, with subscripted variables, is for the observed spectrum of the standard star. In these equations, ϕ is the actual outside-the-atmosphere photon flux, A_{TEL} is the effective light-gathering area of the telescope, $\Delta\lambda$ is the wavelength interval intercepted by the pixel, S is the fraction of the incoming photons absorbed by the atmosphere, Q is the quantum efficiency of the detector, t is the exposure time, and T is the throughput of the spectrograph. The throughput, $T(\lambda)$, is just the fraction of those photons from the source that arrive in the focal plane of the telescope that also arrive at the detector. From Equation (11.15), we define the *sensitivity function* $R(\lambda)$:

$$R(\lambda) \equiv \Delta\lambda A_{\text{TEL}} Q(\lambda) T_{\text{S}}(\lambda) = \left\{ \frac{I_{\text{S}}(\lambda)}{t_s \phi_s(\lambda)} \right\} \frac{1}{S_{\text{S}}(\lambda)}$$

(11.16)

The quantity in the braces you can compute from standard star data and your own observations. The value of the atmospheric absorption must either be estimated from average extinction coefficients for the site, or you must determine the coefficient yourself from one of the methods described in the previous chapter. In either case, if $k(\lambda)$ is the monochromatic extinction coefficient, then at air mass X,

$$S_{\text{S}}(\lambda) = \exp(-0.9208 k(\lambda) X)$$

You usually cannot compute the sensitivity function for every pixel of your data, because the profiles of absorption lines in the standard star spectrum have shapes that depend on the instrumental profile, and because the standard star data are ordinarily tabulated over a rather coarse spacing in wavelength. Therefore, you fit a smooth function for $R(\lambda)$ using only the available calibration points computed from Equation (11.16). As long as the grating tilt is constant, the flat field is good (rendering $Q(\lambda)$ constant pixel-to-pixel), and the throughput is invariant, $R(\lambda)$ should be stationary: independent of the object and time of observation. Observing multiple standards on a photometric night should therefore improve your estimate of $R(\lambda)$. Given a reliable sensitivity function, you may easily compute the flux-calibrated spectrum of the program object from

$$\phi(\lambda) = \frac{I(\lambda)}{tR(\lambda)} \exp(0.9208 k(\lambda) X)$$

(11.17)

There are two obvious impediments to a stationary $R(\lambda)$. First, from Equation (11.16) note that variations in the spectrometer throughput produce variations in $R(\lambda)$. Spectrometer throughput changes if the slit or fiber width is less than the width of the object (the usual case) and if there are any errors in guiding and tip-tilt seeing compensation (also the usual case). Absolute spectrophotometry of stars therefore usually requires a spectrometer slit or fiber opening several (e.g. six) times larger than the FWHM of the seeing disk to capture all significant light from standards and program objects. Fortunately, as long as a narrow slit is aligned perpendicular to the horizon (see Figure 11.15a) variation in throughput due to image motion should be *gray* – independent of wavelength. The variation changes $R(\lambda)$ only by a scale factor from one exposure to another. Therefore, **relative spectrophotometry**, which determines the shape but not the energy scale of the spectrum, should be possible with a narrow slit.

The second impediment to a stationary $R(\lambda)$ is that Equation (11.16) requires an accurate value for the atmospheric extinction. Site-average extinction coefficients are not always available, or they may not be correct for your night. On-the-spot determination of the coefficients requires a stationary spectrometer throughput and observations at multiple air mass. An observing strategy that acquires all exposures at approximately the same air mass minimizes the requirement for accurate extinction coefficients.

11.6.5 Other calibrations

Astronomers assign a program star its temperature and luminosity spectral class by comparing the program's spectrum with that of a **spectroscopic standard star**, noting in particular class-defining features like relative absorption line strengths. The actual appearance of features, and the quantitative relationships among them, depend on the instrumental resolution and its sensitivity function. Precise classification therefore requires observations of spectroscopic standard stars with the same apparatus as the program stars. Phillip Kaler (1963) and James Kaler (1997) discuss the process in detail.

You can determine radial velocities and cosmological redshifts by measuring wavelength shifts in spectra relative to some reference, usually the spectrum of a comparison lamp. Any effects that produce different angles of incidence on the grating for the program and comparison sources will cause systematic errors in your measurements. Similar systematic effects will result if the effective wavelength of a feature changes to an unknown value because of different line blending at different spectrograph resolutions. One precaution against systematic effects for stellar work is to observe a **radial velocity standard** of the same spectral class as the program object. The standards have been cataloged with well-determined velocities in the heliocentric reference frame. If you observe programs and standards under identical conditions, then systematic effects should be apparent in the measured velocities of the standards, and you can therefore eliminate them in the program objects.

11.7 Interpreting spectra

Extracting useful information from the spectrum you have just observed and reduced can require a great deal of analysis, much of it based on astrophysical theory well beyond the scope of our concerns here. However, the analysis almost always begins with some quantitative measurements of the spectrum, a few of which we now examine.

11.7.1 Line strength

The continuum-normalized spectrum as a function of wavelength, $I_N(\lambda)$, when considered only over the region in which a *single* absorption or emission line is present is called the **line profile** of that feature. No line profile is indefinitely narrow, but will extend over a range of wavelengths. The reasons for such line broadening are varied, and we will discuss some of them in the next subsection. You will realize immediately, however, that every line profile must be at least as wide as the spectrometer's resolution. We introduce here the idea of the **equivalent width** as a measure of – not the width – but the **strength** of an emission or absorption line. See Figure 11.20c. We define the equivalent width as

$$
W = \left| \int_{\text{line}} (1 - I_N(\lambda))\mathrm{d}\lambda \right| \tag{11.18}
$$

You can see from this that W will have units of wavelength, and that its meaning is somewhat different for absorption and emission lines. You can also imagine that the equivalent width should be fairly independent of the spectrograph you use to measure it, and indeed, that is part of its appeal: we have catalogs of the equivalent widths of the lines in the solar spectrum, for example, and everyone can agree on these values (the K line, Ca II, has $W = 2.0$ nm, and the Na I D lines have $W = 0.075$ and 0.056 nm). Great care is needed, however. Spectrometer resolution matters tremendously: at low resolution, weak lines disappear because they become too shallow to recognize, and lines can blend together. Any scattered light in a spectrometer will systematically bias equivalent width measurements to lower values.

With a spectrum of sufficiently high resolution, the values of absorption-line equivalent widths (coupled with a good astrophysical model of the stellar or planetary atmosphere, or of the emission nebula) can produce estimates of chemical composition, temperature, and pressure. **Classification of stellar spectra** is traditionally done visually, using criteria like the ratio of critical line strengths, or the presence/absence of certain lines, but always with reference to a set of standard stars. It is possible (and many astronomers have done this) to set completely quantitative classification criteria using equivalent widths, but no scheme has gained widespread acceptance.

11.7.2 Line profiles

The detailed shape of a line profile, $I_N(\lambda)$, depends on a number of physical and observational factors that produce **line broadening**. If you make sufficiently precise measurements of the shape, you can often learn about the physical environment in which the line was formed. We examine the most common line broadening mechanisms:

Natural broadening

The emission or absorption lines of isolated atoms (in a very low-pressure gas, for example) are broadened because of the quantum-mechanical uncertainly in the energies of the quantum states of the transition. Those uncertainties, and the line's half-width, depend directly on transition probability and, therefore, inversely on the lifetimes of the relevant states. The natural widths of most lines are very narrow; so other broadening mechanisms dominate, and, except for interstellar Ly α, natural broadening is not observed in astronomical spectra.

Instrumental broadening

The limited spectral purity of the spectrometer itself will impose an instrumental profile on any infinitely narrow line, and this profile may be convolved with another broadening profile. Although some degree of de-convolution is possible in data analysis, it is best to keep the instrumental profile small compared to other broadening mechanisms under investigation.

Rotational broadening

If a source is spinning on an axis making an angle i to the line of sight, you see, relative to the average, some of its material with positive radial velocities, and some with negative velocities. The Doppler effect means that a line will appear over a corresponding range of wavelengths, with a profile (for absorption) given by

$$I_{NR}(\Delta\lambda) = 1 - I_0\left\{1 - \left[\frac{c}{V\sin i}\frac{\Delta\lambda}{\lambda_0}\right]^2\right\}^{\frac{1}{2}} \tag{11.19}$$

where c is the speed of light and V is the equatorial velocity of rotation of the star. You may recognize the term in braces as the expression for an ellipse. Many

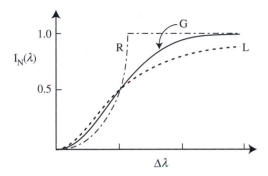

Fig. 11.21 Absorption line profiles produced by different broadening processes. Curves are shown with unit depth and identical FWHM. Profile R is due to solid body rotation. The Gaussian profile, G, is caused by thermal broadening or turbulence. The Lorentzian profile, L, is characteristic of both natural quantum-mechanical broadening, as well as pressure broadening. Observed line profiles will be some convolution of these shapes at different strengths and scales, and will depend as well on the details of radiative transfer.

stars rotate rapidly enough so that this elliptical profile is easy to recognize; see Figure 11.21.

Thermal broadening

On the microscopic level, atoms and molecules will have a spread in velocities that increases with the temperature and decreases with molecular mass. Here again, the Doppler effect means that temperature alters the line profile. The profile for thermal broadening for a molecule of mass m is Gaussian:

$$I_{\rm NT}(\Delta\lambda) = 1 - I_0 \exp\left\{-\frac{mc^2}{2kT}\left(\frac{\Delta\lambda}{\lambda_0}\right)^2\right\} \qquad (11.20)$$

Microturbulence

Microturbulance is small-scale fluid motion caused, for example, by convection in a star's atmosphere. Again, relative velocities between different parts of the source cause line broadening, and microturbulence is expected, like thermal broadening, to produce a Gaussian profile. Since turbulent velocities are independent of molecular mass, the variation in line width as a function of molecular mass in principle allows one to separate the two broadening mechanisms.

Other **organized motions** in a source like expansion, contraction, **macroturbulence** (e.g. rising material might occupy a larger area than falling), and orbital motion can produce asymmetries or secular changes in the line profile.

In emission nebulae, unlike stars, the broadening of profiles is completely dominated by Doppler effects due to large-scale mass motions, temperature or turbulence.

Pressure broadening

For stellar spectra, pressure effects are most often the dominant cause of broadening away from the central part of the line. Physically, broadening arises because nearby charges, especially free electrons, can perturb the energy of the electron states in an atom. Higher pressures mean more frequent encounters and greater broadening. The pressure-broadened profile is Lorentzian, having the form

$$I_{NL}(\Delta\lambda) = 1 - I_{0L}\left[\frac{A}{\Delta\lambda^2 + B}\right] \tag{11.21}$$

Natural broadening also has a Lorentzian profile. You can see from Figure 11.21 that compared to the Gaussian, the Lorentzian has more absorption away from the central parts of the line, and in stellar spectra, lines are very frequently Gaussian in their **cores** and pressure broadened in their **wings**. Indeed, the relative strength of the wings of absorption lines permits estimates of the acceleration due to gravity (which determines pressure) in a stellar photosphere. Since gravitational acceleration depends only on mass and radius, and since stellar masses can be determined from the motions of binary stars, spectroscopic observations of the wings of lines permit estimates of stellar radius, and the recognition that stars of the same temperature exist as dwarves, giants, and supergiants.

An observed line profile is not only a convolution of all the above broadening processes, but also depends on how radiation is transferred through the object. In a stellar atmosphere, for example, once all the light at the core of the line has been absorbed, no more absorption can take place at that wavelength, and the line is saturated.

Summary

- Dispersive spectroscopy relies on optical elements that send light rays in a direction that depends upon wavelength. Concepts:

angular dispersion	*linear dispersion*
spectral purity of optics	*reciprocal linear dispersion, p*
effective spectral purity, $\delta\lambda$	*instrumental profile*
resolving power, R	

- The angular dispersion of a prism varies as λ^{-3}.

- Diffraction gratings depend upon wave interference of diffracted rays. Concepts:

amplitude grating	*transmission grating*	*reflection grating*
grating constant	*groove frequency*	*groove spacing*
order	*free spectral range*	

- The grating equation gives the angle of dispersion as a function of wavelength:

$$\sin\theta + \sin\alpha = \frac{m\lambda}{\sigma}$$

- Phase gratings operate by periodically adjusting the phase of diffracted waves. Concepts:

eschellette	*blazed grating*	*blaze angle*
blaze wavelength	*Littrow configuration*	*zeroth order*
order	*free spectral range*	

- Echelle gratings have steep blaze angles, and usually operate in conjunction with a second (cross-) disperser to separate orders.

- Volumetric phase holographic (VPH) gratings produce phase shifts by periodically adjusting the index of refraction in a transmitting slab. Concepts:

Bragg diffraction	*Bragg angle*	*superblaze*

- Gratings can be produced by scribing lines on a master blank or by holographic techniques. Concepts:

ruled grating	*holographic grating*	*flat-field concave grating*
grating mosaic	*ion etching*	*DCG*

- The objective prism generates a spectrum of every object in the telescopic field of view. The spectra tend to have low resolution, and suffer from high background. Concepts:

non-objective prism	*grism*	*multiplex advantage*

- Spectrometers with slit or fiber inputs restrict incoming light to increase resolution and suppress background. Concepts:

slit width	*collimator*	*anamorphic magnification*
image slicer	*seeing-limited R*	*scaling with telescope diameter*
throughput		

- Resolving power of a conventional slit spectrometer:

$$ R = \frac{\lambda}{r_{an}\phi_s} \frac{D_{COL}}{D_{TEL}} \frac{m}{\sigma \cos\theta} = \frac{1}{D_{TEL}\phi_s} \frac{\lambda m}{\sigma} W $$

- Spectrometers for astronomy have special design requirements. Concepts:

near-Littrow	*off-axis paraboloid*	*quasi-Littrow*
spectrum widening	*folded Schmidt*	*dekker*
multiple-slit mask	*imaging spectrometer*	*multi-object spectrometer*
image slicer	*long-slit spectrometer*	*integral-field spectrometer*
flexure	*wavelength calibration*	*flat-field source*

(*continued*)

Summary (*cont.*)

- Reduction of spectrographic array data varies with one's scientific goals.
  Concepts:

flat field	*comparison source*	*spectrum extraction*
dispersion solution	*spectroscopic standard*	*continuum-normalized spectrum*
linearized spectrum	*photometric standard*	*relative spectrophotometry*
sensitivity function		

- Much of astrophysics concerns the interpretation of spectra. Concepts:

line profile	*line strength*	*equivalent width*
spectra classification	*line broadening*	*natural broadening*
thermal broadening	*rotational broadening*	*instrumental broadening*
microturbulence	*pressure broadening*	*Lorentzian*
line core	*line wings*	*pressure broadening*

Exercises

1. Derive the grating equation for the transmission grating. Clearly state the rule you adopt for measuring positive and negative angles.

2. Explain quantitatively why the free spectral range for a particular order, m, and maximum wavelength, λ_{max}, is not restricted by overlapping light from order $m - 1$.

3. Compute the free spectral range of grating orders 50, 100, and 101 if $\lambda_{max} = 600$ nm in each case.

4. Compare the angular dispersions of a 600 lines per mm amplitude grating at 400 nm and at 900 nm. Assume you are working in first order and the angle of incidence is 25°. Do the same for a 60° prism with angle of incidence $\alpha = 55°$ and $K_2 = 0.01\ \mu m^{-2}$.

5. Manufacturers usually describe blazed gratings by specifying the blaze wavelength in the Littrow configuration and the groove density in lines per mm. (a) Compute the blaze angle for a reflection grating of 1000 lines per mm blazed for a wavelength of 400 nm. (b) Compute the blaze wavelength of this grating when it is used at an angle of incidence of 40° instead of in Littrow.

6. Explain why the self-shadowing of a grating as a function of wavelength is different if the direction of the ray in Figure 11.5b is reversed: i.e. if instead of $\alpha < \theta$ at the blaze wavelength, we have $\alpha > \theta$.

7. A normal VPH grating has an index modulation frequency of 2000 lines per mm. Sketch a spectrograph design (show the relevant angles) that would permit the most efficient observation of spectra near a wavelength of 400 nm in first order. Now sketch how the spectrograph would have to be adjusted to observe efficiently at a wavelength of 600 nm in first order. What is the minimum number of moving parts required for such an adjustable spectrograph?

8. Show that the anamorphic magnification, $d\theta/d\alpha$, of the simple slit spectrograph in Figure 11.12 is $\cos \alpha/\cos \theta$.

9. An astronomer wishes to build a simple fiber-fed spectrometer, using a reflection grating. She will follow the basic plan illustrated in Figure 11.14. She has a CCD detector measuring 1024×1024 pixels, with each pixel 15 µm on a side. An optical fiber with core diameter 100 µm will sample star images in the telescope focal plane and deliver light to the spectrograph. Tests show that the fiber degrades the telescope focal ratio of $f/7.5$ to $f/7.0$. A grating of diameter 50 mm with 600 lines per mm and blaze angle 8.5° is available.

   (a) Compute the first-order blaze wavelength of this grating, and its maximum possible resolving power, R_0.

   (b) The astronomer chooses to illuminate the grating at an angle of incidence, α, of 0°, and to record the first-order spectrum. What is the maximum focal length the collimator mirror must have to avoid light loss at the edge of the grating?

   (c) Compute the value of the anamorphic magnification at the blaze wavelength.

   (d) The astronomer wishes to critically sample the image of the fiber end with her CCD. What is the required focal length for the camera? Compute the resulting plate scale and the wavelength range in nanometers of the complete spectrum the CCD will record.

   (e) Compute the actual resolving power, R, of this spectrometer at the blaze wavelength. If R is less than R_0, describe how the astronomer might improve her value for R.

10. A spectrograph has a very narrow slit, and its CCD is oriented so that the image of the slit (for example, when illuminated by the comparison source) is precisely along the detector's y-axis. Explain why the dispersive effect of atmospheric refraction will tilt the trace of the spectrum of an untrailed exposure so that it is not parallel to the detector x-axis (see Figures 11.15a and 11.18a).

11. Explain the differing meanings of equivalent width for absorption and emission lines.

12. Suggest a method for testing for the presence and severity of scattered light in a spectrometer. Consider two absorption lines each with an identical FWHM: one very strong, the other very weak. In equivalent width measurements, which line is more strongly affected by the presence of scattered light?

13. Compute the FWHM of the line profile for a magnesium absorption line of rest wavelength 500 nm on a star spinning with equatorial velocity 100 km s^{-1} and $\sin i = 1$. Compare this with the FWHM due to thermal broadening if the star's temperature is 8000 K.

Appendix A
General reference data

A1 The Greek alphabet

alpha	A	α	nu	N	ν
beta	B	β	xi	Ξ	ξ
gamma	Γ	γ	omicron	O	o
delta	Δ	δ	pi	Π	π
epsilon	E	ϵ	rho	P	ρ
zeta	Z	ζ	sigma	Σ	σ, ς
eta	H	η	tau	T	τ
theta	Θ	θ, ϑ	upsion	Y	υ
iota	I	ι	phi	Φ	ϕ, φ
kapa	K	κ	chi	X	χ
lambda	Λ	λ	psi	Ψ	ψ
mu	M	μ	omega	Ω	ω

A2 Metric system prefixes and symbols

Use with base unit to indicate decimal multiples. A few units not standard in the SI system are in common use in astronomy. See Tables A3 and A4 below.

Factor	Prefix	Symbol	Factor	Prefix	Symbol
10^{18}	exa	E	10^{-18}	atto	a
10^{15}	peta	P	10^{-15}	femto	f
10^{12}	tera	T	10^{-12}	pico	p
10^{9}	giga	G	10^{-9}	nano	n
10^{6}	mega	M	10^{-6}	micro	μ
10^{3}	kilo	k	10^{-3}	milli	m
10^{2}	hecto	h	10^{-2}	centi	c
10	deca	da	10^{-1}	deci	d

A3 Physical constants

Speed of light in a vacuum	$c = 299792458 \text{ m s}^{-1}$
Planck constant	$h = 6.626075 \times 10^{-34} \text{ J s}$
Gravitational constant	$G = 6.6726 \times \text{ m}^3 \text{kg}^{-1} \text{s}^{-1}$
Mass of the electron	$m_e = 9.10939 \times 10^{-31} \text{ kg}$
Mass of the proton	$m_p = 1.672623 \times 10^{-27} \text{ kg}$
Mass of the neutron	$m_n = 1.674929 \times 10^{-27} \text{ kg}$
Unit elementary charge	$e = 1.6021773 \times 10^{-19} \text{ C}$
	$= 4.803207 \times 10^{-10} \text{ esu}$
Boltzmann constant	$k = 1.380658 \times 10^{-23} \text{ J K}^{-1}$
Stefan–Boltzmann constant	$\sigma = 5.6705 \times 10^{-8} \text{ W m}^{-2} \text{K}^{-4}$
Avogadro number	$N_A = 6.022137 \times 10^{23} \text{ mol}^{-1}$

A4 Astronomical constants

Mass of the Sun	$1.9891 \times 10^{30} \text{ kg}$
Mass of the Earth	$5.975 \times 10^{24} \text{ kg}$
Radius of the Sun	$6.9599 \times 10^8 \text{ m}$
Equatorial radius of the Earth	$6.378140 \times 10^6 \text{ m}$
Tropical year	$365.2421897 \text{ days} = 31{,}556{,}925.19 \text{ sec}$
Solar constant (flux at top of atmosphere)	$1.37 \times 10^3 \text{ W}$
Luminosity of the Sun	$3.825 \times 10^{26} \text{ W}$

A5 Conversions

Length

Angstrom	$1 \text{ Å} = 10^{-10} \text{ m}$
Micron	$1 \mu = 1 \text{ μm} = 10^{-6} \text{ m}$
Astronomical unit	$1 \text{ au} = 1.49598 \times 10^{11} \text{ m}$
Parsec	$1 \text{ pc} = 3.085678 \times 10^{16} \text{ m} = 3.2616 \text{ light years}$
Light year	$1 \text{ lyr} = 9.46053 \times 10^{15} \text{ m}$
Statute mile	$1 \text{ mi} = 1609.344 \text{ m}$
Inch	$1 \text{ in} = .0254 \text{ m}$

Time

Day	$1 \text{ day} = 86{,}400 \text{ sec}$
Tropical year	$1 \text{ yr} = 3.155692597 \times 10^7 \text{ s} = 365.24219 \text{ days}$
Sidereal year	$= 365.25636 \text{ days}$

Mass

Pound (avdp)	$1 \text{ lb} = 0.453592 \text{ kg}$

Pressure

Pascal (SI)	$1 \ Pa = 1 \ N \ m^{-2}$
Bar	$1 \ b = 10^5 \ Pa = 10^3 \ mb$
Atmosphere	$1 \ atm = 1.01325 \times 10^5 \ Pa = 1013.25 \ mb$
Millimeter of mercury	Pressure of 1 mm of Hg = 1 torr = 133.322 Pa
Pound per square inch	$1 \ lb \ in^{-2} = 6894.7 \ Pa$

Energy

Electron volt	$1 \ eV = 1.60218 \times 10^{-19} \ J$
Erg (cgs)	$1 \ erg = 10^{-7} \ J$
Calorie	$1 \ cal = 1.854 \ J$
Kilogram	$1 \ kg \ c^2 = 8.9876 \times 10^{16} \ J$
Kiloton of TNT	$= 4.2 \times 10^{12} \ J$

Monochromatic irradiance (flux density)

Jansky	$1 \ Jy = 10^{-26} \ W \ m^{-2} \ Hz^{-1}$

Velocity

Miles per hour	$1 \ mph = 0.44704 \ m \ sec^{-1}$
au per year	$= 4740.6 \ m \ sec^{-1} = 4.7406 \ km \ sec^{-1}$
Parsec per million years	$10^{-6} \ pc \ yr^{-1} = 977.8 \ m \ s^{-1} = 0.9778 \ km \ s^{-1}$

Appendix B
Light

B1 Photon properties

This table gives the conversion from the photon characteristic in the left-hand column to the corresponding characteristic at the head of each subsequent column. For example, a photon of wavelength of 100 nm has an energy of 1240/100 = 12.4 electron volts.

To From	λ(nm)	λ(m)	ν(Hz)	E(J)	E(eV)
λ(nm)	1	$10^{-9}\lambda$	$2.99729 \times 10^{17}/\lambda$	$1.98645 \times 10^{-16}/\lambda$	$1239.85/\lambda$
λ(m)	$10^7\lambda$	1	$2.99729 \times 10^8/\lambda$	$1.98645 \times 10^{-25}/\lambda$	$1.2985 \times 10^{-6}/\lambda$
ν(Hz)	$2.99792 \times 10^{17}/\nu$	$2.99792 \times 10^8/\nu$	1	$6.62606 \times 10^{-34}\nu$	$4.1357 \times 10^{-15}\nu$
E(J)	$1.98645 \times 10^{-16}/E$	$1.98645 \times 10^{-25}/E$	$1.5092 \times 10^{33}E$	1	$6.2414 \times 10^{28}E$
E(eV)	$1239.85/E$	$1.2985 \times 10^{-6}/E$	$2.4180 \times 10^{20}E$	$1.6022 \times 10^{-19}\,E$	1

B2 The strongest Fraunhofer lines

Telluric lines originate in the Earth's atmosphere, rather than the Sun's. Roman numeral I designates neutral (not ionized) atoms. Wavelengths may vary because of line blending at different spectroscopic resolution.

Designation	Wavelength (Å)	Identification	Comment
A	7593.7	Telluric O_2	Band
a	7160.0	Telluric H_2O	Band
B	6867.2	Telluric O_2	Band
C	6562.8	Hα	
D_1	5895.9	Na I	Doublet
D_2	5890.0	Na I	Doublet
E	5269.6	Fe I	
b_1	5183.6	Mg I	
b_2	5172.7	Mg I	
b_3	(5169.1 + 5168.9)	Fe I	
b_4	5167.3	Mg I	
F	4861.3	Hβ	
G	4314.2	CH	Band
H	3969.5	Ca I	
K	3933.7	Ca I	

B3 Sensitivity of human vision

Extreme range of wavelengths detected by human vision is normally 400–760 nm. There are records of individuals seeing light as long as 1050 nm and as short as 310 nm.

Feature	Photopic vision	Scotopic vision
General illumination level for sole operation	Daylight to twilight	Quarter Moon to darkness
Receptor cells	Cones	Rods
Peak sensitivity	555 nm*	505 nm
10% of peak, blue cut-on	475 nm	425 nm
10% of peak, red cut-off	650 nm	580 nm
Speed of adaptation	Fast	Slow (up to 30 minutes)
Response time	0.02 seconds	0.1 seconds
Color discrimination	Yes	No
Visual acuity	High	Low
Region of retina	Center (fovea)	Periphery
Threshold of detection	High	Low (10^{-4} photopic)

* There are actually three different types of cones, with peak sensitivities at 430, 530 and 560 nm, which permit color discrimination.

B4 The visually brightest stars in the sky

Rank	Classical name	Bayer designation	V
1	Sirius	α CMa	−1.46
2	Canopus	α Car	−0.72
3	Rigel Kent	α Cen	−0.27
4	Arcturus	α Boo	−0.04
5	Vega	α Lyr	0.03
6	Capella	α Aur	0.08
7	Rigel	β Ori	0.12
8	Procon	α CMi	0.38
9	Achernar	α Eri	0.46
10	Betelgeuse	α Ori	0.5 (variable)

Appendix C

C1 The standard normal distribution

The standard normal **probability density function** is defined as

$$G(z) = \frac{1}{\sqrt{2\pi}} e^{-\frac{z^2}{2}}$$

It is related to the Gaussian distribution of mean, μ, and standard deviation, σ, by the transformation

$$z = \frac{x - \mu}{\sigma}$$

Thus, $G(z)$ is simply a Gaussian with a mean of zero and variance of 1. Table C1 gives values for $G(z)$. You can use the table to evaluate any particular Gaussian by applying the desired values of μ, σ and x to the above transformation.

Table C1. *The standard normal distribution*

z	G(z)	P(z)	Q(z)
0.00	0.398942	0.500000	0.000000
0.05	0.398444	0.519939	0.039878
0.10	0.396953	0.539828	0.079656
0.15	0.394479	0.559618	0.119235
0.20	0.391043	0.579260	0.158519
0.25	0.386668	0.598706	0.197413
0.30	0.381388	0.617911	0.235823
0.35	0.375240	0.636831	0.273661
0.40	0.368270	0.655422	0.310843
0.45	0.360527	0.673645	0.347290
0.50	0.352065	0.691462	0.382925
0.55	0.342944	0.708840	0.417681
0.60	0.333225	0.725747	0.451494
0.65	0.322972	0.742154	0.484308
0.70	0.312254	0.758036	0.516073
0.75	0.301137	0.773373	0.546745
0.80	0.289692	0.788145	0.576289
0.85	0.277985	0.802338	0.604675
0.90	0.266085	0.815940	0.631880
0.95	0.254059	0.828944	0.657888
1.00	0.241971	0.841345	0.682689
1.10	0.217852	0.864334	0.728668
1.20	0.194186	0.884930	0.769861

Table C1 (*cont.*)

z	G(z)	P(z)	Q(z)
1.30	0.171369	0.903199	0.806399
1.40	0.149727	0.919243	0.838487
1.50	0.129518	0.933193	0.866386
1.60	0.110921	0.945201	0.890401
1.70	0.094049	0.955435	0.910869
1.80	0.078950	0.964070	0.928139
1.90	0.065616	0.971284	0.942567
2.00	0.053991	0.977250	0.954500
2.10	0.043984	0.982136	0.964271
2.20	0.035475	0.986097	0.972193
2.30	0.028327	0.989276	0.978552
2.40	0.022395	0.991802	0.983605
2.50	0.017528	0.993790	0.987581
2.60	0.013583	0.995339	0.990678
2.70	0.010421	0.996533	0.993066
2.80	0.007915	0.997445	0.994890
2.90	0.005953	0.998134	0.996268
3.00	0.004432	0.998650	0.997300
3.50	8.7268 E-04	0.999767	0.999535
4.00	1.3383 E-04	0.999968	0.999937
4.50	1.5984 E-05	0.9999966	0.9999932
5.00	1.4867 E-06	0.9999997	0.9999994

A related function is the ***cumulative probability***, $P(z)$, also listed in the table:

$$P(z) = \int_{-\infty}^{z} G(t)\,dt,$$

The function $P(z)$ gives the probability that a single sample drawn from a population with a standard normal distribution will measure greater than or equal to z. A second function of interest is

$$Q(z) = \int_{-z}^{z} G(t)\,dt$$

The function $Q(z)$ gives the probability that a single sample drawn from a population with a standard normal distribution will be within z of the mean; Q is also tabulated, although it is also easily computed as

$$Q(z) = 2P(z) - 1$$

Note that P is also related to the ***error function***:

$$\mathrm{erf}(x) = \frac{2}{\sqrt{\pi}} \int_{0}^{x} e^{-t^2}\,dt = 2P(x\sqrt{2}) - 1$$

Appendix D

D1 The nearest stars

Adapted from the table in the RASC Observers Handbook 2001, which incorporates HIPPARCOS data directly. Parallax in mas, μ in mas yr^{-1}, θ is the position angle of the proper motion, radial velocity is in km s^{-1}, and Spt is the spectral type.

D2 The equation of time

Figure D shows the value of the equation of time,

$$\Delta t_{\mathrm{E}} = \text{local apparent solar time} - \text{local mean solar time}$$
$$= \text{RA of the apparent Sun} - \text{RA of the mean Sun}$$

This is the same as 12^{h} minus the mean solar time of transit for the apparent Sun. More precise values for Δt_{E} in a particular year can be obtained from the Astronomical Almanac. Approximate dates for the extrema of Δt_{E} are Feb 11 (minimum), May 14, July 26 and Nov 3 (maximum).

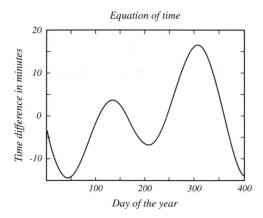

Fig. D2 Equation of time.

Name	Alias	RA 2000	Dec	π	μ	θ	v_R	Spt	V
Sun								G2 V	−26.72
Proxima Cen		14 30	−62 41	772.3	3853	281	−29	M5.5	15.45
α Cen A	Rigel Kent	14 40	−60 50	742.1	3709	277	−32	G2 V	−0.01
α Cen B				742.1	3724	285	−32	K1 V	1.35
Barnard's Star	BD + 4°3561	17 58	4 41	549.0	10358	356	−139	M5 V	9.54
Wolf 359	CN Leo	19 56	7 01	419	4702	235	55	M6.5 V	13.46
BD+36°2147	HD 95735	11 03	35 38	392.4	4802	187	−104	M2 V	7.49
Sirius A	α CMa	6 45	−16 43	379.2	1339	204	−18	A1 Vm	−1.44
Sirius B								DA2	8.44
L726–8 A	LB Cet	1 39	−17 56	373	3360	80	52	M5.5 Ve	12.56
L726–8 B	UV Cet	1 39	−17 56	373	3360	80	53	M5.5 Ve	12.96
Ross 154	V1216 Sag	18 50	−23 50	336.5	6660	107	−10	M3.6 Ve	10.37
Ross 248	HH And	23 42	44 09	316	1588	176	−84	M5.5 Ve	12.27
ε Eri	HD 22049	3 33	−9 27	310.8	977	271	22	K2 V	3.72
CD-36°15693	HD 217987	23 06	−35 51	303.9	6896	79	108	M2 V	7.35
Ross 128	FI Vir	11 48	0 48	299.6	1361	154	−26	M4 V	11.12
L789–6 A	EZ Aqr	22 39	−15 17	290	3256	47	−80	M5 Ve	12.69
L789–6 B									13.6
L789–6 C									
61 Cyg A	HD201091	21 07	38 45	287.1	5281	52	−108	K5 V	5.02
Procyon A	α CMi	7 39	5 13	285.9	1259	215	−21	F5 IV–V	0.40
Procyon B								DF	10.7
61 Cyg B	HD201092	21 07	38 45	285.4	5272	53	−108	K7 V	6.05
BD+59°1915 B	HD173740	18 43	59 38	284.5	2312	323	39	M4 V	9.7
BD+59°1915 A	HD173739			280.3	2238	324	−38	M3.5 V	8.94

D3 Coordinate transformations and relations

To find the angular separation, θ, between two objects having equatorial coordinates (α, δ_1) and $(\alpha + \Delta\alpha, \delta_2)$:

$$\cos\theta = \sin\delta_1 \sin\delta_2 + \cos\delta_1 \cos\delta_2 \cos(\Delta\alpha)$$

With appropriate substitutions, this relation will apply for any similar coordinate system on the surface of a sphere.

To find the altitude, e, and azimuth, a, of an object with equatorial declination, δ, when the object is at hour angle, H, observed from a location with geodetic latitude, β:

$$\sin e = \sin\delta \sin\beta + \cos\delta \cos H \cos\beta$$
$$\sin a = -(\cos\delta \sin H)/\cos e$$

The inverse relationships are

$$\sin\delta = \sin e \sin\beta + \cos e \cos a \cos\beta$$
$$\sin H = -(\cos e \sin a)/\cos\delta$$

D4 Atmospheric refraction

The difference between the true zenith distance, z, and the apparent zenith distance, z', in seconds of arc at visual wavelengths is approximately

$$z - z' = 16.27 \tan(z) \frac{P}{T}$$

where P is the atmospheric pressure in millibars and T is the temperature in kelvins. This formula is reasonably reliable for zenith distances less than 75°. For larger zenith distances, more complex formulae are available – see Problem 3.11. Refraction varies with wavelength.

D5 Days and years

There are several definitions of the length of time the Earth requires to complete one orbit. In all cases, the following hold:

$$
\begin{aligned}
1\ \text{day} &= 24\ \text{hours} = 1440\ \text{minutes} = 86{,}400\ \text{s} \\
1\ \text{Julian calendar year} &= 365.25\ \text{days} \\
&= 8766\ \text{hours} = 525{,}960\ \text{minutes} = 31{,}557{,}600\ \text{s} \\
1\ \text{Julian century} &= 36{,}525\ \text{days} \\
1\ \text{Gregorian calendar year} &= 365.2425\ \text{days}
\end{aligned}
$$

The Julian year (introduced by Julius Caesar) and the Gregorian year (introduced by Pope Gregory XIII in 1582) were each meant to approximate the tropical year. The following values are in units of 1 day of 86,400 SI seconds; T is measured in Julian or Gregorian centuries from 2000.0.

1 tropical year (equinox to equinox)	$= 365^{\text{d}}.242193 - 0^{\text{d}}.000\,0061\ T$
	$= 365^{\text{d}}\ 05^{\text{h}}\ 48^{\text{m}}\ 45^{\text{s}}.5 - 0^{\text{s}}.53\ T$
1 sidereal year (star to star)	$= 365^{\text{d}}.256360 + 0^{\text{d}}.0000001\ T$
	$= 365^{\text{d}}06^{\text{h}}09^{\text{m}}09^{\text{s}}.5 + 0^{\text{s}}.01\ T$
1 anomalistic year (perigee to perigee)	$= 365^{\text{d}}.259635$
	$= 365^{\text{d}}\ 06^{\text{h}}13^{\text{m}}52^{\text{s}}.5$
1 eclipse year (lunar nodes to lunar node)	$= 346^{\text{d}}.620076$
	$= 346^{\text{d}}14^{\text{h}}52^{\text{m}}54^{\text{s}}.6$

Julian dates can be computed from calendar dates by the formulae:

$$
\begin{aligned}
\text{JD} &= 2415020.5 + 365(Y - 1900) - L + d + t \\
&= 2451544.5 + 365(Y - 2000) - L + d + t
\end{aligned}
$$

where Y = current year, L = number of leap years since 1901 or 2001, d = UT day of the year, t = fraction of the UT day.

Appendix E

E1 The constellations

Abbreviation	Nominative	Genitive ending	Meaning	RA	Dec	Area deg^2
And	Andromeda	-dae	Chained princess	1	40 N	722
Ant	Antlia	-liae	Air pump	10	35 S	239
Aps	Apus	-podis	Bird of paradise	16	75 S	206
Aqr	Aquarius	-rii	Water bearer	23	15 S	980
Aql	Aquila	-lae	Eagle	20	5 N	652
Ara	Ara	-rae	Altar	17	55 S	237
Ari	Aries	-ietis	Ram	3	20 N	441
Aur	Auriga	-gae	Charioteer	6	40 N	657
Boo	Böotes	-tis	Herdsman	15	30 N	907
Cae	Caelum	-aeli	Chisel	5	40 S	125
Cam	Camelopardus	-di	Giraffe	6	70 N	757
Cnc	Cancer	-cri	Crab	9	20 N	506
CVn	Canes Venaticium	Canum corum	Hunting dogs	13	40 N	465
CMa	Canis Major	Canis Majoris	Great dog	7	20 S	380
CMi	Canis Minor	Canis Minoris	Small dog	8	5 N	183
Cap	Capricornus	-ni	Sea goat	21	20 S	414
Car	Carina	-nae	Ship's keel	9	60 S	494
Cas	Cassiopeia	-peiae	Seated queen	1	60 N	598
Cen	Centaurus	-ri	Centaur	13	50 S	1060
Cep	Cepheus	-phei	King	22	70 N	588
Cet	Cetus	-ti	Whale	2	10 S	1231
Cha	Chamaeleon	-ntis	Chameleon	11	80 S	132
Cir	Circinus	-ni	Compasses	15	60 S	93
Col	Columba	-bae	Dove	6	35 S	270
Com	Coma Berenices	Comae Berenicis	Berenice's hair	13	20 N	386
CrA	Corona Australis	-nae lis	Southern crown	19	40 S	128
CrB	Corona Borealis	-nae lis	Northern crown	16	30 N	179
Crv	Corvus	-vi	Crow	12	20 S	184
Crt	Crater	-eris	Cup	11	15 S	282
Cru	Crux	-ucis	Southern cross	12	60 S	68
Cyg	Cygnus	-gni	Swan	21	40 N	804
Del	Delphinus	-ni	Dolphin	21	10 N	189
Dor	Dorado	-dus	Swordfish	5	65 S	179
Dra	Draco	-onis	Dragon	17	65 N	1083

(continued)

Abbreviation	Nominative	Genitive ending	Meaning	RA	Dec	Area deg^2
Equ	Equuleus	-lei	Small horse	21	10 N	72
Eri	Eridanus	-ni	River	3	20 S	1138
For	Fornax	-acis	Furnace	3	30 S	398
Gem	Gemini	-norum	Twins	7	20 N	514
Gru	Grus	-ruis	Crane	22	45 S	366
Her	Hercules	-lis	Hero	17	30 N	1225
Hor	Horologium	-gii	Clock	3	60 S	249
Hya	Hydra	-drae	Water snake (F)	10	20 S	1303
Hyi	Hydrus	-dri	Water snake (M)	2	75 S	243
Ind	Indus	-di	Indian	21	55 S	294
Lac	Lacerta	-tae	Lizard	22	45 N	201
Leo	Leo	-onis	Lion	11	15 N	947
LMi	Leo Minor	-onis ris	Small lion	10	35 N	232
Lep	Lepus	-poris	Hare	6	20 S	290
Lib	Libra	-rae	Scales	15	15 S	538
Lup	Lupus	-pi	Wolf	15	45 S	334
Lyn	Lynx	-ncis	Lynx	8	45 N	545
Lyr	Lyra	-rae	Lyre	19	40 N	286
Men	Mensa	-sae	Table	5	80 S	153
Mic	Microscopium	-pii	Microscope	21	35 S	210
Mon	Monoceros	-rotis	Unicorn	7	5 S	482
Mus	Musca	-cae	Fly	12	70 S	138
Nor	Norma	-mae	Square	16	50 S	165
Oct	Octans	-ntis	Octant	22	85 S	291
Oph	Ophiuchus	-chi	Serpent-bearer	17	0	948
Ori	Orion	-nis	Hunter	5	5 N	594
Pav	Pavo	-vonis	Peacock	20	65 S	378
Peg	Pegasus	-si	Winged horse	22	20 N	1121
Per	Perseus	-sei	Champion	3	45 N	615
Phe	Phoenix	-nisis	Phoenix	1	50 S	469
Pic	Pictor	-ris	Painter's easel	6	55 S	247
Psc	Pisces	-cium	Fishes	1	15 N	889
PsA	Piscis Austrinus	-is ni	Southern fish	22	30 S	245
Pup	Puppis	-pis	Ship's stern	8	40 S	673
Pyx	Pyxis	-xidis	Ship's compass	9	30 S	221
Ret	Reticulum	-li	Net	4	60 S	114
Sge	Sagitta	-tae	Arrow	20	10 N	80
Sgr	Sagittarius	-rii	Archer	19	25 S	867
Sco	Scorpius	-pii	Scorpion	17	40 S	497
Scl	Sculptor	-ris	Sculptor	0	30 S	475
Sct	Scutum	-ti	Shield	19	10 S	109
Ser	Serpens	-ntis	Serpent	17	10 N	637

Sex	Sextans	-ntis	Sextant	10	0	314
Tau	Taurus	-ri	Bull	4	15 N	797
Tel	Telescopium	-pii	Telescope	19	50 S	252
Tri	Triangulum	-li	Triangle	2	30 N	132
TrA	Triangulum Australe	-li lis	Southern triangle	16	65 S	110
Tuc	Tucana	-nae	Toucan	0	65 S	295
UMa	Ursa Major	-sae ris	Great Bear	11	50 N	1280
UMi	Ursa Minor	-sae ris	Small Bear	15	70 N	256
Vel	Vela	-lorum	Sails	9	50 S	500
Vir	Virgo	-ginis	Virgin	13	0	1294
Vol	Volans	-ntis	Flying fish	8	70 S	141
Vul	Vulpecula	-lae	Small fox	20	25 N	268

Sources: *Allen's Astrophysical Quantities*, 4th edition, 2000, A. N. Cox, ed., Springer, New York; *The Observer's Handbook*, 2002, Rajive Gupta, ed., The Royal Astronomical Society of Canada, Toronto.

E2 Some named stars

Name	Alternative designation	V	Claim to fame
Albireo	β Cyg	3.08	Telescopic double
Alcor	80 UMa	4.01	Visual double with Mizar
Alcyone	η Tau	2.87	Brightest Pleiad
Aldebaran	α Tau	0.87	Bright red, near ecliptic
Algol	β Per	2.09	Eclipsing variable
Alnilam	ε Ori	1.69	Middle star of Orion's belt
Altair	α Aqi	0.76	
Antares	α Sco	1.06	Very red
Arcturus	α Boo	−0.05	Brightest in northern hemisphere
Barnard's Star	HIP 87937	9.54	Largest proper motion, 2nd nearest system
Bellatrix	γ Ori	1.64	West shoulder of Orion
Betelgeuse	α Ori	0.45	East shoulder. Very red. Variable
Canopus	α Car	−0.62	Second brightest
Capella	α Aur	0.08	
Castor	α Gem	1.58	
Cor Caroli	α CVn	2.90	Undistinguished white star. Named by Halley to mock Charles I
Deneb	α Cyg	1.25	
Denebola	β Leo	2.14	Tail of the lion

(continued)

Name	Alternative designation	V	Claim to fame
Dubhe	α UMa	1.81	Northern of the two pointer stars
Formalhaut	α PsA	1.17	
Kapteyn's Star	HD 33793	8.86	Large proper motion (8.8″ yr^{-1})
Luyten's Star	HIP 36208	9.84	Large proper motion
Merak	β UMa	2.34	Southern of the pointers
Mintaka	δ Ori	2.25	Western end of belt
Mizar	ζ UMa	2.23	Visual double with Alcor
Plaskett's Star	HD 47129	6.05	Most massive binary
Polaris	α UMi	1.97	Pole star
Pollux	β Gem	1.16	
Procyon	α CMi	0.40	
Proxima Centauri	α Cen C	11.01	Nearest star. Member of α Cen system
Regulus	α Leo	1.36	
Rigel	β Ori	0.18	West foot of Orion
Rigel Kent	α Cen A + B	−0.01	Nearest system
Saiph	κ Ori	2.07	East foot
Sirius	α CMa A	−1.44	Brightest. Fourth-nearest system
Sirius B	α CMa B	8.4	Nearest white dwarf
Spica	α Vir	0.98	
Thuban	α Dra	3.65	Former pole star
Vega	α Lyr	0.03	
Zubenelgenubi	α Lib	2.75	

E3 Naming small bodies in the Solar System

Minor planets

The ***provisional designation*** is a four-part name, with all parts related to the date of discovery. Assigned by the Central Bureau for Astronomical Telegrams (CBAT), it combines:

- The year of discovery (all four digits).
- A single upper-case Roman letter, coding the UT half-month of the discovery. Months always divide on the 15th day: e.g. A = Jan 1–15, B = Jan 16–31, D = Feb 16–29. This uses all letters except I and Z.
- A second upper-case letter, indicating the order of discovery within the half-month (A = first, Z = last). The letter I is not used.
- If there are more than 25 discoveries in a half-month (there usually are) append a final number, indicating the number of times the second letter has been recycled. This should be written as a subscript, if practical.

So for example,

- 1999 LA1 is the 26th provisional discovery made during the interval June 1–15 in 1999.
- 2002 WZ_5 is the 150th provisional discovery made during the interval November 16–30 in 2002.

The ***permanent designation***, assigned after observers establish a definitive orbit, consists of a sequential catalog number followed by a name. Names are proposed by the discoverer and approved by the Committee for Small-Body Nomenclature of the IAU. The temporary designation is used if a name is not proposed and approved. Example permanent designations are: (1) Ceres, (2) Pallas, (8) Flora, (88) Thisbe, (888) Parysatis, (8888) 1994 NT_1, (9479) Madres-PlazaMayo, (9548) Fortran, and (9621) MichaelPalin. As of 2010, there were over 200,000 objects with permanent designations, many discovered by auto-mated search programs like **L**incoln **L**aboratory **N**ear **E**arth **A**steroid **R**esearch Project (LINEAR).

Comets

The modern rules for comet designations are similar to those for minor planets. Indeed, the distinction between comets and minor planets is not always clear, and it is helpful that there are similarities in naming schemes. Upon discovery, the Central Bureau for Astronomical Telegrams (CBAT) assigns a candidate new comet a ***provisional designation*** based on the date of the discovery. The designation consists of the four-digit year, a single letter designating the half-month, and a final numeral indicating the order within the half-month. It has been traditional since the eighteenth century that new comets be named after the observer (or group, program, or satellite) who discovers them. Therefore the provisional designation also contains the name of the discoverer, as determined by the IAU committee on Small Body Nomenclature. It may also contain a prefix describing the nature of the orbit, using the codes:

P/, a short-period comet (P < 200 years)
C/, a long-period comet
X/, orbit uncertain
D/, disappeared, destroyed or lost
A/, an object later determined to be an asteroid

Thus, C/2021 E17 (Smith) would be the provisional designation for the seven-teenth cometary discovery announced in the interval March 1–15, 2021. The (fictional) object is in a long-period orbit, and Smith is credited with its discov-ery. D/1993 F2 (Shoemaker-Levy 9) was an actual comet discovered by Eugene and Carolyn Shoemaker and David Levy in the second half of March 1993. It was destroyed in a spectacular collision with Jupiter.

Most newly discovered comets are in such large orbits that a reliable ephemeris cannot be computed for the comet's next perihelion passage. The "periodic" (i.e. short period) and destroyed comets are the exception, and these are given a ***permanent designation*** that prefixes a catalog number, assigned in order of orbit discovery or comet destruction to the provisional name. For most references, the date segment of the name can be dropped. An example, undoubtedly the most famous, is 1P/1682 Q1 (Halley) = 1P/Halley. As of 2010, there were 227 comets with permanent catalog numbers.

Natural satellites of the major and minor planets

Again, designation of these objects parallels the practice for minor planets. A provisional designation consists of (1) the prefix S/, (2) the year of discovery, (3) a roman letter coding the planet, or a parenthetical numeral of the numbered asteroid, and (4) a numeral giving order of announcement within the year. For example:

> S/2000 J 7 is a satellite of Jupiter
> S/2002 (3749) 1 is a satellite of the minor planet (3749) Balam

Once an orbit is well defined, the temporary designation is replaced with a sequential roman numeral affixed to the planet name, and a permanent name whose selection is based on mythological or literary themes. For example:

> Jupiter I = Io
> Jupiter XVI = Metis
> Uranus XIII = Belinda
> Neptune VI = Galatea

Appendix F

F1 A timeline for optical telescopes

See also: http://amazing-space.stsci.edu/resources/explorations/groundup/

c. 3500 BC	Invention of glass in Egypt and Mesopotamia.
c. 2000 BC	Lenses fashioned from rock crystal in Ionia. Use unknown.
424 BC	Aristophanes (*The Clouds*, Act II, Scene 1) describes the focusing power of a glass globe filled with water.
c. 300 BC	Euclid gives a rudimentary treatment of the ray theory of light and of refraction at a plane interface. Euclid, following Plato, believed rays moved from the eye of the beholder to the object beheld.
212 BC	Archimedes is reported to have used curved mirrors to focus sunlight and set fire to the sails of Roman ships during the siege of Syracuse. It is not reported how proponents of Euclid's ray theory accounted for this effect.
c. 300 BC– AD 170	Hellenistic astronomers make naked-eye observations with the armillary sphere and the mural quadrant, predecessors of the equatorial mount and the transit telescope. Ptolemy's catalog of 1000 stars, with positions precise to about 15 minutes of arc, becomes a standard for the next millennium.
c. AD 1000	Ibn al-Haitham (Latin: Alhazen), in Egypt, conducts experiments in optics and writes on spherical mirrors, lenses, and refraction.
c. 1275	Roger Bacon, English philosopher, conducts optics experiments and describes, for example, the magnifying power of a plano-convex lens. Vitello of Silesia publishes a large volume on optics, founded upon and advancing Alhazen.
c. 1285	Spectacles invented (first examples appear in northern Italy, manufactured from high-quality glass produced in Venice).
1565–1601	Tycho designs a spectacular series of instruments that permit naked-eye observations to the unprecedented accuracy of one minute of arc.
1608	Hans Lippershey, a spectacle-maker, petitions the States-General of the Netherlands for a patent on his invention of the spy-glass (Galilean telescope). The patent is denied because "many other persons have a knowledge of the invention."
1609	Galileo, on hearing rumors of Lippershey's device, constructs one for himself. He uses subsequent models to observe the night sky, with momentous consequences.
1611	After acquiring a telescope, Kepler, familiar with the work of Vitello, writes a treatise on optics, *Dioptrice*. This includes the first description of spherical aberration and of the Keplerian telescope (objective and ocular both convex lenses).
1612–1690	Era of very long focal length refractors, constructed with single-lens objectives and limited by both chromatic aberration and SA. Large focal ratios minimize both aberrations.

(continued)

1655	Invention of the pendulum clock by Huygens makes positional astronomy with transit instruments a much more precise enterprise. At about the same time, Huygens also introduces a micrometer ocular, and "divided instruments" of the sort used by Tycho begin to appear equipped with telescopic sights.
1663–1674	First reflecting telescopes designed by Gregory, Newton, Cassegrain, and Hooke. A number of large (up to 80-cm aperture) speculum-metal reflecting telescopes appear over the next century, but most of the advances in astronomy come from refractors.
1688–1720	Flamsteed observes the positions of 3000 stars with a 7-foot quadrant equipped with telescopic sights. Positional accuracy is about 10 seconds of arc in declination, and about 1 second of time (15 seconds of arc at the equator) in RA.
1721	John Hadley produces a 6-inch $f/10.3$ Newtonian reflector, which rivals the performance of the "long" refractors.
1728	Bradley discovers the velocity aberration of starlight using a zenith sector – a telescope suspended vertically – to measure changes in the apparent declination of stars transiting near the zenith. Bessel (in 1818) shows that Bradley's fundamental catalog, derived from observations with a quadrant and a transit telescope, have a positional accuracy of 4 seconds of arc in declination and 1 second of time in RA.
1729	Chester Moor Hall works out the theory of the achromatic doublet. Doublets will not be used in telescopes until the work of Dolland in the 1750s.
1733–1768	James Short in England manufactures a number of excellent speculum-metal reflectors, many in the Gregorian configuration.
c. 1760	The first achromatic objectives begin to appear. These are of small aperture (< 10 cm) owing to the difficulty of casting blanks of flint glass that is free from optical flaws.
1761–1764	Clairaut gives a rigorous explication of the achromatic doublet, and uses ray tracing to characterize most of the third-order aberrations.
1781	William Herschel discovers the planet Uranus with a 16-cm Newtonian reflector. Herschel's fame and productivity, especially with his 45-cm reflector, led to increasing popularity of speculum-metal reflectors.
1812–1826	Fraunhofer combines an ability to produce large flint blanks with practical methods of optical testing, and produces a number of excellent achromatic refractors. The tide at observatories turns away from reflectors. With the "Great Dorpat Refractor" (aperture 24 cm), Fraunhofer introduces the German equatorial mount.
1848	Fighting the tide, William Parsons, Earl of Rosse, builds a Herschel-style reflector with an aperture of 1.8 meters. This telescope is heavy, unwieldy, and located at a poor site. Refractors with 0.4-meter apertures, installed at Pulkova and at Harvard at about this time, prove more productive. The refractor is supreme.
1852	William Lassell applies the equatorial mount to large reflectors. He moves his 0.6-meter telescope from England to the island of Malta in order to obtain better seeing. It will be 50 years before astronomers fully recognize the importance of site selection.
1856	Leon Foucault, von Steinheil, and others introduce methods for depositing silver on glass mirrors.

1892	The refractor reaches its technological limit with the Alvan Clark 1.0-meter Yerkes Observatory telescope. At about this time, there is a flurry of construction of astrographs – refractors optimized for photographic work.
1895	Edward Crossley donates his private telescope, a silver-on-glass 0.9-meter reflector, to the Lick Observatory, because he recognizes the telescope is wasted at its site at Halifax in northern England. The remounted telescope demonstrates the suitability of large reflectors for photographic work.
1917	With the commission of the 2.5-meter (100-inch) reflector on Mt. Wilson, it is clear that telescopes with silvered-glass mirrors have surpassed refractors in light-gathering power and cost effectiveness. From now on, all new major optical telescopes will be reflectors. Because of the world wars, new telescope construction is greatly curtailed in Europe for 60 years.
1948	Reflector technology reaches a plateau with the 5.0-meter on Palomar Mountain. Over the next 50 years, gradual technological advances result in 24 large optical–infrared telescopes with apertures between 2.5 and 4.2 meters. All are reflectors. Most have equatorial mounts, although altazimuths become common after 1985.
1953	US astronomer Horace Babcock lays out the principle of adaptive optics. No practical system will be developed for another 20 years.
1974	The Anglo-Australian Telescope, a 3.9-meter equatorial, is the first large telescope designed with computer-controlled pointing and tracking.
1975	The 6.0-meter Bolshoi Teleskop Azimutal'ny is installed in the Caucus Mountains. Generally regarded as only partially successful, it demonstrates the superiority of the altazimuth for large telescopes.
1979	The Multiple-Mirror Telescope (MMT) on Mt. Hopkins, Arizona, uses six 1.8-meter mirrors to bring light to a common focus, and demonstrates several concepts employed by future large-aperture systems.
1982	The Advanced Research Projects Agency (ARPA; US Department of Defense) demonstrates the Compensated Imaging System, a practical adaptive optics system for imaging artificial Earth satellites, culminating ten years of development effort.
1984	The 2.6-meter Nordic Optical Telescope on La Palma is the first to use effective climate control and active primary mirror support.
1989	The first astronomical adaptive optics images are obtained – diffraction-limited K-band images with the 1.5-meter telescope at the Observatoire de Haute-Provence in France. The successor systems, COME-ON/ADONIS, produce a steady stream of practical results from the ESO 3.6-meter telescope at la Silla, Chile. The 3.5-meter New Technology Telescope (NTT) is a highly developed active optics system that achieves seeing discs as small as 0.3 seconds of arc.
1990	The Hubble Space Telescope (HST) is launched. After repair of residual SA in the optics in 1993, this 2.4-meter telescope achieves a resolution of 0.1 arcsec in the visible. The HST was followed by three other large space telescopes in the NASA great observatories program: the Compton Gamma-Ray Observatory in 1991, the Chandra X-Ray Observatory in 1999, and the Spitzer Infrared Observatory in 2003.

(continued)

1993	Keck I 9.8-m aperture telescope with an active optics segmented primary is installed on Mauna Kea.
1997–2008	A half-dozen large optical telescopes with light-gathering power equivalent to apertures in the 6.5–16.4 meter range see first light (see Appendix A6).
1999	Adaptive optics system operational on the second Keck telescope (Keck II), achieving resolutions of 0.02 arcsec for bright stars in K band.
2001	Tests successfully combine beams of the Keck I and II telescopes for operation as an interferometer with a resolution of 1 mas. Similar tests combine beams of the ESO VLT telescopes in Chile for a similar resolution.
2013	Projected launch of the James Webb Space Telescope (6.6-m aperture).
2018–2020	Projected first lights for the Thirty-Meter Telescope, the Giant Magellan Telescope (24.5-meter aperture), and the European Extremely Large Telescope (E-ELT, 42-meter aperture).

Appendix G

G1 Websites

AO (ESO): http://www.eso.org/projects/aot/introduction.html
AO (Keck): http://www2.keck.hawaii.edu:3636/realpublic/inst/ao/ao.html
HET: http://www.as.utexas.edu/mcdonald/het
HST: http://www.stsci.edu/hst/
JWST: http://www.ngst.nasa.gov/
Optical glass (Schott, Inc): http://www.schott-group.com/english/
company/us.html
Spin casting mirrors: http://medusa.as.arizona.edu/mlab
Subaru telescope: http://SubaruTelescope.org/index.html

G2 Largest optical telescopes (2009)*

Name	Organization location	Aperture, mirror type	Focal ratios	Year	Comments
VLT (Very Large Telescope)	European South Observatory (ESO),	8.2 m × 4,	R:13.4,	1999–2001	Can combine four large beams equivalent to 16.4 m aperture. Four additional 1 m telescopes improve interferometric resolution
	Cerro Paranal, Chile	monolithic zerodur meniscus	N:15, C:47.3		
Keck I & II	University of California;	9.82 m × 2,	P:1.75,	1993–1996	Interferometric combination of two beams (2001) gives 13.9 m equivalent aperture
	California Institute of Technology;	1.8 m × 36 (each), hexagonal zerodure segments	R:15		

(continued)

Name	Organization location	Aperture, mirror type	Focal ratios	Year	Comments
	Mauna Kea, Hawaii				
LBT (Large Binocular Telescope)	LBT	8.4 m × 2,	P:1.14,	1999–2001	Major partners are University of Arizona and the Vatican Observatory. Combined beams give 11.9 m aperture
	Consortium, Mt. Graham, Arizona	spin-cast ribbed borosilicate	5.4 R:15		
GTC (Gran Telescopio Canarias)	Spain + others,	10.4 m	P:1.75	2002	
	La Palma, Canary Islands	1.9 m × 36, hexagonal segments	R:15 N:25		
HET (Hobby-Eberly Telescope)	University of Texas, Pennsylvania State University, Stanford, München, Gottingen,	1.0 m × 91, hexagonal zerodure segments	P:1.42	1997	Spherical primary is stationary during observations, with tracking done in the focal plane. 9.2 m entrance pupil on 11 m primary
	McDonald Observatory				
Magellan	Consoritum Las Campanas, Chile;	2 × 6.5 m, spin-cast honeycomb borosilicate	P:1.25, N:11, C:15	1999 2003	
Subaru	National Observatory of Japan; Mauna Kea, Hawaii	8.3 m Corning ULE thin meniscus	P:2.0, R:12.2, N:12.6	1998	

Gemini North	USA; UK; Canada; Chile; Australia; Argentina; Brazil;	8.1 m,	R:16	1998 (N)	Optimized for near infrared
Gemini South	Mauna Kea, Hawaii Cerro Pachon, Chile	Corning ULE meniscus		2000 (S)	
Smithsonian Astrophysical Observatory, Mt. Hopkins, Arizona (MMT)		6.5 m, spin-cast ribbed borosilicate	P:3.0	1978	

* Focal ratios for prime focus (P), RC (R), Naysmith (N), and coudé (C). Corning ULE is an ultra low thermal expansion coefficient glass, zerodur and borosilicate are types of glass.

G3 Large Schmidt telescopes

Name	Location	Diameter: corrector / mirror	Focal ratio	Year	Comments
Tautenberg	Tautenberg, Germany	1.34/2.00	f/3.00	1960	Equipped with a Nasmyth and coudé focus, multi-object spectrograph
Oschin	Palomar Mt., California	1.24/1.83	f/2.47	1948	Important because it was the first very large Schmidt. Produced the Palomar Sky Survey (PSS), a basic reference tool
UK Schmidt	Siding Spring Mt., Australia	1.24/1.83	f/2.5	1973	Collaborated on the ESO-Science and Engineering Research Council (UK) survey, extending the PSS project to the southern hemisphere
Kiso	Kiso, Japan	1.05/1.5	f/3.1	1975	
Byurakan	Mt. Aragatz, Armenia	1.0/1.5	f/2.13	1961	Conducted objective prism survey for galaxies with ultraviolet bright nuclei
Uppsala	Kvistaberg, Sweden	1.0/1.35	f/3.00	1963	Surveys for near-Earth objects.
ESO	La Silla, Chile	1.0/1.62	f/3.06	1972	Decommissioned in 1998
Venezuela	Merida, Venezuela	1.0/1.52	f/3.0	1978	

Appendix H

H1 Some common semiconductors

Forbidden band-gap energies and cutoff wavelengths are given at room temperature, except where noted. Band-gap data from Section 20 of Anderson (1989) or from Kittel (2005).

Material		Band gap (eV)	λ_c (μm)
IV			
Diamond	C	5.48	0.23
Silicon	Si	1.12	1.11
	Si (4 K)	1.17	1.06
	Si (700 K)	0.97	1.28
Germanium	Ge	0.67	1.85
	Ge (1.5 K)	0.744	1.67
Gray tin	αSn	0.0	
Silicon carbide	SiC	2.86	0.43
III–V			
Gallium arsenide	GaAs	1.35	0.92
Gallium antimonide	GaSb	0.68	1.83
Indium phosphide	InP	1.27	0.98
Indium arsenide	InAs	0.36	3.45
Indium antimonide	InSb	0.18	6.89
	InSb(77K)	0.23	5.39
Boron phosphide	BP	2.0	0.62
II–VI			
Cadmium sulfide	CdS	2.4	0.52
Cadmium selenide	CdSe	1.8	0.69
Cadmium teluride	CdTe	1.44	0.86
Mercury cadmium teluride	$Hg_xCd_{1-x}Te$	0.1 to 0.5 (x = 0.8 to 0.5)	12.4–2.5
IV–VI			
Lead sulfide	PbS	0.42	2.95

Appendix I

I1 Characteristics of some commercial CCDs for astronomy

These are advanced devices nevertheless within budgetary reach of a small observatory. See manufacturers' websites (I2) for additional examples.

Device	e2v CCD42–90	Kodak KAF4301E	SITe SI242A
Type	Three-phase BCCD, backthinned, polysilicon gates, three-side buttable	Two-phase, front-illuminated ITO gates, can switch to low (LR) or high (HR) responsivity amplifier	Three-phase BCCD, backthinned, polysilicon gates
Dimensions, $C \times R$	2048 × 4096	2084 × 2084	2048 × 2048
Pixel size, μm	13.5 × 13.5	24 × 24	24 × 24
Output amplifiers	Two parallel	Two (select either LR or HR)	Two parallel + two serial
Amplifier responsivity, μV electron^{-1}	4.5	2 (LR) or 11.5 (HR)	1.3
Pixel full well (electrons)	150,000	570,000	200,000
Serial full well (electrons)	600,000	na	na
Summing full-well or amplifier saturation (electrons)	900,000	1,500,000 (LR) 150,000 (HR)	na
CTE	0.999 995	0.999 990	0.999 990
Read noise, rms	3 (at 20 kHz)	13 (HR) 22 (LR) (at 1 Mhz)	7 (at 45 kHz)
Dark current, electrons pixel^{-1} s^{-1}	0.0003 (at –100 °C)	85 (at 20 °C)	1800 (at 20 °C)
AR coating: QE with coating, % at wavelength (nm):	Blue, visual	Red	Ultraviolet, std, frontside
350	50, 17	10	65, 25, 2
400	80, 52	35	70, 63, 3
500	85, 92	50	72, 75, 22
650	80, 93	65	71, 86, 38
900	50, 55	32	40, 53, 25

I2 Manufacturers of sensors and cameras for astronomy

Fairchild Imaging: http://www.fairchildimaging.com

Teledyne Scientific and Imaging (SITe, HAWAII and PICNIC arrays): http://www.teledyne-si.com/

DALSA Corporation: http://www.dalsa.com

Hamamatsu Corporation (photo-emissive devices): http://jp.hamamatsu.com/products/sensor-etd/pd007/index_en.html.

Hamamatsu Photonics: http://www.hamamatsu.com

HORIBA Jobin Yvon Inc.: http://www.jobinyvon.com

Kodak: http://www.kodak.com/global/en/business/ISS/Applications/scientific.jhtml

Santa Barbara Instrument Group: http://www.sbig.com/

Apogee: http://www.ccd.com/

Finger Lakes Instruments: www.flicamera.com/

Pan-STARRS: http://pan-starrs.ifa.hawaii.edu/public/design-features/cameras.html

e2v: http://www.e2v.com

Appendix J

J1 The point-spread function

The point-spread function (PSF) is the two-dimensional brightness distribution produced in the plane of the detector by the image of an unresolved source, such as a distant star. A real detector has finite pixel size, so it records a matrix of pixel-sized samples of the PSF. (see, for example, Figures 9.12 and 9.20). Astronomers often face the problem of determining the PSF that best fits the pixilated detector image of one or more point sources. There are two different approaches:

(1) One approach is to assume some analytical function, $P(x-x_0, y-y_0)$, describes the brightness distribution as a function of the displacement from the image centroid at (x_0, y_0). In this case you select a likely function with a sufficient number of free parameters, and select the parameter values that best fit the one or more observed stellar images. For example, images dominated by seeing are often thought to assume the **Moffat** profile:

$$P = B_0 \left[1 + \left(\frac{r}{\alpha} \right)^2 \right]^{-\beta}$$

where

$$r^2 = (x - x_0)^2 + (y - y_0)^2$$

The parameters α and β determine the PSF shape, while B_0 sets the scale, and a least-squares fit can estimate their values. The above profile has circular isophotes. You would require additional parameters to fit the more irregular profiles often seen in practice. For example, you could fit profiles that had elliptical isophotes of ellipticity ε, elongated direction θ_0, by setting

$$\alpha^2 = \alpha_0^2 \left[\cos^2 \left\{ (\tan^{-1} \frac{y}{x}) - \theta_0 \right\} + \varepsilon \sin^2 \left\{ (\tan^{-1} \frac{y}{x}) - \theta_0 \right\} \right]$$

in the Moffat formula. **Gaussian** functions are also popular choices for analytical PSFs.

(2) A second approach is to use bilinear interpolation to estimate the brightness values at fractional pixel positions in an observed stellar image (or in several). This produces a completely empirical PSF, and has the advantage of coping nicely with very irregular profiles. It has difficulty near the centroid, where interpolated values are unstable, especially in under-sampled images.

A frequent strategy is to meld the two approaches: First fit the data with an analytical function, then fit the residuals using the empirical method, so the final PSF is the sum of the two fits.

Appendix K

K1 Intrinsic broadband colors for various spectral types

Calibration of MK spectral types with temperatures, absolute magnitudes, bolometric corrections (magnitudes) and UBV colors is from Table 15.7 in *Allen's Astrophysical Quantities* (Cox, 1999). The JHK colors are from Ducati *et al.* (2001). Cousins R and I colors are an average of the computed values from Bessel (1990) and the values in the Johnson system given by Ducati *et al.* (2001) transformed to the Cousins system. Data for spectral types later than M5 are from Bessel (1991), Barsi (2000) and Chabrier and Baraffe (2000) and should be regarded as preliminary. Tables K1 and K2 are inconsistent by about one spectral type.

Table K1. *Main sequence dwarves (luminosity class V)*

Spectral type	T_{eff}	M_V	BC	$B - V$	$U - B$	$(V - R)_C$	$(V - I)_C$	$V - J$	$V - H$	$V - K$
O5	42000	−5.7	4.40	−0.33	−1.19	−0.18	−0.33			
B0	30000	−4.0	−3.16	−0.30	−1.08	−0.16	−0.26	−0.80	−0.92	−0.97
B2	20900	−2.45	−2.35	−0.24	−0.84	−0.11	−0.24	−0.67	−0.79	−0.89
B5	15200	−1.2	−1.46	−0.17	−0.58	−0.07	−0.17	−0.51	−0.62	−0.71
B8	11400	−0.25	−0.80	−0.11	−0.34	−0.05	−0.11	−0.36	−0.45	−0.49
A0	9790	+0.65	−0.30	−0.02	−0.02	−0.02	−0.02	−0.16	−0.19	−0.17
A2	9000	+1.3	−0.20	+0.05	+0.05	0.01	0.02	−0.07	−0.04	−0.05
A5	8180	+1.95	−0.15	+0.15	+0.10	0.05	0.12	0.09	0.19	0.15
F0	7300	+2.7	−0.09	+0.30	+0.03	0.16	0.35	0.37	0.57	0.52
F2	7000	+3.6	−0.11	+0.35	0.00	0.19	0.41	0.48	0.71	0.66
F5	6650	+3.5	−0.14	+0.44	−0.02	0.24	0.49	0.67	0.93	0.89
F8	6250	+4.0	−0.16	+0.52	+0.02	0.28	0.58	0.79	1.06	1.03
G0	5940	+4.4	−0.18	+0.58	+0.06	0.31	0.63	0.87	1.15	1.14
G2	5790	+4.7	−0.20	+0.63	+0.12	0.33	0.65	0.97	1.25	1.26
G5	5560	+5.1	−0.21	+0.68	+0.20	0.35	0.68	1.02	1.31	1.32
G8	5310	+5.5	−0.40	+0.74	+0.30	0.38	0.72	1.14	1.44	1.47
K0	5150	+5.9	−0.31	+0.81	+0.45	0.44	0.80	1.34	1.67	1.74
K2	4830	+6.4	−0.42	+0.91	+0.64	0.52	0.95	1.60	1.94	2.06
K5	4410	+7.35	−0.72	+1.15	+1.08	0.71	1.28	2.04	2.46	2.66
M0	3840	+8.8	−1.38	+1.40	+1.22	0.8	1.71	2.49	3.04	3.29
M2	3520	+9.9	−1.89	+1.49	+1.18	0.91	2.07	2.74	3.42	3.67
M5	3170	+12.3	−2.73	+1.64	+1.24	1.30	3.18			

Table K2. *Lower main sequence dwarves*

Spectral type	T_{eff}	M_V	M_I	$(V-I)_C$	$R-I$	$I-J$	$J-H$	$H-K$
M6	2800	16.6	12.5	4.13	2.22	2.96	0.66	0.38
M8	2400	14.7	14	4.5	2.99	2.90	0.68	0.431
L0	2200	19.6:	15	4.6	3.1	3.0	0.82	0.58
L2	2000	21.8	17	4.8	3.22	3.2	1.02	0.70
L5	1700	24.7?	18.5:	6.25?	3.94	3.3	1.17	0.70
T2	1400		>19.		5.47	4.8	0.84	0.15
T6	1200					6.3	0.0	−0.03

Table K3. *Giant stars (luminosity class III)*

Spectral type	T_{eff}	M_V	BC	$B-V$	$U-B$	$(V-R)_C$	$(V-I)_C$	$V-J$	$V-H$	$V-K$
G5	5050	+0.9	−0.34	+0.86	+0.56	0.47	0.93	0.94	1.44	1.53
G8	4800	+0.8	−0.42	+0.94	+0.70	0.50	0.47	1.11	1.61	1.77
K0	4660	+0.7	−0.50	+1.00	+0.84	0.49	1.00	1.23	1.72	1.94
K2	4390	+0.5	−0.61	+1.16	+1.16	0.84	1.11	1.56	2.08	2.39
K5	4050	−0.2	−1.02	+1.50	+1.81	0.60	1.53	2.25	2.87	3.14
M0	3690	−0.4	−1.25	+1.56	+1.87	0.88	1.78	2.55	3.23	3.46
M2	3540	−0.6	−1.62	+1.60	+1.89	0.92	1.97	2.99	3.76	3.89
M5	3380	−0.3	−2.48	+1.63	+1.58	1.30	3.04	3.95	4.98	4.73
M6	3250						3.80	4.34	5.50	5.04

Table K4. *Supergiant stars (luminosity class I)*

Spectral type	T_{eff}	M_V	BC	$B-V$	$U-B$	$(V-R)_C$	$(V-I)_C$	$V-J$	$V-H$	$V-K$
O9	32000	−6.5	−3.18	−0.27	−1.13	−0.08	−0.16	−0.57	−0.75	−0.84
B2	17600	−6.4	−1.58	−0.17	−0.93	−0.06	−0.09	−0.43	−0.56	−0.63
B5	13600	−6.2	−0.95	−0.10	−0.72	−0.01	−0.07	−0.28	−0.34	−0.39
B8	11100	−6.2	−0.66	−0.03	−0.55	0.03	0.07	−0.12	−0.14	−0.15
A0	9980	−6.3	−0.41	−0.01	−0.38	0.04	0.09	0.02	0.06	0.08
A2	9380	−6.5	−0.28	+0.03	−0.25	0.05	0.11	0.11	0.17	0.21
A5	8610	−6.6	−0.13	+0.09	−0.08	0.08	0.20	0.20	0.29	0.35

(*continued*)

Table K4 (*cont.*)

Spectral type	T_{eff}	M_V	BC	$B - V$	$U - B$	$(V - R)_C$	$(V - I)_C$	$V - J$	$V - H$	$V - K$
F0	7460	−6.6	−0.01	+0.17	+0.15	0.13	0.28	0.36	0.51	0.60
F2	7030	−6.6	−0.00	+0.23	+0.18	0.18	0.33	0.44	0.62	0.73
F5	6370	−6.6	−0.03	+0.32	+0.27	0.21	0.42	0.57	0.79	0.91
F8	5750	6.5	0.09	+0.56	+0.41	0.27	0.52	0.87	1.17	1.34
G0	5370	−6.4	−0.15	+0.76	+0.52	0.36	0.66	1.14	1.52	1.71
G2	5190	−6.3	−0.21	+0.87	+0.63	0.40	0.75	1.35	1.80	1.99
G5	4930	−6.2	−0.33	+1.02	+0.83	0.41	0.76	1.61	2.13	2.32
G8	4700	−6.1	−0.42	+1.14	+1.07	0.56	1.03	1.83	2.41	2.59
K0	4550	−6.0	−0.50	+1.25	+1.17	0.61	1.12	2.01	22.64	2.80
K2	4310	−5.9	−0.61	+1.36	+1.32	0.65	1.22	2.20	2.87	3.01
K5	3990	−5.8	−1.01	+1.60	+1.80	0.86	1.72	2.74	3.55	3.59
M0	3620	−5.6	−1.29	+1.67	+1.90	0.97	2.04	3.07	3.97	3.92
M2	3370	−5.6	−1.62	+1.71	+1.95	1.11	2.39	3.45	4.45	4.28
M5	2880	−5.6	−3.47	+1.80	+1.60:	2.7		5.26	6.68	5.73

References

Allen, R. H., *Star Names and their Meanings* (Original edition, Stechert, New York, NY, 1899; reprinted: Dover, Washington, DC, 1963). Online: http://penelope.uchicago.edu/ Thayer/E/Gazetteer/Topics/astronomy/home.html

Anderson, H. L., ed., 1989. *A Physicist's Desk Reference*, New York, American Institute of Physics.

Barsi, G. 2000. *Annual Review of Astronomy and Astrophysics*, **38**, 485.

Bessell, M. S., 1990. *Publications of the Astronomical Society of the Pacific*, **102**, 1181.

Bessell, M. S., 1991. *Astronomical Journal*, **101**, 662.

Bessell, M. S., 1992. *IAU Colloquium*, **136**, 22.

Bessell, M. S., 2005. *Annual Review of Astronomy & Astrophysics*, **43**, 293.

Bessell, M. S., Caselli, F. and Plez, B., 1998. *Astronomy and Astrophysics*, **333**, 231

Bevington, P. R., 1969. *Data Reduction and Error Analysis for the Physical Sciences*, New York, NY, McGraw-Hill.

Birney, D. S., Gonzalez, G., and Oesper, D., 2006. *Observational Astronomy,* 2nd. edn., Cambridge, Cambridge University Press.

Blundell, S. J., 2009. *Superconductivity: A Very Short Introduction*, New York, NY, Oxford University Press.

Boyle, W. and Smith, G. E., 1971. *IEEE Spectrum*, **8**(7), 18.

Butler, R. P., Marcy, G. W., Williams, E., *et al.*, 1996. *Publications of the Astronomical Society of the Pacific*, **108**, 500.

Chabrier, G. and Baraffe, I., 2000. *Annual Review of Astronomy and Astrophysics*, **38**, 337.

Colina, L., Bohlin, R., and Castelli, F., 1996. Absolute flux calibration of Vega. Space Telescope Instrument Science Report CAL/SCS-008. See: www.stsci.edu/instruments/ observatory/PDF/scs8.rev.pdf

Cox, A., 1999. *Allen's Astrophysical Quantities*, New York, NY, AIP Press.

Dekker, H., D'Odorico, S., Kaufer, A., Delabre, B., and Kotzlowski, H., 2000. *Proceedings of the Society of Photo-optical Instrumentation Engineers*, **4008**, 534.

Ducati, J. R., Bevilacqua, C. M., Rembold, S. B., and Ribeiro, D., 2001. *Astrophysics Journal*, **558**, 309.

Glass, I. S., 1999. *Handbook of Infrared Astronomy*, Cambridge, Cambridge University Press.

Hardy, J. W., 1998. *Adaptive Optics for Astronomical Telescopes*, New York, NY, Oxford University Press.

Harris, R., 1998. *Nonclassical Physics*, New York, NY, Addison Wesley.

Hearnshaw J. B., 1986. *The Analysis of Starlight*, Cambridge, Cambridge University Press.

Hearnshaw, J. B., 1996. *The Measurement of Starlight*, Cambridge, Cambridge University Press.

Hearnshaw J. B., 2009. *Astronomical Spectrographs and their History*, Cambridge, Cambridge University Press.

Hogg, D. W., Baldry, I. K., Blanton, M. R., and Eisenstein, D. J., 2002. eprint arXiv:astro-ph/0210394.

Howell, S. B., 2006. *Handbook of CCD Astronomy*, 2nd edn., Cambridge, Cambridge University Press.

Jaschek, C. and Murtagh, F., 1990. *Errors, Bias and Uncertainties in Astronomy*, Cambridge, Cambridge University Press.

Johnson, H. L. and Harris, D. L., 1954. *Astrophysical Journal*, **120**, 196.

Kaler, J. B., 1997. *Stars and their Spectra*, Cambridge, Cambridge University Press.

Kaler, P. C., 1963. *Basic Astronomical Data*, ed. Strand, K. Aa., Chicago, University of Chicago Press, ch. 8.

King, Henry C., 1979. *The History of the Telescope*, New York, NY, Dover. (Originally published by Griffin & Co., 1955.)

Kitchin, C. R., 1995. *Optical Astronomical Spectroscopy*, Bristol, Institute of Physics.

Kitchin, C. R., 2008. *Astrophysical Techniques*, 5th edn., Boca Raton, FL, CRC Press.

Kittel, C., 2005. *Introduction to Solid State Physics*, 8th edition, New York, NY, John Wiley and Sons.

Landolt, A. U., 1983. *Astronomical Journal*, **88**, 439.

Landolt, A. U., 1992. *Astronomical Journal*, **104**, 340, 436.

Lasker, B. M. and Lattanzi, Mario G., 2008. *Astronomical Journal*, **236**, 735.

Lyons, L., 1991. *Data Analysis for Physical Science Students*, Cambridge, Cambridge University Press.

Martinez, P. and Klotz, A., 1998. *A Practical Guide to CCD Astronomy*, Cambridge, Cambridge University Press.

McLean, I. S., 2008. *Electronic Imaging in Astronomy*, 2nd edn., Chichester, Springer-Praxis.

Menzies, J. W., Cousins, A. W. J., Banfield, R. M., Laing, J. D., and Coulson, I. M., 1989 *SAAO Circular*, **13**, 1.

Menzies, J. W., Marang, F., Laing, J. D., Coulson, I. M., and Engelbrecht, C. A., 1991. *Monthly Notices of the Royal Astronomical Society*, **248**, 642.

Möller, K. D., 1988. *Optics*, Mill Valley, CA, University Science Books

Monet, D. G., 1988. *Annual Reviews of Astronomy and Astrophysics*, **26**, 413.

Owens, J. C., 1967. *Applied Optics*, **6**, 51.

Palmer, C., 2002. *Diffraction Grating Handbook*, 5th edn., Rochester, NY, Thermo RGL (widely available online).

Peacock, T., Verhoeve, P., Rando, *et al.*, 1997. *Astronomy and Astrophysics Supplement Series*, **123**, 581.

Pedrotti, F. S., Pedrotti, L. M., and Pedrotti, L. S., 2006. *Introduction to Optics*, 3rd edn., San Francisco, CA, Benjamin Cummings.

Rieke, G., 2003. *Detection of Light: From the Ultraviolet to the Submillimeter*, 2nd edn., Cambridge, Cambridge University Press.

Rieke, G. H. and Lebofsky, M. J., 1985. *Astrophysical Journal*, **288**, 618.

Rieke, G. H., Lebofsky, M. J., and Low, F. J., 1985. *Astronomical Journal*, **90**, 900.

Roddier, F., 1999. *Adaptive Optics in Astronomy*, Cambridge, Cambridge University Press.

Rutten, H. G. J. and van Venrooij, M. A. M., 1988. *Telescope Optics, A Comprehensive Manual for Amateur Astronomers*, Richmond, VA, Willmann-Bell, Inc.

Schaefer, B., 2006. *Scientific American*, **295**, 5, 70.

Schroeder, D. J., 1987. *Astronomical Optics*, London, Academic Press.

Sterken, C. and Manfroid, J., 1992. *Astronomical Photometry*, Dordrecht, Kluwer.

Thoren, V. E., 1990. *The Lord of Uraniborg: a Biography of Tycho Brahe*. Cambridge, Cambridge University Press, 1990.

Tonry, J., Burke, B. E., and Schechter, P. L., 1997. *Publications of the Astronomical Society of the Pacific*, **109**, 1154.

Verhoeve, P., Martin, D., Brammetz, G., Hijmering, R., and Peacock, A., 2004. *Proceedings of the 5th International Conference on Space Optics*, ESA SP-554.

Wall, J. V. and Jenkins, C. R., 2003. *Practical Statistics for Astronomers*, Cambridge, Cambridge University Press.

Weaver, H., 1946. *Popular Astronomy*, **24**, 212; 287; 339; 389; 451; 504.

Wilson, R. N., 1996. *Reflecting Telescope Optics*, vol. I, New York, Springer.

Wilson, R. N., 1999. *Reflecting Telescope Optics*, vol. II, New York, Springer.

Young, A., 2006. See http://mintaka.sdsu.edu/GF/bibliog/overview.html.

Index